Pathways to a Carbon-Free Future Through Advanced Nuclear Systems

Shrikaant Kulkarni
Sanjivani University, India & Victorian Institute of Technology, Australia

Susanta Das
Ajeenkya D.Y. Patil University, India

IGI Global
Scientific Publishing
Publishing Tomorrow's Research Today

Vice President of Editorial	Melissa Wagner
Director of Acquisitions	Mikaela Felty
Director of Book Development	Jocelynn Hessler
Production Manager	Mike Brehm
Cover Design	Jose Rosado

Published in the United States of America by
IGI Global Scientific Publishing
701 East Chocolate Avenue
Hershey, PA, 17033, USA
Tel: 717-533-8845
Fax: 717-533-7115
Website: https://www.igi-global.com E-mail: cust@igi-global.com

Library of Congress Cataloging-in-Publication Data

LCCN: 2025005841 (CIP Data Pending)
ISBN13: 9798337306858
Isbn13Softcover: 9798337306865
EISBN13: 9798337306872

British Cataloguing in Publication Data
A Cataloguing in Publication record for this book is available from the British Library.

This book, *Pathways to a Carbon-Free Future Through Advanced Nuclear Systems*, is dedicated to all researchers, educators, and students who have been relentlessly working toward a sustainable society and a healthier environment.

Your unwavering commitment to knowledge, innovation, and responsible stewardship of our planet is the foundation upon which a carbon-free future can be built.

Table of Contents

Preface xiii

Acknowledgement xvi

Introduction xvii

Chapter 1
Nuclear Energy in the Context of Global Climate Objectives 1
Adil Aknouch, LICPM Laboratory, Faculty of Sciences and Techniques, Sultan Moulay Slimane University, Morocco
Youssef El-Ouardi, LIMAS Laboratory, Department of Physics, Faculty of Science, Sidi Mohamed Ben Abdellah University, Morocco
Hasna Belcadi, LAOC Laboratory, Department of Chemistry, Faculty of Sciences, Abdelmalek Essaadi University, Morocco
Mohammed Yjjou, LPTPME Laboratory, Department of Physics, Faculty of Sciences, Mohammed 1st University, Morocco
Lhoucine Oufni, LICPM Laboratory, Faculty of Sciences and Techniques, Sultan Moulay Slimane University, Morocco
Mohammed Talbi, Faculty of Science, University Ibn Tofail, Kenitra, Morocco
M'hamed El Mansouri, Faculty of Science, University Ibn Tofail, Kenitra, Morocco
Slimane Semghouli, Higher Institute of Nursing Professions and Health Techniques, Agadir, Morocco
Halima Soussi, Mohammed V University, Rabat, Morocco
Khawla Rachmoune, ILRBREM, Higher Normal School, Cadi Ayyad University, Marrakech, Morocco

Chapter 2
Nuclear Energy's Role in Global Energy Transition 25
Amitava Podder, Brainware University, India
Piyal Roy, Brainware University, India
Shivnath Ghosh, Brainware University, India
Anupama Sen, Brainware University, India
Subrata Paul, Brainware University, India

Chapter 3
Introduction to Advanced Nuclear Systems 63
M. D. Mathew, Saintgits College of Engineering, India
Bhanu Sankara Rao, Indian Institute of Technology, Hyderabad, India

Chapter 4
Advanced Nuclear Systems and Climate Resilience 93
Ranjit K. Parmar, Government Polytechnic Junagadh, India

Chapter 5
Innovations in Nuclear Fuel Technology: Accident Tolerance, Efficiency, and Sustainability.. 129
J. Mangaiyarkkarasi, Nadar Mahajana Sangam S. Vellaichamy Nadar College, India
J. Shanthalakshmi Revathy, Velammal College of Engineering and Technology, India

Chapter 6
Small Modular Reactor: Road to Carbon Free and Sustainable Energy............ 153
Kashvi Chaturvedi, Ajeenkya D.Y. Patil University, India
Sunil Sankathala, Ajeenkya D.Y. Patil University, India
Aditya Shrivastav, Ajeenkya D.Y. Patil University, India
Yadnyesh Khapekar, Ajeenkya D.Y. Patil University, India
Atharva Saraf, Ajeenkya D.Y. Patil University, India
Shaurya Talewar, Ajeenkya D.Y. Patil University, India
Susanta Das, Ajeenkya D.Y. Patil University, India
Shrikaant Kulkarni, Sanjivani University, India

Chapter 7
Integrating AI and Nuclear Technology for Environmental Sustainability 195
Usharani Bhimavarapu, Department of Computer Science and Engineering, Koneru Lakshmaiah Education Foundation, Vaddeswaram, India

Chapter 8
The Pertinency of Cyber Security Towards Critical Infrastructure 215
Shaurya Talewar, Ajeenkya D.Y. Patil University, India
Kashvi Chaturvedi, Ajeenkya D.Y. Patil University, India
Sunil Sankathala, Ajeenkya D.Y. Patil University, India
Yadnyesh Khapekar, Ajeenkya D.Y. Patil University, India
Aditya Shrivastav, Ajeenkya D.Y. Patil University, India
Atharva Saraf, Ajeenkya D.Y. Patil University, India
Susanta Das, Ajeenkya D.Y. Patil University, India

Chapter 9
Technological Advancement of Space Explorations With Nuclear Energy 239
Shaurya Talewar, Ajeenkya D.Y. Patil University, India
Kashvi Chaturvedi, Ajeenkya D.Y. Patil University, India
Sunil Sankathala, Ajeenkya D.Y. Patil University, India
Yadnyesh Khapekar, Ajeenkya D.Y. Patil University, India
Aditya Shrivastav, Ajeenkya D.Y. Patil University, India
Atharva Saraf, Ajeenkya D.Y. Patil University, India
Susanta Das, Ajeenkya D.Y. Patil University, India
Shrikaant Kulkarni, Sanjivani University, India

Chapter 10
Adoption Challenges of Hydrogen Energy: A Critical Analysis 263
Suchismita Satapathy, KIIT University, India

Conclusion 283

Compilation of References 285

About The Contributors 317

Index 327

Detailed Table of Contents

Preface xiii

Acknowledgement xvi

Introduction xvii

Chapter 1
Nuclear Energy in the Context of Global Climate Objectives 1
Adil Aknouch, LICPM Laboratory, Faculty of Sciences and Techniques, Sultan Moulay Slimane University, Morocco
Youssef El-Ouardi, LIMAS Laboratory, Department of Physics, Faculty of Science, Sidi Mohamed Ben Abdellah University, Morocco
Hasna Belcadi, LAOC Laboratory, Department of Chemistry, Faculty of Sciences, Abdelmalek Essaadi University, Morocco
Mohammed Yjjou, LPTPME Laboratory, Department of Physics, Faculty of Sciences, Mohammed 1st University, Morocco
Lhoucine Oufni, LICPM Laboratory, Faculty of Sciences and Techniques, Sultan Moulay Slimane University, Morocco
Mohammed Talbi, Faculty of Science, University Ibn Tofail, Kenitra, Morocco
M'hamed El Mansouri, Faculty of Science, University Ibn Tofail, Kenitra, Morocco
Slimane Semghouli, Higher Institute of Nursing Professions and Health Techniques, Agadir, Morocco
Halima Soussi, Mohammed V University, Rabat, Morocco
Khawla Rachmoune, ILRBREM, Higher Normal School, Cadi Ayyad University, Marrakech, Morocco

This chapter examines the key role of nuclear energy in achieving global climate goals, such as carbon neutrality by 2050 and the reduction of greenhouse gas emissions. It highlights nuclear energy's potential to address growing energy demands while reducing carbon footprints, especially by complementing renewable sources like wind and solar. Despite its benefits, nuclear power faces challenges such as safety concerns, waste management, high infrastructure costs, and public opposition. The chapter explores how innovations like Small Modular Reactors (SMRs) and nuclear fusion could overcome some of these challenges and improve nuclear energy's sustainability. It also discusses the integration of nuclear energy into global decarbonization strategies, citing case studies from countries like France, Sweden, and Japan. In conclusion, nuclear energy is essential for meeting climate targets, but its effective role in a low-carbon future requires a balanced approach, including supportive policies, technological advancements, and international collaboration.

Chapter 2
Nuclear Energy's Role in Global Energy Transition 25
Amitava Podder, Brainware University, India
Piyal Roy, Brainware University, India
Shivnath Ghosh, Brainware University, India
Anupama Sen, Brainware University, India
Subrata Paul, Brainware University, India

As more people across the world gain access to electricity then these sustainable energy models need to be managed effectively. This chapter presents an overview of common modern trends, technologies, and heutagogy in energy optimization based on fundamentals of power minimization and incorporation of renewable energy sources such as solar, wind, biomass, and geothermal. It gives importance to combined and comprehensive systems for better energy availability and productivity plus smart grid and energy storage in energy supply and utilization. Optimization techniques, such as deterministic and stochastic approaches together with multi-objective optimization are provided for enhanced performance. Many aspects of technical, economical, policy, and environmental nature of sustainable energy optimization are divided and discussed, giving a more extensive perspective. AI and machine learning represent further perspectives as well as decentralized energy systems, energy as a service, and the circular economy.

Chapter 3
Introduction to Advanced Nuclear Systems 63
M. D. Mathew, Saintgits College of Engineering, India
Bhanu Sankara Rao, Indian Institute of Technology, Hyderabad, India

Nuclear energy is clean energy. It does not produce any greenhouse gases. Hence, it is a game-changing technology in addressing global energy demands, climate change mitigation, and sustainable development. This chapter describes the principles of nuclear fission, and demonstrates that nuclear fission is a mature and efficient method of large-scale electricity production. The different types of commercial thermal reactors such as Pressurized Water Reactors, Boiling Water Reactors, and Pressurized Heavy-Water Reactors, as well as Advanced Gas-Cooled Reactors are discussed and evaluated based on their moderator type, fuel enrichment levels, coolant mechanisms, and safety features. Development of advanced nuclear technologies like Small Modular Reactors and Generation IV reactors is also discussed. Fast reactors and emerging fusion reactor technologies hold great promise to supplement thermal reactors and renewable energy sources to replace fossil fuels to address the challenges of global warming.

Chapter 4
Advanced Nuclear Systems and Climate Resilience 93
Ranjit K. Parmar, Government Polytechnic Junagadh, India

Due to global warming and other factors, the frequency of climate induced calamities like flood, tsunami, draught are increasing. They damage the energy infrastructure a lot. To combat these challenges, existing infrastructure must be strengthened for energy resilience. Renewable energy sources are gaining momentum as they are non-polluting unlike fossil fuels, but their energy output is intermittent, unpredictable and depends on weather conditions. In this context, advanced nuclear reactors may play an important role to supply stable base load power. Small modular reactors with passive safety features can help in meeting the energy, heat and drinking water demand even in remote area where traditional rectors are unpractical. Integrating renewable energy sources with advanced nuclear reactors though smart grid can show the path for a carbon free future. This chapter discusses the role of SMRs in climatic adaptation along with related opportunities and challenges.

Chapter 5
Innovations in Nuclear Fuel Technology: Accident Tolerance, Efficiency, and Sustainability.. 129
J. Mangaiyarkkarasi, Nadar Mahajana Sangam S. Vellaichamy Nadar College, India
J. Shanthalakshmi Revathy, Velammal College of Engineering and Technology, India

This chapter highlights the development of nuclear fuel and the current status with special reference to accident tolerable fuels and efficiency and sustainability. Low-gamma materials like chromium-coated zirconium and FeCrAl alloys enhance the safety by operation at harsh conditions and long-lived harmless materials; sophisticated fuels like HALEU and thorium-based fuels increase energy extraction efficiency, decrease radioactive waste, and help in long reactor cycles. Technologies in recycling also advance sustainability since areas of uranium usage are reduced and unsafe burial also eliminated by pyroprocessing. The chapter has devoted due attention to the economic and regulation factors with concern to the fact that the overall costs are high in compare to the perceived long term advantages. The book examines the roles of advanced fuels in new generations of reactors, including SMRs and Gen IV reactors, and how they improve safety, efficiency, and sustainability to guarantee nuclear energy's competitive and sustainable future.

Chapter 6
Small Modular Reactor: Road to Carbon Free and Sustainable Energy............ 153
Kashvi Chaturvedi, Ajeenkya D.Y. Patil University, India
Sunil Sankathala, Ajeenkya D.Y. Patil University, India
Aditya Shrivastav, Ajeenkya D.Y. Patil University, India
Yadnyesh Khapekar, Ajeenkya D.Y. Patil University, India
Atharva Saraf, Ajeenkya D.Y. Patil University, India
Shaurya Talewar, Ajeenkya D.Y. Patil University, India
Susanta Das, Ajeenkya D.Y. Patil University, India
Shrikaant Kulkarni, Sanjivani University, India

Small Modular Reactors (SMRs) are safer, more flexible, & more affordable than big reactors. They are becoming popular as an affordable, scalable, & adaptable substitute for conventional large-scale nuclear power. SMRs are positioned to be a key component of low-carbon, sustainable energy systems due to growing international interest & regulatory support. Rapid deployment in remote locations & integration with renewable energy sources are made possible by their factory-based modular construction. SMRs contribute to clean energy objectives by offering dependable electricity & purification. Their scalability & passive safety features lower capital risks & improve resilience. However, there are still issues such as public awareness, economic viability, & licensing barriers. This study investigates how SMRs can work together to create a future energy landscape that is secure, varied, & sustainable.

Chapter 7
Integrating AI and Nuclear Technology for Environmental Sustainability 195
Usharani Bhimavarapu, Department of Computer Science and Engineering, Koneru Lakshmaiah Education Foundation, Vaddeswaram, India

The integration of nuclear technology, ecosystem services, and sustainable development is faced with numerous challenges, including regional heterogeneities, environmental risks, and socio-economic constraints. There exist no models particularly to tackle localized sustainability change, transfer of technology, and long-term ecological impacts. The study discusses a more intricate framework that enhances the sustainability indicators by considering region-specific drivers, advanced risk assessment, and adaptive policy advice. The proposed algorithms improve upon the shortcomings of past studies by including predictive modeling of long-term material stability, improving life-cycle analysis of greenhouse gas emissions, and improving isotopic analysis for nutritional and environmental research. Through the integration of dynamic modeling techniques, multi-proxy data techniques, and case studies in real-world environments, this study improves the usability and applicability of nuclear-based environmental and energy solutions.

Chapter 8
The Pertinency of Cyber Security Towards Critical Infrastructure 215
Shaurya Talewar, Ajeenkya D.Y. Patil University, India
Kashvi Chaturvedi, Ajeenkya D.Y. Patil University, India
Sunil Sankathala, Ajeenkya D.Y. Patil University, India
Yadnyesh Khapekar, Ajeenkya D.Y. Patil University, India
Aditya Shrivastav, Ajeenkya D.Y. Patil University, India
Atharva Saraf, Ajeenkya D.Y. Patil University, India
Susanta Das, Ajeenkya D.Y. Patil University, India

Protection of digital facilities operating nuclear power plants constitutes a vital infrastructure concern that demands immediate attention because of digital era. Implementation of connections of various systems & modern technologies by nuclear facilities lead to efficient operations but create risk for cyber related threats. ICS systems together with SCADA networks & safety mechanisms face high exposure to attacks which became apparent due to the damaging Stuxnet incident. The research examines cybersecurity problems that affect nuclear power plants before evaluating current protocols & suggesting potential remedial options. This investigation depends on a defence system based on AI & blockchain technology alongside ML capabilities for it to detect threats instantly & trigger decisions beforehand. Examination of regulatory aspects & policy needs to emphasize global cybersecurity standards requirements. The analysis integrates recommendations developed by three expert fields - cybersecurity experts, nuclear engineers, along with policymakers to safeguard nuclear infrastructure.

Chapter 9
Technological Advancement of Space Explorations With Nuclear Energy 239
Shaurya Talewar, Ajeenkya D.Y. Patil University, India
Kashvi Chaturvedi, Ajeenkya D.Y. Patil University, India
Sunil Sankathala, Ajeenkya D.Y. Patil University, India
Yadnyesh Khapekar, Ajeenkya D.Y. Patil University, India
Aditya Shrivastav, Ajeenkya D.Y. Patil University, India
Atharva Saraf, Ajeenkya D.Y. Patil University, India
Susanta Das, Ajeenkya D.Y. Patil University, India
Shrikaant Kulkarni, Sanjivani University, India

The quest for deep space exploration requires increasingly efficient, powerful, & enduring energy solutions. This chapter discusses some of the most significant developments in space nuclear technologies, starting with the application of RTGs in Apollo & Voyager missions. It then goes on to the modern systems such as SMRs for lunar and Martian settlements, and NTP systems for high-speed interplanetary travel. Advances in fuel types, miniaturization, shielding, & thermal control have greatly improved system viability in space. Major institutional roles, NASA's

Kilopower, DARPA's DRACO, & international collaborations, are discussed. The chapter is also critically evaluating ethical, environmental, & geopolitical issues such as launch hazards, planetary protection, and space militarization. Looking ahead, it discusses next-generation ideas such as DFD, nuclear-powered in-situ resource utilization, & self-sustaining off-Earth colony designs. From a multi-disciplinary perspective, this chapter demonstrates how nuclear systems are transforming the future of space exploration.

Chapter 10
Adoption Challenges of Hydrogen Energy: A Critical Analysis 263
Suchismita Satapathy, KIIT University, India

Globally, hydrogen energy is considered as a clean energy carrier, which will not impact on climate change and reduce environmental pollution. Hydrogen energy is also being called as a potential solution for reducing greenhouse gas emissions. Hydrogen can bring a revolution to vehicles and to the grid system. Still, there are some challenges creating barriers to adoption of hydrogen energy in residential, commercial and industrial sectors. In this chapter, the challenges of hydrogen energy adoption are found in ARAS(Additive ratio assessment)method followed by ISM(Interpretative structural modelling) and QFD(Quality function deployment is used such that the stakeholders can frame new policies and develop new methodologies with advanced technology to mitigate these challenges and familiarize hydrogen use as sustainable energy source in India.

Conclusion 283

Compilation of References 285

About The Contributors 317

Index 327

Preface

The intensifying urgency of the global climate crisis, coupled with rising energy demands and the limitations of traditional fossil fuels, has created a pivotal moment for reimagining our energy systems. In this context, nuclear energy stands at a transformative crossroads—offering both a proven foundation and a springboard for innovation as humanity strives toward a carbon-free future. This edited volume, *Pathways to a Carbon-Free Future Through Advanced Nuclear Systems*, brings together critical perspectives, pioneering research, and practical frameworks to examine how advanced nuclear technologies can serve as a keystone in the decarbonization of global energy systems.

CHAPTER OVERVIEW

Chapter 1, **"Nuclear Energy in the Context of Global Climate Objectives,"** sets the stage by framing nuclear energy as a critical player in meeting international climate goals, such as carbon neutrality by 2050. It underscores nuclear's potential to complement intermittent renewables like wind and solar while addressing systemic challenges through innovations such as Small Modular Reactors (SMRs) and nuclear fusion.

Expanding on this foundation, Chapter 2, **"Nuclear Energy's Role in Global Energy Transition,"** provides a holistic view of how nuclear fits within broader sustainable energy models. It highlights optimization strategies and smart integration with renewables, emphasizing the role of artificial intelligence and decentralized systems in building robust, future-ready energy frameworks.

Chapter 3, **"Introduction to Advanced Nuclear Systems,"** offers readers a detailed grounding in nuclear fission and contemporary reactor technologies, ranging from Pressurized Water Reactors to next-generation concepts such as Generation IV reactors. It presents a compelling case for nuclear as both a mature and evolving solution to energy and climate challenges.

In Chapter 4, **"Advanced Nuclear Systems and Climate Resilience,"** the conversation shifts toward the adaptability and resilience of nuclear technologies in the face of climate-induced disruptions. SMRs, in particular, are shown to provide stable, scalable energy solutions for disaster-prone or remote regions where traditional infrastructure is vulnerable.

Chapter 5, **"Innovations in Nuclear Fuel Technology: Accident Tolerance, Efficiency, and Sustainability,"** explores cutting-edge advances in nuclear fuel materials and cycles. By focusing on safety enhancements, fuel efficiency, and sustainability through recycling and novel fuel types, this chapter demonstrates how material science is redefining the future of nuclear energy.

Chapter 6, **"Small Modular Reactor – Road to Carbon-Free & Sustainable Energy,"** delves further into SMRs as a cornerstone of modern nuclear strategy. Their modular design, economic scalability, and suitability for remote or decentralized deployment position them as a practical path toward a cleaner and more resilient energy mix.

Emerging technologies and interdisciplinary approaches are further explored in Chapter 7, **"Integrating AI and Nuclear Technology for Environmental Sustainability."** This chapter proposes new models that harness artificial intelligence and data science to optimize long-term sustainability, regional adaptation, and environmental risk management in nuclear applications.

Recognizing the increasing digitization of energy infrastructure, Chapter 8, **"The Pertinency of Cyber Security Towards Critical Infrastructure,"** addresses the escalating threat of cyberattacks on nuclear systems. It proposes an integrated cybersecurity framework using AI, blockchain, and machine learning to safeguard operational integrity and national security.

Chapter 9, **"Technological Advancement of Space Explorations with Nuclear Energy,"** broadens the horizon by examining how nuclear technologies are revolutionizing deep space exploration. From powering extraterrestrial habitats to enabling interplanetary propulsion, nuclear systems are shown to be vital for the next era of space exploration, while also raising ethical and geopolitical considerations.

Finally, Chapter 10, **"Adoption Challenges of Hydrogen Energy: A Critical Analysis,"** examines hydrogen energy not in opposition to nuclear, but as a complementary clean energy vector. Using advanced decision-making tools, it identifies key barriers and solutions to hydrogen adoption, particularly in the Indian context, where integrated energy policies may benefit from nuclear-hydrogen symbiosis.

Together, the chapters in this volume form a multidimensional map of nuclear energy's evolving role in global sustainability. From terrestrial grids to extraterrestrial missions, from AI-powered optimization to fuel innovation, this book aims to equip policymakers, engineers, researchers, and energy strategists with insights and tools to accelerate the transition to a carbon-free world. We hope the ideas presented here will inspire cross-disciplinary collaboration, informed policy, and continued innovation at the intersection of nuclear science and sustainable development.

Shrikaant Kulkarni
Sanjivani University, India & Victorian Institute of Technology, Australia

Susanta Das
Ajeenkya D.Y. Patil University, India

Acknowledgement

We are thankful to all the contributors for their valuable input and for reviewing the chapters. We greatly appreciate the continuous support and help received from our family members, friends, and colleagues—from the submission of the proposal to the completion of the project—without which it would have been impossible for us to embark on this journey. Last but not least, we are grateful to IGI Global for providing us with the opportunity to publish this edited book.

Shrikaant Kulkarni
Sanjivani University, India & Victorian Institute of Technology, Australia

Susanta Das
Ajeenkya D.Y. Patil University, India

Introduction

It is with great anticipation and purpose that we present *Pathways to a Carbon-Free Future Through Advanced Nuclear Systems*, a comprehensive exploration of how advanced nuclear energy can redefine the global energy landscape. In the face of an escalating climate crisis, there is growing recognition that no single solution can singularly decarbonize the world's energy systems. Yet, nuclear power—particularly in its modern, advanced forms—stands out as one of the most viable, scalable, and reliable options for achieving net-zero goals. This volume seeks to capture the technological progress, policy frameworks, and integration strategies that position nuclear energy at the forefront of climate and energy solutions.

This book brings together leading experts, researchers, and thought leaders to explore the full spectrum of next-generation nuclear systems—from small modular reactors (SMRs) and microreactors to Generation IV designs, fusion breakthroughs, and AI-enhanced safety protocols. The chapters delve into the science and engineering behind these innovations, contextualizing them within global sustainability goals and climate policy imperatives.

What distinguishes this volume is its focus on the *real-world applications* of nuclear technologies. The transition to clean energy must go beyond theoretical potential; it must demonstrate how nuclear systems can be deployed in harmony with renewables, support energy-intensive industries like hydrogen production, and deliver reliable power to regions facing energy poverty. The book thus serves not only as a technical reference but also as a strategic guide for stakeholders involved in energy planning, investment, regulation, and public engagement.

We have also made it a priority to address key societal and regulatory challenges. Misinformation and legacy perceptions of nuclear risk continue to pose barriers to progress. By presenting evidence-based analyses of modern safety features, accident-tolerant fuels, waste management strategies, and cyber-resilience for nuclear infrastructure, this book helps dismantle outdated narratives and promotes informed discourse.

Our goal with *Pathways to a Carbon-Free Future Through Advanced Nuclear Systems* is multifaceted:

- To educate current and future nuclear professionals on the latest technologies and practices.
- To highlight nuclear energy's role as a backbone in global decarbonization strategies.
- To promote integration with renewable energy for robust and flexible power grids.
- To inform public perception and regulatory development through transparency and facts.
- To inspire innovation and investment across nations and economic contexts.

This book is crafted for a broad yet specialized audience—nuclear engineers, researchers, policymakers, environmental advocates, investors, and students alike. Whether you are designing reactors, setting energy policy, leading a classroom, or simply curious about clean energy solutions, we hope this book empowers your work and enriches your understanding.

We would like to express our gratitude to the contributing authors who have brought depth, diversity, and rigor to this volume. Their expertise has shaped a resource that reflects the interdisciplinary and global nature of the clean energy challenge. We also thank the peer reviewers and collaborators who supported the development of this work with critical insights and technical refinement.

As we look toward a carbon-neutral future, this book invites readers to envision nuclear energy not as an outdated relic of the past, but as a dynamic and indispensable component of the clean energy era ahead.

Chapter 1

Nuclear Energy in the Context of Global Climate Objectives

Adil Aknouch
https://orcid.org/0000-0003-4340-6687
LICPM Laboratory, Faculty of Sciences and Techniques, Sultan Moulay Slimane University, Morocco

Youssef El-Ouardi
https://orcid.org/0000-0002-5430-374X
LIMAS Laboratory, Department of Physics, Faculty of Science, Sidi Mohamed Ben Abdellah University, Morocco

Hasna Belcadi
LAOC Laboratory, Department of Chemistry, Faculty of Sciences, Abdelmalek Essaadi University, Morocco

Mohammed Yjjou
https://orcid.org/0000-0002-1255-0785
LPTPME Laboratory, Department of Physics, Faculty of Sciences, Mohammed 1st University, Morocco

Lhoucine Oufni
LICPM Laboratory, Faculty of Sciences and Techniques, Sultan Moulay Slimane University, Morocco

Mohammed Talbi
Faculty of Science, University Ibn Tofail, Kenitra, Morocco

M'hamed El Mansouri
Faculty of Science, University Ibn Tofail, Kenitra, Morocco

Slimane Semghouli
https://orcid.org/0000-0002-5256-2836
Higher Institute of Nursing Professions and Health Techniques, Agadir, Morocco

Halima Soussi
https://orcid.org/0009-0003-0735-0097
Mohammed V University, Rabat, Morocco

DOI: 10.4018/979-8-3373-0685-8.ch001

Khawla Rachmoune
https://orcid.org/0009-0001-6483-1060
ILRBREM, Higher Normal School, Cadi Ayyad University, Marrakech, Morocco

ABSTRACT

This chapter examines the key role of nuclear energy in achieving global climate goals, such as carbon neutrality by 2050 and the reduction of greenhouse gas emissions. It highlights nuclear energy's potential to address growing energy demands while reducing carbon footprints, especially by complementing renewable sources like wind and solar. Despite its benefits, nuclear power faces challenges such as safety concerns, waste management, high infrastructure costs, and public opposition. The chapter explores how innovations like Small Modular Reactors (SMRs) and nuclear fusion could overcome some of these challenges and improve nuclear energy's sustainability. It also discusses the integration of nuclear energy into global decarbonization strategies, citing case studies from countries like France, Sweden, and Japan. In conclusion, nuclear energy is essential for meeting climate targets, but its effective role in a low-carbon future requires a balanced approach, including supportive policies, technological advancements, and international collaboration.

1. INTRODUCTION

Nuclear energy plays a positive role in achieving global climate goals, including carbon neutrality by 2050 and reducing greenhouse gas emissions (Liu et al., 2023). The Intergovernmental Panel on Climate Change (IPCC) states that for addressing energy demand and reducing global greenhouse gas emissions required to cap global warming to 1.5°C, nuclear power is a key decarbonized source (IPCC, 2022). Furthermore, nuclear power can be a supplement to other sources of energy, such as wind and solar, particularly in managing the intermittent nature and production fluctuations of these variable sources (El-Emam et al., 2024).

But for all its advantages, nuclear power has a number of barriers to overcome, from safety concerns to nuclear waste disposal to the high costs of infrastructure. Its own problems may also be compounded by popular worries that are unlikely to be helpful to its advancement. Research has found that people's attitudes towards nuclear power are affected by historical accidents especially the Fukushima accident causing a fear of the nuclear power in many countries (Visschers & Wallquist, 2013). This identifies the importance of a multicomponent effort to negotiate these barriers.

New technologies like SMRs and nuclear fusion provide new potential solutions to some of these challenges. For instance, SMRs are supposed to be safer and less expensive to construct while being more flexible and easier to deploy (Black et al., 2021; Rowinski et al., 2015). In addition, while nuclear fusion is still in its early stages, it holds the potential to provide a large-scale, long-term, and nearly inexhaustible energy source, without the disadvantages typically associated with fission reactors (Reinders, 2021).

The integration of nuclear energy into global decarbonisation strategies is a major challenge for current energy policies. Countries such as France, Sweden and Japan are showing how nuclear energy can be integrated into decarbonisation strategies while guaranteeing energy security. France, for example, has made nuclear power a pillar of its energy policy, contributing to a significant reduction in carbon emissions, with around 70% of its electricity generation coming from nuclear sources (Soto & Martinez-Cobas, 2024). Similarly, after the Fukushima disaster, Japan had to review its energy policy, but continues to regard nuclear power as a central element in achieving its long-term climate objectives (Suzuki, 2019).

To understand how important is the contribution that nuclear energy can make to the world energy revolution, some relevant aspects are presented in this chapter. Let's start by looking at why nuclear power is a necessary provider of low-carbon electricity and grid stability. This section will emphasize the continuous generation potential of nuclear power and its role in mitigating greenhouse gases. We also cover the key challenges and controversies that have long encircled nuclear power, from disposing of waste to concerns about safety and the public's perception of the process. These concerns emerged and the obstacles to nuclear power, will be discussed within global climate objectives. We will also examine where nuclear power fits in the decarbonization plans of the world, considering the varying ways in which nuclear power is being incorporated into the energy plans of various countries. We will also discuss technological advancements, including the development of small modular reactors (SMRs) and nuclear fusion, that could help to increase the future application of nuclear power. These advances offer solutions to the problems of today, in which nuclear power will be safer, less expensive and even more sustainable.

2. NUCLEAR ENERGY: A KEY SOLUTION FOR LOW-CARBON ENERGY AND GRID STABILITY

It is commonly regarded that nuclear power is an essential weapon in this battle, as it produces electricity without emitting greenhouse gases during operation. This is because, unlike coal, oil or natural gas power generation, nuclear power stations do not emit CO_2, which must be reduced to meet global emission reduction targets.

It already is, says the International Energy Agency (IEA, 2020), saving about 2.5 billion tonnes of CO_2 per year, a significant strike against climate change. What's more, its energy density is breathtaking: A small bit of nuclear fuel yields far more energy than coal or oil. For the record, 1 kilogram of uranium produces the equivalent energy of 15,000 tonnes of coal (World Nuclear Association, 2021). Nuclear energy is therefore not only more economical, but is also less resource- and space- dependent compared to other energy options.

One of the biggest challenges we face today is the transition to renewable energy sources such as solar and wind power, which are intermittent by nature. Indeed, these energy sources are dependent on weather conditions and the time of day, which makes it impossible to guarantee a constant and reliable supply of electricity. This is where nuclear energy comes in. Unlike weather-dependent renewables, nuclear power plants operate without interruption, providing a stable and reliable source of electricity regardless of external conditions. Combined with renewables, nuclear power plays a vital role in providing what is known as base load energy - a constant and reliable supply that powers homes for everyday needs, whether that's running a cooker or charging a phone. Recognizing this complementary role is crucial to maintaining the stability of electricity networks, especially as the share of renewable energies increases.(Gross et al., 2003; Nelson, 2009; Sayed et al., 2023).

Then there's the issue of energy security. In today's world marked by growing geopolitical and economic uncertainties, we are beginning to see fossil fuel supplies become less reliable, making it more important than ever to secure a reliable and locally controlled energy source. One of the solutions to lower this dependency is "Nuclear Energy". It makes countries less dependent on volatile world energy prices, and it guarantees a more reliable, secure energy supply. The World Energy Council (2019) report states that diversification of energy sources, types of which include nuclear, enhances energy security and resiliency of countries (Ould Amrouche et al., 2016; Sayed et al., 2023).

Ultimately, nuclear power has an important part to play in the energy transition. With low CO_2 emissions, high energy density and ability to deliver firm power, it is an essential part of the solution. It is not only crucial to meet our electricity needs and reduce our carbon emissions, but it is also vital to making our transition to an energy system that is both seamless and secure a reality. Nuclear power when combined with renewable energy, plays a major role in decarbonizing the energy industry (Lenzen, 2008).

3. CHALLENGES AND CONTROVERSIES OF NUCLEAR ENERGY IN ACHIEVING CLIMATE GOALS

One of the central topics in the debate on climate change mitigation strategies is the role of nuclear power. As countries strive to meet their net-zero carbon emission targets, the discussion around nuclear energy remains contentious. Although nuclear power is recognized for its ability to deliver substantial amounts of low-carbon energy, it is also challenged by significant economic, environmental, and social obstacles. Moreover, there is ongoing skepticism regarding its potential to effectively support global climate initiatives. These issues are reflected in the main advantages of nuclear energy in climate mitigation, as illustrated in Figure.1.

Figure 1. Advantages of nuclear energy

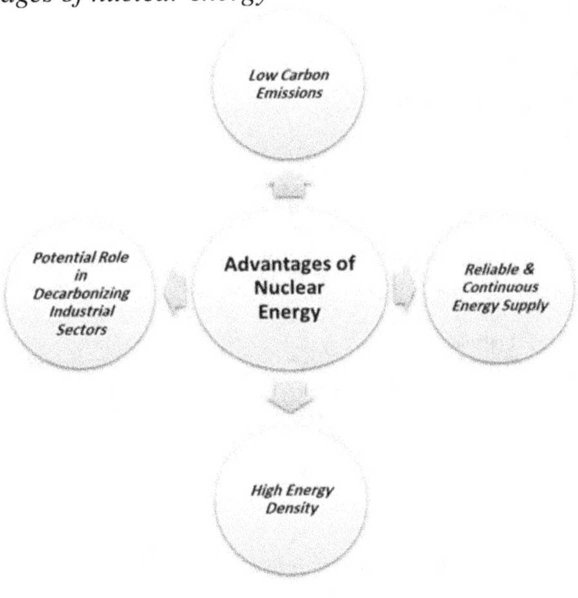

Low Carbon Emissions: Large-scale nuclear power plants have relatively low greenhouse gas (GHG) emissions during operation. As a result, they can generate significant amounts of low-emission energy. According to the IPCC, nuclear power ranks among the lowest in CO_2 emissions per kilowatt-hour, comparable to renewable sources such as wind and solar energy(Lights and Institut, 2022).

Reliable and Continuous Energy Supply: Nuclear power provides a stable and continuous base-load electricity supply, unlike intermittent sources such as solar and wind. This consistent base-load is essential for maintaining grid stability and ensuring the uninterrupted availability of electricity (Drouineau et al., 2014).

High Energy Density: Nuclear energy has an energy density that is several orders of magnitude higher than that of fossil fuels or renewable sources. A small quantity of uranium can generate a substantial amount of power, significantly reducing both the land footprint and the extent of mining required to obtain the fuel (Harris, 2022).

Potential Role in Decarbonizing Industrial Sectors: Advanced nuclear technologies, including SMRs, offer a viable solution for supplying energy to highly energy-intensive industries—such as steel and cement production—that require high-temperature heat and are otherwise challenging to decarbonize (Jiang et al., 2016).

3.a. Challenges of Nuclear Energy

Although nuclear energy can play a significant role in addressing climate change, several challenges must be overcome before it can be scaled up effectively. These challenges are summarized in Figure 2:

Figure 2. Challenges of nuclear energy

High Costs and Long Development Timelines: Constructing nuclear power plants involves substantial financial investment and long lead times. For instance, the Vogtle Electric Generating Plant in the United States experienced major cost overruns and delays, which have undermined the competitiveness of nuclear energy compared to more affordable renewable alternatives (Jiang et al., 2016).

Nuclear Waste Management: The safe management and disposal of radioactive waste remain long-standing challenges associated with nuclear power. Spent nuclear fuel remains hazardous for thousands of years, and its permanent storage in deep geological repositories presents both political and technical difficulties (Carter and Stillerman, 2020).

Safety Concerns and Accidents: Major nuclear accidents such as the Chernobyl disaster in 1986 and the Fukushima incident in 2011 have significantly shaped public perception of nuclear energy. While modern reactors are designed with enhanced safety features, the risk of catastrophic accidents though extremely rare remains a serious concern (Connolly et al., 2010).

Proliferation Risks: The expansion of nuclear power carries the risk of nuclear proliferation. Byproducts of nuclear reactors, such as highly enriched uranium and plutonium, could potentially be used by terrorists or rogue states for military purposes, posing international security threats (Fukami, 2024).

Finite Uranium Resources: Although current uranium supplies are sufficient for operating existing reactors, the future scalability of nuclear energy may be limited unless alternative fuel cycles such as thorium reactors or breeder reactors are developed. (Harris, 2022).

3.b. Controversies Surrounding Nuclear Energy

Nuclear energy is a contentious issue, raising concerns about safety, environmental impact, and social equity. This complex debate reflects the tension between the potential for hope and progress and the looming threat of disaster.

Public Opposition: Public opinion on nuclear power is divided in many regions. For some, it is seen as a necessary step to save the planet, while others oppose it due to concerns about safety, environmental risks, and a deep skepticism toward the reassurances of government and industry (Drouineau et al., 2014).

Competition with Renewable Energy: Opponents argue that investments in nuclear power could divert funding from more affordable and faster methods of integrating renewable technologies, such as wind, solar, and energy storage, into the grid. Given the urgency of the climate crisis, there is increasing pressure to prioritize renewable energy sources (Carter and Stillerman, 2020).

Environmental Impacts: Although nuclear power is low-carbon, it still has environmental impacts. Uranium mining and processing pose significant threats to habitats, leading to degradation or destruction. Additionally, thermal pollution from nuclear facilities can alter aquatic ecosystems, further affecting the environment (Harris, 2022).

Equity and Justice Issues: Nuclear facilities are often situated near marginalized communities, raising concerns about environmental justice. These populations tend to face disproportionate risks while reaping few benefits (Carter et al., 2021).

3.c. Path Forward

For nuclear energy to make its fullest contribution to climate goals, it must overcome existing challenges through a combination of innovative solutions and policy measures:

Investing in Research and Development: Continuous advancements in reactor technologies, such as SMRs and fusion energy, present opportunities to reduce costs, improve safety, and expand the applications of nuclear energy (Yang & Khan, 2022).

Enhancing Safety and Transparency: Public confidence can be strengthened through stringent safety standards, transparency in decision-making, and effective communication about the risks and benefits of nuclear energy (Fukami, 2024).

Integrating with Renewable Energy Systems: Rather than competing, the integration of flexible nuclear technologies with renewable energy sources can help create a balanced and resilient energy mix (Carter et al., 2021).

Developing Sustainable Waste Management Strategies: Implementing permanent solutions for the disposal of radioactive waste, such as deep geological repositories, is crucial for addressing long-term environmental and safety concerns (Bohrson et al., 2020).

4. NUCLEAR ENERGY IN GLOBAL DECARBONIZATION STRATEGIES

The majority of major international reports on climate change and the transition to a clean energy economy agree on one point: nuclear energy could play a decisive role in slowing global warming and decarbonizing the world's energy systems. Leading organizations such as the International Atomic Energy Agency (IAEA), the IPCC and the International Renewable Energy Agency (IRENA) have published in-depth studies highlighting the importance of nuclear power in achieving goals such as carbon neutrality by 2050. These institutions point out that electricity generated by nuclear power is not only very low in carbon emissions, but is also constant and predictable. This is a major advantage in a context where renewable energies such as wind and solar power are still dependent on weather conditions.

According to the IAEA (2020), nuclear energy can play a key role in the global energy transition, thanks in particular to technological advances such as SMRs and advanced fuel cycles. These innovations offer greater flexibility in the deployment of nuclear power plants, reducing initial costs and making it easier to adapt them to the specific needs of developing countries or isolated regions. The IAEA also emphasizes that integrating nuclear power into modern energy systems can significantly reduce dependence on fossil fuels, thereby contributing to energy security and the resilience of power grids.

In its report Climate Change 2021: The Physical Science Basis, the IPCC stresses the urgency of drastically reducing CO_2 emissions if we hope to contain global warming to 1.5°C. This presupposes massive use of low-emission energy production technologies. Among these, nuclear power plays a central role. Its integration into the global energy mix is often presented as complementary to the development of renewables, such as wind and solar power. The scenarios studied by the IPCC predict an increase in the use of nuclear power between now and 2050, which is considered necessary to reconcile the reduction of greenhouse gases with the growth in energy demand, in a sustainable way.

The IRENA, in the 2020 report Global Energy Transformation: A Roadmap to 2050, emphasizes a global energy transformation of mainly renewable energies. IRENA asserts that renewable energy, including solar and wind power, needs to constitute the "backbone" of decarbonization in order to reduce the amount of greenhouse gases released in the atmosphere. At the same time, however, the agency acknowledges that nuclear power could feature prominently as an ancillary power source in such circumstances, especially in areas with either climatic, geographical, or economic constraints on achieving a stable and fit energy power supply from renewable sources.

While not the only solution, the future of low-carbon power generation depends on nuclear. Its contribution is not limited to reducing greenhouse gases, it also helps to stabilize electricity grids and guarantee sufficient production to meet the growing demand of populations, while respecting international climate commitments. The challenges inherent in this technology, including high costs, safety concerns and radioactive waste management, can be overcome through technology advances, increased international cooperation and long-term strategic planning (Addo et al., 2023).

The nuclear power's integration in the world energy mix varies according to strategic priorities, available natural resources and the energy policies of individual nations (Davidson, 2022; El-Emam et al., 2024). Certain countries have emphasized nuclear power as an essential lever for decarbonization and energy security, while others have preferred to reduce or gradually eliminate their current dependence on this technology. This scientific analysis explores different nuclear integration models around the world (Mathew, 2022).

French model: Historical commitment and controlled transition

France is a world pioneer in the use of nuclear energy, with about 70% of its electricity coming from this source. This policy, initiated in the 1970s in response to the oil crisis, was aimed at ensuring the country's energy independence while limiting CO_2 emissions (*Nuclear Power in France - World Nuclear Association*, s. d.). According to Pervès (2022), nuclear power has enabled France to maintain a comparatively low carbon emissions footprint compared with other developed nations. Nevertheless, issues such as aging reactors and radioactive waste management have led France to adopt an energy diversification strategy. The country aims to gradually reduce the nuclear power share to 50% by 2035, while increasing the contribution of renewable energies.

Sweden: Complementarity between Nuclear and Renewable Energy

Sweden has a hybrid approach, integrating both nuclear power and renewable energies into its energy mix. Although the country has massively developed wind power and hydroelectricity, nuclear power remains an essential component in ensuring the stability of the power grid, particularly during variations in renewable energy production (Ćetković & Stockburger, 2024). According to Mikovits et al. (2021), the Swedish approach demonstrates that a balance between nuclear and renewables can effectively contribute to a decarbonization strategy, therefore reducing greenhouse gas emissions while guaranteeing energy security.

Japan: A Complex Transition after Fukushima

Prior to the Fukushima accident in 2011, nuclear power accounted for around 30% of Japan's overall energy mix. Following the disaster, the country significantly reduced its nuclear production, resulting in increased imports of fossil fuels. However, to meet its climatic targets, Japan has gradually brought certain reactors back on stream, while improving their safety standards (Thiri & Borsi, 2024). A study by Ohta & Barrett (2023) highlights the challenges of this transition and Japan's desire to combine nuclear and alternative energy sources to secure its energy supply.

United Kingdom: Nuclear Renewal with SMRs

The UK regards nuclear power as a cornerstone of its decarbonization strategy. The UK authorities have announced investments in the construction of new reactors, as well as in the development of SMRs, which offer advantages in terms of cost and flexibility (Bodel et al., 2021). In its Clean Growth Strategy (2017), the UK identified nuclear power as a strategy for reducing emissions while ensuring stable, reliable energy production.

Germany: Nuclear Phase-Out with Contrasting Consequences

Germany, through its Energiewende policy, decided to phase out nuclear power after Fukushima, with a total closure of its reactors by 2022. Yet this transition to a fully renewable mix has led to increased dependence on coal and gas, temporarily compromising CO_2 emission reduction targets. According to (Kemmerzell, 2022), the German example highlights the challenges of a swift nuclear phase-out without large-scale alternative infrastructure (*Germany's Energiewende - World Nuclear Association*, n. d.).

The United States: A Heterogeneous Evolution

In the United States, nuclear energy has long been a mainstay of electricity generation, but its expansion is being held back by high costs and rising competition from renewables. Some states, however, such as California and Michigan, are maintaining nuclear power plants to complete their energy transition (Leppert & Kennedy, 2024). The German federal government is banking on SMRs as a solution to reintroduce competitive, low-carbon nuclear generation (*Nuclear Power in the USA - World Nuclear Association*, n. d.).

Political and technological choices to integrate nuclear power into the global energy mix are varied. While some governments rely on nuclear power to guarantee stable, low-carbon electricity production, others have opted to abandon it, raising challenges in terms of energy efficiency and CO_2 emissions (Gungor & Sari, 2022). The future of nuclear power will depend on technological developments, waste management and global energy policies.

5. TECHNOLOGICAL INNOVATIONS TO ENHANCE NUCLEAR ENERGY'S ROLE

5.a. SMRs

SMRs represent a significant evolution in nuclear reactor design. Unlike traditional nuclear power plant with large scale, SMRs as the new generation nuclear power plant, are designed to generate electricity with a lower power output, between 50 and 300 megawatts per module (Zarębski & Katarzyński, 2023). They are thus well-suited for areas with less demand for ener of further-to-reach areas where construct relatively small plants is prohibitively expensive.

SMRs widely regarded as the future of nuclear power, are claimed to provide cost and safety advantages over traditional gigawatt-scale light water reactors (LWRs). Nevertheless, little attention has been paid to their back-end effect on the nuclear fuel cycle. Quantitatively comparing low, intermediate, and high waste streams, SMRs produce larger amounts of more chemically and physically reactive waste than LWRs, exacerbating the challenges with waste management and disposal (Krall et al., 2022).

The modularity of SMRs allow for critical components to be built in factories with controlled settings, lowering costs and reducing risks associated with onsite construction. Advanced reactors of this sort frequently offer enhanced safety features, such as passive cooling systems that limit the need for additional power supplies to control levels of heat. It helps to lessen the chances of accidents in cases of emergency (Vijayan & Gangwar, 2024).

Additionally, SMRs offer enhanced flexibility. Their smaller footprint allows integration into existing infrastructure, and they can be deployed incrementally to match growing energy demands. However, challenges remain. The development of regulatory frameworks, public perception, and the scaling of manufacturing processes to achieve economic competitiveness are critical hurdles to widespread adoption.

5.b. Nuclear Fusion

Nuclear fusion, often described as the 'holy grail' of energy, has the potential to revolutionize global power generation. Unlike fission, where heavy atomic nuclei are split, fusion involves combining light atomic nuclei, such as isotopes of hydrogen, to form a heavier nucleus, releasing immense amounts of energy in the process (Reinders, 2021).

The primary advantage of fusion is its virtually limitless fuel supply. Hydrogen, derived from water, is abundant and easily accessible. Moreover, fusion generates minimal radioactive waste compared to fission and does not produce greenhouse gases (Nicholas et al., 2021). It is also inherently safer, as it requires extreme conditions to sustain the reaction, and any disruption would naturally terminate the process.

Ongoing initiatives like ITER and progress in magnetic confinement techniques, including tokamaks and stellarators, are driving efforts toward sustainable fusion energy (Dahal, 2023).

Meanwhile, laser-based inertial confinement approaches are also under investigation. Despite these advancements, critical challenges persist, such as attaining net energy gain, creating materials capable of withstanding the extreme conditions within reactors, and overcoming the substantial upfront costs of research and development.

5.c. Advanced Nuclear Technologies

The next generation of nuclear technologies is focused on improving the efficiency, sustainability, and safety of nuclear power. Fast reactors, for example, utilize high-energy neutrons to convert nuclear waste into additional fuel, thereby reducing the volume and toxicity of radioactive waste. This capability addresses one of the primary criticisms of nuclear energy—long-term waste management.

Thorium reactors are another promising innovation. Thorium, a more abundant and widely distributed element than uranium, offers several advantages (Manchanda, 2023).

Reactors fueled by thorium are less prone to proliferation risks, as they produce minimal quantities of plutonium. Additionally, thorium-based systems operate at higher thermal efficiencies, making them more economical.

Improved fuel recycling methods are also at the forefront of research. These methods aim to extend the life of nuclear fuel by extracting usable isotopes from spent fuel. Combined with enhanced reactor designs, these innovations promise to make nuclear power a more sustainable and environmentally friendly energy source.

5.d. Safety Enhancements

Safety is the utmost importance considering the most recent advances in reactor technology. Passive safety systems is a major advantage, using physical principle such as gravitational and natural connection and no any active mechanical or electrical system. For instance, if a blackout happens, the systems may shut down a reactor while keeping it cooled without further human input.

There has also been a focus on building resilience to natural disasters. The new reactors are constructed to resist extra ordinal conditions, such as earthquakes, tsunamis and hurricanes and hurricanes (Eidemüller, 2021).

For example, more robust containment facilities and installations less susceptible to flooding are elements that make reactors secure in worst case scenarios. Furthermore, progress in the field of real-time monitoring and diagnostic tools make it possible to detect early malfunctions and address a possible problem in due time. Together, these innovations, along with an unwavering safety culture, are essential to restoring the public's confidence in nuclear as a safe, reliable, and clean source of power.

6. RECOMMENDATIONS

At the end of this analysis, it is appropriate to propose a set of recommendations based on the elements discussed above. These aim to respond to the main challenges identified, to support the current dynamics and to guide future policies and actions for a more controlled and sustainable development of nuclear energy as part of the global energy transition.

Expeditious development of SMR

SMRs are a potentially groundbreaking technology for the nuclear industry. They represent an ideal solution, being safer to construct, cheaper to construct, and a valid solution for smaller and less developed countries and areas. Governments should pursue research-supportive policies, while loosening safety regulations and facilitating faster certification. This will facilitate more countries in gaining from this energy source, which will greatly enhance nuclear power's application in a broader set of markets.

Nuclear fusion research investment

Nuclear fusion is the future of nuclear power in the long term. At present, this is still considered an experimental energy source. Yet, it offers to the world an energy source with the least amount of greenhouse gas emissions, and in the greatest quantity. To realize this goal, further investments in fundamental research and facilities at large scale like ITER are needed. Last but not least is the urgent need for international collaboration to address scientific and economic imperatives. In addition, both public and private spending needs to be stepped up in order to accelerate the translation of research into a clean and viable energy source by the close of the current century.

Using cutting-edge technology to tackle nuclear waste

Management of nuclear waste is one of the most challenging issues preventing nuclear industry. One way to tackle this challenge is to develop new solutions for the long-term storage and disposal of radioactive waste that are inherently safer and sustainable. Other technologies including closed fuel cycle and fast neutron reactors could largely decrease the volume and lifetime of waste. Similarly, large-scale underground nuclear waste storage is one such solution that has to be made efficient as well. But it is crucial that local communities also are engaged in choosing storage sites to promote social acceptance and transparent waste management.

The growing social acceptance of nuclear power

Social acceptability of nuclear power in many areas is the most important issue. To meet this challenge, there has to be better communication regarding any progress made in the field of nuclear security. Governments should engage in open dialogue with societies in close cooperation with nuclear industry representatives, pointing out the benefits of nuclear power in climate protection and providing clear safety guarantees. At the same time, it is also important to promote educational programs when children are still young, in order to deepen the understanding of the characteristics of nuclear energy. An openness in involving local communities in projects will help establish trust and acceptance of this energy source.

Set up clear and consistent policy and legal structures

Nuclear energy needs public policy to be developed in a sustainable way. It is crucial to have a supportive and transparent legislative framework in place that encourages investment in nuclear infrastructure and enforces a rigorous approach to safety. Leadership in the form of financial and fiscal incentives, as well as appropriate regulation that can keep up with new technology, would create a fertile environment for innovation and competition. Just as important is incorporating nuclear in a coherent national strategy for energy that focuses on renewables of all types to reach the goal of a balanced, low-carbon energy mix.

Promoting international cooperation in the nuclear field

Nuclear energy will never fulfill its promise without being shared globally, particularly with developing countries that have ever-increasing energy needs. For this to occur, international coordination is necessary to promote a secure and equitable technology transfer. Advanced countries have an important part to play in providing assistance to emerging countries in building nuclear facilities, as long as the technologies are employed in a peaceful manner in a safe environment. A stronger global regime, such as the Non-Proliferation Treaty (NPT), will be important to ensure that nuclear technology is used exclusively for peaceful purposes.

Using data and AI to run nuclear plants more efficiently

Digitization and AI offer great potential to increase the efficiency of nuclear power plants. Some of the ways AI could ultimately help monitor and optimize performance of nuclear reactors in real time, so that potential issues might be foreseen and addressed well before they occur and maintenance schedules might also be more smoothly coordinated. Investing in these technologies is not only about improving the performance of current plants, it also means lower operating costs, while maintaining safety as the priority. Better still, these advances could lead to less risk and optimal energy output, which would help to make sure a nuclear reactor is working at its peak.

Guaranteeing a fair job transition for nuclear workers

As we transition to cleaner energy, it is essential we do not leave behind workers in fossil fuels and nuclear. A just transition involves retraining those who have worked in traditional energy sources, enabling them to find new roles in renewable energy sources or the rapidly expanding field of advanced nuclear technologies. But it doesn't end there, communities that have depended on nuclear plants for their payrolls also need that financial marching order. We can guarantee that the move to a greener future is fair, inclusive and good for everyone by offering both new job training and regional assistance.

7. CONCLUSION

Nuclear power is a key element in the fight against climate change and the worldwide goal of decarbonization. It is a big part of the solution to the climate emergency, because it allows us to generate large amounts of electricity without emitting greenhouse gases. The significance of nuclear in global energy transition and its critical challenges limiting its development across the globe are the focus of this chapter. Among these challenges are the question of nuclear waste, the issue of safety, the high infrastructure costs and that there is strong public resistance to nuclear energy in worst case. The barriers hindering nuclear development including costs, threats to population health and the environment, the long-term unsolved issue for the storage of spent fuel, etc. There are a number of solutions to overcome these challenges, including the development of nuclear fusion, and recent technological advances, such SMRs.

Renewable power from sources including solar and wind is expanding fast and is seen as a keystone of the energy transition. They're good alternatives for generating electricity without emitting greenhouse gases, much like nuclear energy. But there is a problem with renewable energies: their intermittency. Indeed, solar and wind power are reliant on weather that can't be predicted for all time, and thus make it difficult to produce stable and constant amounts of energy. By contrast, nuclear provides a reliable and weather-independent energy production that ensures an energy supply sufficient to the needs at all times when the input from renewables falls short. Nuclear and renewables are therefore complementary solutions for a cleaner global energy mix which will curb the emission of greenhouse gases.

In order for nuclear energy to become a widely used and deployed source of electricity generation in the world, as well as to be embraced as a key solution to mitigate greenhouse gas emissions and help decarbonize the globe, a multi-fold approach is required. Part of this strategy will be the further development of already existing technologies and, most importantly, new developments in security and nuclear fuel efficiency. Fusion reactors and small modular reactors are significant developments that could transform how nuclear energy is viewed and utilized. They are also among the ways to mitigate the issues of conventional reactors, more flexible and less costly, and can be scaled to the size of the energy capacity.

As much as technological upgrade, a consistent, viable and perceptive public policy is required for successful energy transition to nuclear power. Under this policy, governments need to provide robust policy resilience and enabling regulation to drive innovation while keeping citizens safe. There are also policies to incentivize research and development of advanced nuclear reactors and to train workers for a new industry. The social legitimacy of nuclear energy plays a big role. Nuclear power remains an emotive subject in many countries, linked to human health and environmental dangers from past catastrophes such as Chernobyl and Fukushima. Only public information on this issue, the transparency of the production of nuclear energy, and by providing clear information on the actual risks and benefits can social acceptance be secured.

The management of nuclear waste is one of the main problems faced by the nuclear industry. Although a deep geo-repository can be a solution in the short term, the management of nuclear waste is still an issue for the long term and the overall safety. And on this note, it is also crucial to further investigate the optimal way to deal with nuclear waste with an absolute certainty that future generations will be safe.

International cooperation is also key to the nuclear energy making a fair and sustainable contribution to a global energy transition. With the developments in the world regarding climate change, the world cannot afford to live its own life, so countries have to learn and exchange best practice and technologies related to nuclear power generation. Innovation cannot be the privilege of a couple of countries only, but must be the heritage of an international community sharing common knowledge to tackle the challenges connected with nuclear energy. Collaboration between countries also bolsters global nuclear security and promotes responsible operation of nuclear facilities, especially in less economically developed countries with limited technical abilities.

The future of nuclear energy lies in this comprehensive and integrated vision. This vision requires technology, political support and international cooperation, as well as an appropriate social basis for the world to have a secure, equitable, and low-carbon energy system. This guaranteed vision will ensure that nuclear energy, in synergy with renewable energies, contributes to a transition to a prosperous global climate economy. This long-term energy mix will ensure that the world can meet growing global energy consumption while respecting the environment and climate growth.

In the near future, it is not out of the question that nuclear energy will become one of the pillars in creating a resilient and sustainable energy system. This would contribute to global energy security and significantly reduce greenhouse gas emissions. Ultimately, for a cleaner and more sustainable energy future for all, technology, societal concerns, and related policies will each play a central role. As such, nuclear energy should not be viewed solely as an energy solution, but as a strategic tool towards a low carbon and environmentally friendly future.

REFERENCES

Addo, E. K., Kabo-bah, A. T., Diawuo, F. A., & Debrah, S. K. (2023). The Role of Nuclear Energy in Reducing Greenhouse Gas (GHG) Emissions and Energy Security : A Systematic Review. *International Journal of Energy Research, 2023*(1), 8823507. DOI: 10.1155/2023/8823507

Black, G., Shropshire, D., & Araújo, K. (2021). 22 - Small modular reactor (SMR) adoption : Opportunities and challenges for emerging markets. In Ingersoll, D. T., & Carelli, M. D. (Eds.), *Handbook of Small Modular Nuclear Reactors* (2nd ed., pp. 557–593). Woodhead Publishing., DOI: 10.1016/B978-0-12-823916-2.00022-9

Bodel, W., Butler, G., & Matthews, J. (2021). *Nuclear energy for net zero—Dalton Nuclear Institute—The University of Manchester.* https://www.dalton.manchester.ac.uk/nuclear-energy-for-net-zero/

Bohrson, W. A., Spera, F. J., Heinonen, J. S., Brown, G. A., Scruggs, M. A., Adams, J. V., Takach, M. K., Zeff, G., & Suikkanen, E. (2020). Diagnosing open-system magmatic processes using the Magma Chamber Simulator (MCS) : Part I—major elements and phase equilibria. *Contributions to Mineralogy and Petrology, 175*(11), 104. DOI: 10.1007/s00410-020-01722-z

Carter, J., & Stillerman, K. P. (2020). *Rebuilding Science and Scientific Integrity at the US Department of Agriculture.* Union of Concerned Scientists. https://www.jstor.org/stable/resrep26328

Carter, R., Choularton, R., Ferdinand, T., Ding, H., Ginoya, N., & Preethan, P. (2021). *Food Systems at Risk : Transformative Adaptation for Long-Term Food Security.* https://www.wri.org/research/food-systems-risk

Ćetković, S., & Stockburger, J. (2024). Hydrogen Strategy of Sweden : Unpacking the Multiple Drivers and Potential Barriers to Hydrogen Development. In R. Quitzow & Y. Zabanova (Éds.), *The Geopolitics of Hydrogen : Volume 1 : European Strategies in Global Perspective* (p. 191-211). Springer Nature Switzerland. DOI: 10.1007/978-3-031-59515-8_10

Connolly, D., Lund, H., Mathiesen, B. V., & Leahy, M. (2010). A review of computer tools for analysing the integration of renewable energy into various energy systems. *Applied Energy, 87*(4), 1059–1082. DOI: 10.1016/j.apenergy.2009.09.026

Dahal, S. (2023). Plasma Physics and Fusion Energy : A Promising Future for Sustainable Power Generation. *Koshi Pravah: Multidisciplinary Peer Reviewed Journal, 2*(1), 1. Advance online publication. DOI: 10.3126/koshipravah.v2i1.67130

Davidson, A. J. (2022). *The Role of Nuclear Energy in the Global Energy Transition*. Oxford Institute for Energy Studies. https://www.oxfordenergy.org/publications/the-role-of-nuclear-energy-in-the-global-energy-trasition/

Drouineau, M., Maïzi, N., & Mazauric, V. (2014). Impacts of intermittent sources on the quality of power supply : The key role of reliability indicators. *Applied Energy, 116*, 333–343. DOI: 10.1016/j.apenergy.2013.11.069

Eidemüller, D. (2021). Radioactive Incidents and Disasters. In Eidemüller, D. (Ed.), *Nuclear Power Explained* (pp. 195–240). Springer International Publishing., DOI: 10.1007/978-3-030-72670-6_10

El-Emam, R. S., Constantin, A., Bhattacharyya, R., Ishaq, H., & Ricotti, M. E. (2024). Nuclear and renewables in multipurpose integrated energy systems : A critical review. *Renewable & Sustainable Energy Reviews, 192*, 114157. DOI: 10.1016/j.rser.2023.114157

(2022). Framing and Context. InIntergovernmental Panel on Climate Change (IPCC) (Ed.), *Global Warming of 1.5°C: IPCC Special Report on Impacts of Global Warming of 1.5°C above Pre-industrial Levels in Context of Strengthening Response to Climate Change, Sustainable Development, and Efforts to Eradicate Poverty* (pp. 49–92). Cambridge University Press., DOI: 10.1017/9781009157940.003

Fukami, T. (2024). Enhancing Healthcare Accountability for Administrators : Fostering Transparency for Patient Safety and Quality Enhancement. *Cureus, 16*. DOI: 10.7759/cureus.66007 PMID: 39221336

Germany's Energiewende—World Nuclear Association. (s. d.). Consulté 30 janvier 2025, à l'adresse https://world-nuclear.org/information-library/energy-and-the-environment/energiewende

Gross, R., Leach, M., & Bauen, A. (2003). Progress in renewable energy. *Environment International, 29*(1), 105–122. DOI: 10.1016/S0160-4120(02)00130-7 PMID: 12605943

Gungor, G., & Sari, R. (2022). Nuclear power and climate policy integration in developed and developing countries. *Renewable & Sustainable Energy Reviews, 169*, 112839. DOI: 10.1016/j.rser.2022.112839

Harris, W. (2022). *Developments in Safety Culture—Views by the World Nuclear Association and the International System of Occupational Exposure*. 399-403.

IEA. (2020). *World Energy Outlook*. https://iea.blob.core.windows.net/assets/a72d8abf-de08-4385-8711-b8a062d6124a/WEO2020.pdf

Jiang, Y., Oron, T. R., Clark, W. T., Bankapur, A. R., D'Andrea, D., Lepore, R., Funk, C. S., Kahanda, I., Verspoor, K. M., Ben-Hur, A., Koo, D. C. E., Penfold-Brown, D., Shasha, D., Youngs, N., Bonneau, R., Lin, A., Sahraeian, S. M. E., Martelli, P. L., Profiti, G., & Radivojac, P. (2016). An expanded evaluation of protein function prediction methods shows an improvement in accuracy. *Genome Biology*, *17*(1), 184. DOI: 10.1186/s13059-016-1037-6 PMID: 27604469

Jin, T., & Kim, J. (2018). What is better for mitigating carbon emissions – Renewable energy or nuclear energy? A panel data analysis. *Renewable & Sustainable Energy Reviews*, *91*(C), 464–471. DOI: 10.1016/j.rser.2018.04.022

Kemmerzell, J. (2022). Energy Governance in Germany. In Knodt, M., & Kemmerzell, J. (Eds.), *Handbook of Energy Governance in Europe* (pp. 667–708). Springer International Publishing., DOI: 10.1007/978-3-030-43250-8_11

Krall, L. M., Macfarlane, A. M., & Ewing, R. C. (2022). Nuclear waste from small modular reactors. *Proceedings of the National Academy of Sciences of the United States of America*, *119*(23), e2111833119. DOI: 10.1073/pnas.2111833119 PMID: 35639689

Lenzen, M. (2008). Life cycle energy and greenhouse gas emissions of nuclear energy : A review. *Energy Conversion and Management*, *49*(8), 2178–2199. DOI: 10.1016/j.enconman.2008.01.033

Leppert, R., & Kennedy, B. (2024). Growing share of Americans favor more nuclear power. *Pew Research Center*. https://www.pewresearch.org/short-reads/2024/08/05/majority-of-americans-support-more-nuclear-power-in-the-country/

Lights, Z., & Institut, E. I. for M. S. on H. R. and S.-K. (2022). The (Missing) Role of Nuclear Energy in the Sustainable Development Goals. *Journal of the European Institut for Multidisciplinary Studies on Human Rights & Science*, *3*(1). Advance online publication. DOI: 10.5281/zenodo.6632131

Liu, L., Guo, H., Dai, L., Liu, M., Xiao, Y., Cong, T., & Gu, H. (2023). The role of nuclear energy in the carbon neutrality goal. *Progress in Nuclear Energy*, *162*, 104772. DOI: 10.1016/j.pnucene.2023.104772

Long, X., Naminse, E. Y., Du, J., & Zhuang, J. (2015). Nonrenewable energy, renewable energy, carbon dioxide emissions and economic growth in China from 1952 to 2012. *Renewable & Sustainable Energy Reviews*, *52*(C), 680–688. DOI: 10.1016/j.rser.2015.07.176

Manchanda, V. K. (2023). Thorium as an abundant source of nuclear energy and challenges in separation science. *Radiochimica Acta, 111*(4), 243–263. DOI: 10.1515/ract-2022-0006

Mathew, M. D. (2022). Nuclear energy : A pathway towards mitigation of global warming. *Progress in Nuclear Energy, 143*, 104080. DOI: 10.1016/j.pnucene.2021.104080

Menyah, K., & Wolde-Rufael, Y. (2010). CO2 emissions, nuclear energy, renewable energy and economic growth in the US. *Energy Policy*, *38*(6), 2911–2915. DOI: 10.1016/j.enpol.2010.01.024

Mikovits, C., Wetterlund, E., Wehrle, S., Baumgartner, J., & Schmidt, J. (2021). Stronger together : Multi-annual variability of hydrogen production supported by wind power in Sweden. *Applied Energy*, *282*, 116082. DOI: 10.1016/j.apenergy.2020.116082

Nelson, V. (2009). *Wind Energy : Renewable Energy and the Environment.* CRC Press., DOI: 10.1201/9781420075694

Nicholas, T. E. G., Davis, T. P., Federici, F., Leland, J., Patel, B. S., Vincent, C., & Ward, S. H. (2021). Re-examining the role of nuclear fusion in a renewables-based energy mix. *Energy Policy*, *149*, 112043. DOI: 10.1016/j.enpol.2020.112043

Nuclear Power in France—World Nuclear Association. (s. d.). Consulté 30 janvier 2025, à l'adresse https://world-nuclear.org/information-library/country-profiles/countries-a-f/france

Nuclear Power in the USA - World Nuclear Association. (s. d.). Consulté 30 janvier 2025, à l'adresse https://world-nuclear.org/information-library/country-profiles/countries-t-z/usa-nuclear-power

Ohta, H., & Barrett, B. (2023). Politics of climate change and energy policy in Japan : Is green transformation likely? *Earth System Governance*, *17*(100187), 100187. Advance online publication. DOI: 10.1016/j.esg.2023.100187

Ould Amrouche, S., Rekioua, D., Rekioua, T., & Bacha, S. (2016). Overview of energy storage in renewable energy systems. *International Journal of Hydrogen Energy*, *41*(45), 20914–20927. DOI: 10.1016/j.ijhydene.2016.06.243

Pervès, J.-P. (2022). *What are the CO2 emissions of nuclear power?* Polytechnique Insights. https://www.polytechnique-insights.com/en/columns/energy/carbon-footprint-of-nuclear-reactors-in-france/

Reinders, L. J. (2021). What is Nuclear Fusion? In Reinders, L. J. (Ed.), *Sun in a Bottle?... Pie in the Sky!: The Wishful Thinking of Nuclear Fusion Energy* (pp. 1–11). Springer International Publishing., DOI: 10.1007/978-3-030-74734-3_1

Rowinski, M. K., White, T. J., & Zhao, J. (2015). Small and Medium sized Reactors (SMR) : A review of technology. *Renewable & Sustainable Energy Reviews, 44*, 643–656. DOI: 10.1016/j.rser.2015.01.006

Saidi, K., & Omri, A. (2020). Reducing CO2 emissions in OECD countries : Do renewable and nuclear energy matter? *Progress in Nuclear Energy, 126*, 103425. DOI: 10.1016/j.pnucene.2020.103425

Sayed, E. T., Olabi, A. G., Alami, A. H., Radwan, A., Mdallal, A., Rezk, A., & Abdelkareem, M. A. (2023). Renewable Energy and Energy Storage Systems. *Energies, 16*(3), 3. Advance online publication. DOI: 10.3390/en16031415

Siegrist, M., & Visschers, V. H. M. (2013). Acceptance of nuclear power : The Fukushima effect. *Energy Policy, 59*, 112–119. DOI: 10.1016/j.enpol.2012.07.051

Soto, G. H., & Martinez-Cobas, X. (2024). Nuclear energy generation's impact on the CO2 emissions and ecological footprint among European Union countries. *The Science of the Total Environment, 945*, 173844. DOI: 10.1016/j.scitotenv.2024.173844 PMID: 38871309

Suzuki, T. (2019). Nuclear Energy Policy after the Fukushima Nuclear Accident : An Analysis of "Polarized Debate" in Japan. In *Energy Policy*. IntechOpen. DOI: 10.5772/intechopen.83435

Thiri, M. A., & Borsi, M. T. (2024). From Fukushima to fossil fuels : Carbon emissions, climate narratives, and grassroots movements in Japan's energy transition. *Energy Research & Social Science, 112*, 103520. DOI: 10.1016/j.erss.2024.103520

Vijayan, P. K., & Gangwar, S. (2024). Intrinsically Safe Thermohydraulic Designs for SMRs : Design advantages and challenges. *IEEE Electrification Magazine, 12*(4), 75-83. *IEEE Electrification Magazine*. Advance online publication. DOI: 10.1109/MELE.2024.3473332

Visschers, V. H. M., & Wallquist, L. (2013). Nuclear power before and after Fukushima : The relations between acceptance, ambivalence and knowledge. *Journal of Environmental Psychology, 36*, 77–86. DOI: 10.1016/j.jenvp.2013.07.007

Yang, X., & Khan, I. (2022). Dynamics among economic growth, urbanization, and environmental sustainability in IEA countries : The role of industry value-added. *Environmental Science and Pollution Research International, 29*(3), 4116–4127. DOI: 10.1007/s11356-021-16000-z PMID: 34402019

Zarębski, P., & Katarzyński, D. (2023). Small Modular Reactors (SMRs) as a Solution for Renewable Energy Gaps : Spatial Analysis for Polish Strategy. *Energies*, *16*(18), 18. Advance online publication. DOI: 10.3390/en16186491

Chapter 2
Nuclear Energy's Role in Global Energy Transition

Amitava Podder
https://orcid.org/0009-0004-9268-3781
Brainware University, India

Piyal Roy
https://orcid.org/0009-0007-5378-2789
Brainware University, India

Shivnath Ghosh
https://orcid.org/0000-0001-7368-7139
Brainware University, India

Anupama Sen
https://orcid.org/0009-0008-4806-0738
Brainware University, India

Subrata Paul
https://orcid.org/0000-0001-5920-0654
Brainware University, India

ABSTRACT

As more people across the world gain access to electricity then these sustainable energy models need to be managed effectively. This chapter presents an overview of common modern trends, technologies, and heutagogy in energy optimization based on fundamentals of power minimization and incorporation of renewable energy sources such as solar, wind, biomass, and geothermal. It gives importance to combined and comprehensive systems for better energy availability and productivity plus smart grid and energy storage in energy supply and utilization. Optimization techniques,

DOI: 10.4018/979-8-3373-0685-8.ch002

such as deterministic and stochastic approaches together with multi-objective optimization are provided for enhanced performance. Many aspects of technical, economical, policy, and environmental nature of sustainable energy optimization are divided and discussed, giving a more extensive perspective. AI and machine learning represent further perspectives as well as decentralized energy systems, energy as a service, and the circular economy.

1. INTRODUCTION

The dynamism in the energy systems is one of the most critical drivers in the increasing global push for sustainability. Considering the global emergencies of climate change and depletion of resources, seeking for efficient solutions for sustainable energy supply is one of the ultimate goals of political leaders, scientists and engineers (Prasad, Venkatramanan, & Singh, 2020). It is of great significance that extant literature has identified the optimization of these sustainable energy systems as critical in meeting energy demands in the most efficient, reliable, and sustainable fashion possible. This chapter specifically focuses on the detailed analysis of the best practices concerning the sustainable energy models: its essence, problems, and prospects.

1.1 Sustainable Energy: An Overview

Sustainable energy is the energy that is produced and consumed in a manner that does not limit the ability of the next generation from generating and using energy in the same or even better way (United Nations, 2015). This concept is rather a quite wide one and can be adjusted to such renewable power supplies as solar, wind, biomass, and geothermal power. Unlike the traditional source of energy which are the fossil fuels which are limited in supply and which cause pollution of the environment, renewable energy sources are vast they are also replenishable and they least impact on the environment. The global energy mix has been changing over the years due to innovation in technology and increase in awareness to environment and cost of power. Renewable energy as of this decade has been growing at a very fast rate and largely driven by the solar and wind energy. The IEA stated that the share of renewables that have increased in the last few years and made up 29% of the global power generation in 2021 as compared to 24% in the year 2018. Hence, this transition is not done out of concern with environment issues only but more under energy security and economic stability (International Energy Agency, 2022). However, the shift to renewable energy is not without its problems as noted below. These are technical ones, whereby the renewable energy sources are stochastic, in the sense that they may be available at one point in time, and not available at another; economic barriers, where renewable energy technologies are capital-intensive; policy barriers, given that energy policies in different jurisdictions are complicated. These are not easy tasks to address but they have to be done through the use of technology, incentives, and policy changes. This is why it is possible to refer to optimization as such crucial and vital notion.

1.2 Importance of Optimization in Energy Systems

In the context of energy systems, optimization is defined as the search for the best strategies of using energy resources with the least costs and hassles, and in the most sustainable manner. Application of the optimization in relation to energy can be defined in the context of sustainable design as being the allocation of available resources in a way that results in the production of high levels of energy from renewable resources at small cost and harm to the earth (Anastasovski, 2021). Optimization is needed because of the nature of energy systems and its objectives where its rather hard and also costly to achieve more than one tangible goal without detriments to the other goals such as cost, emission or the need for security. The other reason why optimization plays a crucial role in sustainable energy system is due to the uncertainty of renewable power sources. For instance, solar and wind energy depend on the weather conditions, which means that the two energy sources cannot be relied on to produce electricity throughout the day and night in equal measure Zhang et al., 2020. There is normally variability in energy supply, and thus, to cater for this, there is the use of procedures including the stochastic programming and machine learning algorithms so as to meet the demand for energy regardless of the prevailing uncertainty (Wang & Zhao, 2023). The next key area where optimization plays a crucial importance is the improvement of sustainability economics of energy systems. One of the major challenges linked with the deployment of renewable energy systems include the high capital costs that come with installation of renewable energy systems like photovoltaic systems or wind turbines (Anastasovski, 2021). Coordination is a major factor where optimization is also well applied in the grid integration of renewable energy. As it is now clear that the intermittent nature of renewable energy sources is some of the most pressing issues affecting grid stability and reliability (Lund & Mathiesen, 2009). Optimal resource sharing can make the energy system support the variable nature of renewable resources through demand side management, storage and the efficient grid (Aghaei, & Alizadeh, 2013). In addition, these strategies contribute the grid stability as well as increase the share of renewables within the grid in order to provide the shift towards sustainable future.

1.3 Chapter Objectives and Structure

This chapter seeks to guide the reader through most of the significant models for optimisation of sustainable energy while outlining the highlights, principles, difficulties, ideas, approaches, and methods connected with the problem field. The chapter is divided into sections that enable the reader to follow a flow of understanding the principles of energy optimization, status of sustainable energy systems, and the techniques used to improve the energy systems. The initial part of the chapter begins with presenting the fundamentals of optimization, kinds of optimization problems existed in the energy systems, and the mathematical notion of the varying optimization approaches. This section also covers the objective functions that are used in energy optimization namely cost, emission and energy efficiency functions as well as the constraints that are considered in sustainable energy models which include technical constraints, economic constraints and environmental constraints. The second division of the chapter gives more information about sustainable energy systems including details on renewable energy sources, solar energy, wind energy, biomass energy, and geothermal energy. Here, we specifically discuss merits and issues attached to each renewable power source and ways in which optimization can be used to increment their validity. The third section of the chapter is the Hybrid and Integration system where several renewable energy sources are integrated for the development of more secure energy systems.The fourth section of the chapter focuses in smart grids and energy storage as the key players in efficiency and management of sustainable energy systems. However, this section focuses on the applications of smart grids for dealing with variability and uncertainty of RE sources and the various kinds of energy storage systems that can be used for optimizing supply and demand balances in sustainable energy systems. The fifth section of the chapter concerns advanced methods of optimization and these are deterministic, stochastic, and heuristic. This section gives a brief information on the new improvements in optimization algorithms and strategies used in sustainable energy systems; how these can be helpful in solving complicated energy optimization issues. The sixth sub section of the chapter describes and discusses some case studies in optimization in sustainable energy systems for elucidating real world applications of the optimization techniques in the solar, wind and hybrid energy systems. These case allows understanding the real life issues and potential of improving sustainable energy systems. The last part of this chapter analyzes the problems and prospects of further development in the field of sustainable energy optimization. This section looks at the opportunities and barriers, technical, economic and policy, that need to be addressed in order to harness sustainable energy, and the trends and technologies that can be expected in the next few years.

2. KEY CONCEPTS IN ENERGY OPTIMIZATION

Energy management is among the most significant factors for the creation of the more sustainable future energy systems. Optimization strategies are used as a means of improving the performance, reliability and unit costs of energy systems with the view of making energy production, distribution and utilization as efficient as is practically feasible. In this section, the basic concept of optimization is looked at, the more general types of objective functions utilized in energy models are discussed, and the constraints to be placed on these models are also described. It is important for the management of energy to have an understanding of the above noted key concepts in order to formulate and apply efficient energy optimization plans.

2.1 Fundamentals of Optimization

Optimization is a complex mathematical and computational procedure that aims at altering value of one or more variables so as to get the best solution in regards to a specified objective function. In terms of energy systems, optimisation has to do with finding how the structures, functions and use of energy resources should be setup with a view to achieving specific goals including reduction of costs, enhancement of efficiency or reduction of impacts on the environment among others (Financial Ratio Analysis in Launchpad Learning Labs Pvt Ltd, 2019). To understand the underpinning of optimization, it is helpful to give credit to operations research and mathematical programming where one can find all the tools and methods to address the optimization issue at hand or other related optimization problems. At its core, an optimization problem consists of three main components: The objective function, decision variables and constraints form the basis of optimization problem according to Boyd and Vandenberghe (2004). This function is used to capture the optimization goal or the aim of the optimization exercise such as the lowest cost of energy generation or the highest energy utilization. Decision variables are the factors which can be controlled in the optimization process such as the amount of energy produced by various types of resources or at given intervals of time. Constraints are the conditions which should be met during the optimization process, for example available resources, technical characteristics, or legal provisions (Financial Ratio Analysis in Launchpad Learning Labs Pvt Ltd, 2019). There are quite several classification of the optimization process depending on the nature of the objective, the decision variable, and constraints and among them are as follows: There are two main categories of optimization; linear optimization where there are linear relationships between the decision variables and the objective function and nonlinear optimization where; there are non-linear relationships between the decision variables and the objective function (Bazaraa, Sherali, & Shetty, 2005). Another categorization is deterministic and stochastic which is based on understanding that all parameters and variables are well-known in deterministic optimization and no account is taken of uncertainty and variability in the system in the stochastic optimization (Birge & Louveaux, 2011). Optimisation methods can also be grouped into classical and those that are the new generation known as heuristic. Some of the traditional approaches include linear programming which aims at solving optimization problems through mathematical algorithms, quadratic programming, dynamic programming among others; (Boyd & Vandenberghe 2004). These methods are particularly applicable to well formulated problems characterized by pre-defined parameters and conditions. Nondhowever, an number of real-word energy systems subject the optimization proglems to be much more complicated, nonlinear and, involving a number of decision variable and constraint within the system. However, in such cases, several heuristic techniques of contemporary optima search such as genetic algorithm, simulated annealing, and particle swarm optimization (Financial Ratio Analysis in Launchpad Learning Labs Pvt Ltd, 2019). These methods do not always help to reach the global optimum of the problem but are able to provide solution of adequate quality in terms of reasonable amount of computational time, which makes them useful when the problem is large-scale and complex (Nature-Inspired Optimization Algorithms, 2014). In the given context of sustainable energy systems, optimization is related to all aspects of energy production, distribution as well as consumption. For example, in renewable energy systems, optimization is used in selecting the location and dimensions of renewable energy resources likes solar power and wind generators that may produce the highest amount of energy and at the same time with the least cost and environmental effects. In smart grid system application, optimization methods are used for controlling and balancing electricity flow throughout the grid to meet consumers' demands at minimal energy loss and keeping the grid stability (Aghaei & Alizadeh, 2013). Furthermore, optimization helps energy storage systems to decide the charging and discharging policies of power storage equipment including the batteries, in order to attain adequate energy supply and minimize the utilization of fossil fuels.

Optimization is not confined to the technical aspect, but it include the economic and policy optimization as well. For example, optimization can be applied in determining policies of charging for energy that will encourage people to use energy during times of little demand thus reducing the peak loads which are not efficient in the provision of energy (Fleten & Pettersen, 2005). Optimization models, for example, in policy-making enable the decision makers to compare actual trade-offs between for instance, the subsidy to renewable energy or carbon price, in order to establish the most economical and socially acceptable solution (Loulou et al., 2005).

2.2 Common Objective Functions

In any energy optimization model the 'objective function' is simply another term for the formula or equation that defines what the optimization procedure is designed to accomplish. It is understood that different objective functions could be employed depending on the general system interests and necessities of the energy system that is being optimised. Nonetheless, there are several objective functions that are quite popular when used in sustainable energy models These are identified by the major objectives of cost, maximum utilization of system's and environmental impact. The most widely used objective function in energy optimization is cost function where the aim is to minimize cost. Cost and the management of energy production, distribution and usage is a main factor of concern to energy providers consumers as well as policy makers (Bazaraa, et al., 2013). In renewable energy systems the objectives may include cost control which can be a minimization of capital costs, operational costs and costs of maintenance (Anastasovski, 2021). For example, the objective function in a solar energy system could be to minimise the cost per kilowatt hour of electricity generated depending on the geographic position and orientation of the solar panels as well as their size (Sonar, 2021). Likewise, in a wind energy system, cost minimisation may mean finding the number, capacity and locations of turbines that would provide the least LCOE (Zhang et al., 2020).

Another example of an objective function is efficiency maximization that defines the idea to enhance the given energy systems as much as possible with the specific goal of achieving the maximum energy output to a given input (Boyd & Vandenberghe, 2004). It is most relevant in renewable energy systems because the sources of energy like solar and wind cannot produce energy at desired rates thereby resulting in lower efficiency if the systems are not well designed. For instance, in a system that integrates solar and wind power, parameters of the objective function may involve increasing the total efficiency of the system through effective interaction between the two energy sources and based on the weather conditions, demand for energy or, perhaps power storage capacity (Deb 2019). refers to adjustment of the grids to get the most out of them in terms of energy flow, for example, in smart grids efficiency may mean minimizing energy losses while delivering energy to the consumers in the most efficient manner possible (Aghaei & Alizadeh, 2013). Environmental loss reduction is an another significant objective function which is emphasizing the environmental factor and its necessity for sustainable energy optimization in the planning as well as policy formulation stage (Loulou et al., 2005). This objective function is intended to avoid environmental impacts on energy systems by minimizing the emission of green house gases, strike a balance between land and water consumption, and minimization of other unfavorable environmental impacts of energy production and energy use (Zhang et al., 2020). Minimizing the environmental impacts of renewable energy systems may entail course of action and adjustments to the structure and management of the energy resources to reduce CO2 emissions, the use of non-renewable resources, and impact on the local ecosystem (Sonar, 2021). For instance, in the biomass energy system, the objective may include minimizing the samples of the GHG emission which could be a function of feedstock type, conversion efficiency and byproducts management (Anastasovski, 2021).

Multi-objective optimization is a quite effective methodology which permits the formulation and solution of problems with several objective functions, which is quite natural in the context of sustainable energy systems (Financial Ratio Analysis in Launchpad Learning Labs Pvt Ltd, 2019). In multi-objective optimization, instead of establishing a detailed solution it aims on discovering a set of Pareto solution where no one of the solution can be improved without reducing other. For instance, objective functions in a multiple objective optimization model of wind energy system could be to minimize the cost of produced energy, maximize the efficiency of energy produced and minimize the environmental cost of wind turbines (Bazaraa et al., 2013). The optimization process would then try and find a way of achieving all these objectives – essentially this means trying to find trade-offs between the above objectives with a view of presenting decision makers with a number of solutions from which they can make their choices depending on their priorities (Birge & Louveaux, 2011).

Besides these standard objective functions there may be other particular objectives that have to be considered as objective functions depending on the general purpose of the certain energy system. For instance, in the energy storage systems, the objective function might focus on enhancing reliability and robustness of energy distribution by properly solving charge-discharge control problems of the energy storage devices (Yu Ru, Kleissl, & Martinez, 2012). In DSM management, the objective function could be to work the energy consumption profile in order to reduce the power during periods of peak load and enhance total system performance.

2.3 Constraints in Energy Models

Constraints are an essential part of energy optimization models as these are the conditions which should be met and limitations which have to be complied with during the optimization process. Of these, the following constraints are considered as the ones pertaining to the technical, economic, environmental and regulatory aspects of the systems that are applied in the energy sector. Real-life engineering problems are always bounded by constraints, and these form an essential part of the feasible solution set that the optimization models need to produce in order to come up with realistic solutions that are implementable. Resource availability constraints are often the most conventional among the constrains applied to energy optimization models.

This constraint can be associated with the availability of energy reserves including fossil and renewable, energy generation and storage capabilities (Aghaei & Alizadeh, 2013). In renewable energy systems, the resource availability constraints may include; limits of the amount of solar radiation or wind speed or the biomass feedstock that is available for energy conversion (Anastasovski, 2021). For instance in a solar energy optimization model, the resource availably may restrict the amount of the sunlight that can be harnessed by the solar panels depending on the location of the solar panel or the direction, position and orientation of the panel, or even climatic factors such as weather (Zhang et al., 2020). In the same manner, in a wind energy optimization model the resource availably constraint may restrain the amount of wind energy that is available for extraction through the use of wind turbines subjected to factors such as wind speed, turbine capacity, and site circumstances (Sonar, 2021). About technical constraints, they are also defining an integral part of the optimization of energy models, which takes into account the characteristics of physical and technical impossibility of energy systems (Financial Ratio Analysis in Launchpad Learning Labs Pvt Ltd, 2019). Examples of technical constraints are the limit of the capacity on the generation units, the energy conversion efficiency rate and the steadiness and reliability of the energy network (Birge & Louveaux, 2011). For instance, technical constraints of a smart grid optimization model can be the allowable limit of power flow in the transmission line, the voltage and frequency control of the grid, and capability of storage systems (Aghaei & Alizadeh, 2013). These constraint sets help in filtering the optimal solutions and only those solutions are considered which are technically plausible and can maintain the safety and reliability of the energy system. Other constraints include environmental constraints which are incorporated into the energy optimization models due to the environmental effects that are associated with energy production and use (Loulou et al., 2005). Limitation could be in the form of environmental where there are restrictions on Green house gas emissions, the use of land and water as well as other environmental effects related to energy systems. For instance, in a biomass energy optimization model, constraints can be; the level of GHG emissions in the conversion process, the use of non-renewable source and effects on the ecosystem (Anastasovski, 2021). The constraints guarantee that the optimization can be conducted within the economic, technical and environmental realms in order to arrive at sound solutions. Cost considerations are also other essential elements in energy models of optimization because of the demands that need to be met at minimal costs by producers, distributors and consumers of energy (Bazaraa et al., 2013). The available financial resources, cost of energy and bearing market demands are some of the examples of economical limitations to growth (Boyd & Vandenberghe, 2004). For instance in renewable energy optimization model, will constraints related to capital and operational costs of energy resources, availability of subsidies and incentives, effects of energy price on consumers etc., may act as constraint (Fleten & Pettersen, 2005). It is essential for these constraints to limit the solutions obtained by optimization process, not only for technical and environmental acceptability but for economical too. It is also important in energy optimization models that a unit and its related equipment have to adhere to the laws and policies of the state and/or country (Loulou et al., 2005). The regulatory constraints may be requirements in emissions, renewable energy integration, and grid connection among others (Sonar, 2021).

3. RENEWABLE ENERGY SYSTEMS

Renewable energy systems form one of the most critical catalysts to the realization of sustainable energy society. They use natural occurrences and materials to produce power which makes them environmentally friendly as compared to the fossil energy sources. In this section the basics of solar power, wind power, bio mass, and geothermal energy are also discussed. Each subsection outlines general and detailed information on the basics of the specific form of renewable energy sources with challenges and opportunities presented by their implementation dealt with as well.

3.1 Solar Energy: Basics and Challenges

Solar energy is widely used of all of the renewable energy sources, owing to its availability and the trends that have emerged, making solar energy more affordable. Solar power is one of the renewable energy sources which entails harnessing of solar power through the use of solar panels and then converting the solar power into electricity or heat. Solar photovoltaic (PV) systems employ semiconductor materials to generate electricity whereas solar thermal system converts sunlight directly into heat that can for instance be used for water and space heating (Green et al., 2018). The basic component and working principle of solar PV system can be understood with reference to photovoltaic cells formed from materials like silicon. Interconnection of these cells occurs in a way that when the cell gets exposed to sunshine, it generates electricity since the sunshine causes the electron to behave in a certain manner. New innovations in the PV technology have however seen the introduction of different types of solar cells such as mono, multi, and thin film, every one of them possessing its usefulness. Monocrystalline cells have one of the highest efficiencies and lasting effects while thin film cells are light weight and versatile for use in various applications. Nevertheless, solar energy has the following challenges; One major disadvantage of solar energy is that power generation from solar panels is not constant as it relies on the availability of sunlight of which it is influenced by time of the day, weather or geographical location. This intermittency has implications on the need for efficient energy storage options and all round balancing of the energy grid. The one weakness that is connected with the use of solar energy is the cost that is needed to put up the solar panels. The cost of a solar PV system has moderately declined in the last one decade and therefore the upfront cost involved in the purchase and installation of solar panels remains high. Such a cost may be prohibitive and a hindrance especially for the use in residential areas and for small-scale developers (Financial Ratio Analysis in Launchpad Learning Labs Pvt Ltd, 2019). Further, the fabrication of solar panel consumes materials that have effects on the environment for instance in the extraction of raw materials for use in the production of solar panels as well as the manufacturing processes result in production of wastes.

These problems are however being tackled progressively by technological innovation. New technologies in pv modules, for example, bifacial modules that are sensitive to the light on both surfaces and improved energy storage technologies including coupled batteries and thermal energy storage are raising the efficacy and stability of solar electricity systems. In addition, established policies and another form of monetary incentives like tax credits and rebates are aiding in breaking the financial thresholds for solar energy use and thus fostering its utilization.

3.2 Wind Energy: Key Factors

Wind energy is another well-known renewable energy source which harness the kinetic energy of wind in to electricity production. Wind turbines harness mechanical energy of wind through the creation of a rotary motion that can mechanically turn a generator and generate electricity. Wind power has become a more popular source of energy in the last few years due to the opportunities it offers to supply cheap, clean and abundant electricity (Mehta, Jha, & Liang, 2020).

Some of the important factors affecting the wind power comprise of wind velocity, type of turbine and the area chosen. Another characteristic is the wind speed because the capacity of the wind turbine to produce energy is dependent on the cube of the wind speed. Thus, it can be seen that power output increases exponentially with wind speeds. Wind turbines are usually fixes in places with stable and fast wind speed levels like near the seashore, mountain ridges or anywhere that is flat (Martin & Rice, 2021). Several factors relating to the wind turbine structure or configuration including; the blade dimensions, the tower heighth the generator type also influence the ability of the wind turbine in generating electrical power (Hamilton, 2014). However, one of the significant issues that are frequently linked to the use of wind energy is that wind is a rather unpredictable resource. As with the case of solar energy, wind energy has discrete nature, as the availability of wind varies depending on weather conditions and the time of day. This volatilities necessitates the need for adequate grid integration and energy storage systems for management of supply and demand as well as to maintain a reliable power supply (Current issues in operational planning (of power systems), 1992). However, wind farming also brings about environmental and social effects including; noise, visual effects and impact on wildlife. That is why the appropriate assessment of the site and measures to reduce the impact on populations and the environment are needed (Pereira, Salgueiro, & Mesquita, 2017). The following are other factors involved in wind energy development; Economic factors Another factor that is involved in wind energy development is economic factors. One of the largest drawbacks is the fact that the initial cost of establishing turbines which convert wind into electricity is high while in the long run the benefits which include minor maintenance costs, and electrical energy make up for the initial outlay many folds. Tax credits including production tax credits as well as the renewable energy certificates is a means through which the cost threshold can be decreased and hence promoting the wind energy sector (Kruger & Eberhard, 2018).

3.3 Biomass and Geothermal: Opportunities

Biomass and geothermal energy are two other categories of renewable energy and offering good prospects and uses. Bio energy may also be known as biomass energy which entails the usage of biological materials in order to generate heat, electrical energy or bio fuels. Geothermal power utilizes heat occurring from within the earth to produce electricity and as a direct heat source. The following are the benefits of biomass energy; Biomass materials can be transformed into bio-fuels like ethanol and biodiesel which are green substitutes to gasoline and diesel in transport. Biomass can also be utilized in generation of electricity by either direct combustion, gasification or through the process of anaerobic digestion. The generation of electricity using biomass also has the advantage in that biomass feedstocks can be stockpiled and used as and when required (Jingjing, Xing, DeLaquil, & Larson, 2001). There is also an ability of biomass energy to lessen greenhouse gases and in turn enforce carbonless energy or neutrality. This how biomass is said to be a renewable source of energy because for each unit of carbon dioxide that is emitted by combustion of biomass, an equivalent unit was absorbed by plants while in the process of growth and development. However, the usability of biomass energy is very sensitive to the source of the feed, as well as, the social costs of sourcing it and processing. For instance, biomass crops used for feedstock may result in land Use conversion and competition with food crops, while the burning of biomass maybe result in emissions of air pollutants together with particulate matter as analyzed (Du et al., 2023). The technologies identified in the assessment have the potential of producing both electricity and useful heat. Geothermal power stations harness the heat of the ground in the creation of steam to turn the turbines to generate electricity. Geothermal energy is available in areas where there are high geothermal activities like volcanic areas and in areas of overlying of tectonic plates (Lund et al., 2010). Direct use applications include; heating of buildings, plant cultivation under greenhouses, and industrial uses of the heat .

Certainly an attraction of geothermal energy is that of reliability and predictability. The energy characteristics of geothermal energy differ from solar and wind energies which are invariable and generate power in fixed amount irrespective of the weather conditions and time of day. Also, the cost of generating power by geothermal power plants is relatively small, and most of these power plants last for a long time (Lund et al., 2010). But know that geothermal energy has its own limitations as well. Geothermal power entails large capital investment and involves tasks such an drilling of geothermal wells, constructing power stations, among others. As the sources of geothermal energy, heat resources are limited to geographical basin and not all countries can exploit geothermal energy due to prevailing conditions (Lund et al., 2010). Secondly, there are negative environmental effects which may arise as a result of the extraction of the geothermal fluids such as subsidence and emission of trace gases and minerals.

4. HYBRID AND INTEGRATED ENERGY SYSTEMS

Microgrids and assimilated energy systems are gradually considered significant solutions in the context of the reliable and sustainable energy projects. These systems involve integration of various sources of energy and technologies in a manner that gives maximum output, reliability and respect to the environment. This section aims at introducing basic notions on hybrid energy systems and discussing possibilities of integration and presents examples of hybrid models.

4.1 Understanding Hybrid Energy Systems

Integrated Renewable Energy Systems or Integrated Hybrid Energy Systems are intended to incorporate both the conventional and renewable resources to improve the performance of the system. Hybrid systems, unlike single-source systems, utilize the best in multiple source of energy to overcome their shortcoming and offer the society a more effective means of energy supply .For instance, the combined solar-wind system combines the solar photovoltaic (PV) panels and the wind turbines because the two are examples of intermittent resources. Solar energy will be expensive during sunny periods while wind energy will be expensive when there isfavorable wind conditions. When used together, these two types of sources can ensure a more constant and continuous supply of power benefitting the system against cases of power failure that comes with using a single type of source (Muhammad Lawan & Azlan Wan Zainal Abidin, 2020). A similar type of system is the combined use of energy storage systems with renewable energy systems. BM can use resources like battery storage devices, battery or pumped hydro storage to store energy produced in time with high renewable output and utilize it in time with low generation. This integration makes renewable energy sources to be complementarity in ways that reduce reliance on renewable power production thus increasing the reliability and flexibility of the energy systems (Veerendra et al., 2024).

Further, hybrid energy systems may also comprise of integration of various forms of renewable energy sources. For example, a solar-biomass system combines solar PV with biomass energy so as to attain a steady and reliable source of energy with less reliance on the conventional sources of energy. Biomass energy can help to support solar energy by power generation during the night or when solar power insufficient because of bad weather (Caliskan, 2015). The use of hybrid energy systems brings out some considerations such as the type of energy sources and their characteristics, the design of the system and its interaction with other energy systems. The detailed modeling and optimization procedures are used to study the efficiency of the hybrid systems and to determine the optimum configuration for the intended application of the hybrid system (Xie et al., 2019).

4.2 Benefits of Integration

Hybrid energy systems where more than one energy systems are incorporated into one system, has the following advantages that make energy supply efficient, reliable and sustainable. The most important and obvious advantage of integration is in the sphere of the energy security, or rather, the stability of the energy supply. System integration can be achieved by building different systems that have varying generation patterns so that variations in production from one or several systems can be compensated by other systems. For instance, although solar and wind energy source are stochastic and their generation is irregular, careful amalgamation of the stochastic renewable sources with power storage systems can help to level the irregular method of power generation (Veerendra et al., 2024).

The last advantage of using hybrid systems is the increased effectiveness of energy usage, Since the hybrid systems use both renewable and non-renewable energy sources, the consumption of energy is optimized in the hybrid systems. Hybrid systems are found to be efficient than single-source systems since they make use of more than one energy source in a way that this one is optimized to the other. For instance, a solar-wind that is an integrated system can be able to benefit from situations whereby, solar power is being generated in abundance while the wind is not blowing, and the reverse can be true as well. The optimization of energy flow in complex societies could mean better utilization of the resources available as well as decrease of the frequency in the relying on the backup power supplies; (Muhammad Lawan & Azlan Wan Zainal Abidin, 2020). Renewable energy sources form a part of hybrid systems and such a system can also enhance environmental sustainability. Hybrid systems can lower greenhouse gas emission and the effects of energy generation to the environment if integrated with renewable sources of energy storage or other low-emission technologies. For instance, an integrated solar-wind-battery system helps the elimination of the standard fossil fuel-dependent peaking power plant and decreases the carbon intensity of the energy generation (Caliskan, 2015).

There are also economic advantages of Hybrid energy systems. This is because the utilization of a number of technologies for generating energy in a renewable power system can be cheaper as compared to the single technology. For instance, integration of storage with renewable sources can eliminate the need for costly upgrades of grid network and bring down the costs for energy thus providing cheaper solutions (Xie et al., 2019). Also, hybrid systems make energy security more effective through the availability of a variety of energy sources and widgets.

4.3 Case Examples of Hybrid Models

There are some general case studies given below, which depicts the successful integration of hybrid energy systems in various applications and geographical parts of the world. These examples show that the best and worst outcomes of multisourcing the energy electricity or utilizing renewable and conventional sources of energy technology. The latter one can be illustrated by the combined solar, wind and battery power of the Samsø island, Denmark. This system includes solar photovoltaic plant, wind power plant wind and battery power storage system to constitute a secure and renewable energy platform to the island populace. Incorporation of the solar and wind systems capture energy from both sources, while the battery storage enables provision of energy from the sources at the time of low generation (Sperling, 2017). The presented profile of the Samsø demonstrates that hybrid systems can successfully contribute to high rates of renewable power plant applications as well as minimal use of fossil fuel sources.

An example is the hybrid solar-diesel system that has been implemented in the off-grid remote communities across the India. This system uses combine solar photovoltaic with diesel generation for efficient power generation for places that are not connected to the main power supply. The solar PV generates power in the daytime while the diesel generators are used at night time or in the time of high power consumption (Elhadidy, 2002). Integrating solar and diesel supply systems has led to enhancement of fuel utilization and overall costs, besides addressing issues to do with reliability of the power supply in the areas of demand. In the USA, the integration of both the solar PV and the natural gas in the hybrid power plants has emerged as a mechanism of increasing the flexibility of the conventional power generation. The combined solar gas system includes the integration of solar PV modules with natural gas turbines in order to generate dispatchable and largely clean power. In situations when the amount of solar power is adequate or high, the solar PV panel delivers electric power to the electricity grid while the natural gas turbines can be idle. With flexibility taken to mean variable operation the natural gas turbines can always be called upon to serve customer demand during low solar generation (Gambini & Vellini, 2019). Accounting for all of these significantly contributes to getting the best out of renewable energy with due regard to the resiliency in the grid and the levels of emissions in the environment. The combined geothermal biomass power plant which has been put into practice in Iceland is another good example of the integration of decoupled systems. This system integrates geothermal heat to heat only biomass combustion processes for district heating and power generation. The geothermal has thermal energy input which is constant and could be relied upon unlike the biomass which can be utilized during peak demands as it act as a back up to the geothermal systems (Harrison, Mortimer, & Smarason, 1990). However integration of geothermal and biomass technologies is an effective way of improving the overall performance and reliability of the energy system as well as decreasing the reliance on imported fuel. As illustrated by above discussed case examples, hybrid energy systems have a wide variety of uses and can provide numerous advantages. Both systems demonstrate how the successful incorporation of several energy types is possible and can solve certain problems in order to provide an effective, stable, and renewable power supply. Such case studies will still be useful as it allows for the identification of important lessons that can be applied to further development of hybrid energy systems.

5. SMART GRIDS AND ENERGY STORAGE

Change in energy systems has brought in new technologies and approaches aimed at improving the processes, availability and sustainability. That is why smart grids and energy storage systems have become the keystone of this process, providing superior potential for the effective control of energy supply and demand. This section focuses on smart grid and discusses possibilities of energy storage, separate methods of optimization for grid management.

5.1 Role of Smart Grids

Smart grids can be described as a major improvement in the generation, distribution and consumption of electricity. Smart grids are distinguished from the conventional electrical power grid, organized as a one-way broadcast system from electric generation facilities to consumers, in that smart grid systems provide two-way interactive communications with all components of the relevant electrical power system. The incorporation of information and communication technology (ICT) with the power grid enables monitoring, control and optimization of energy flows in real-time Zhang et al., 2020. Smart grid is another essential part of the whole utility, which main purpose is to increase the reliability of the energy delivery process. Smart grids employ intelligent sensors and communication to keep record of the state of the grid in real-time manner. This capability helps utilities to be able to identify and act timely on events like equipment malfunctions, supply disruptions, and mismatch of demand and supply. Smart grid, which relay information occurring on the grid in real-time, offers proper management of grids and quick maintenance and getting over the breakdowns (Miller et al., 2018). There are other functions of smart grid: one of them is to increase the share of RES in energy consumption. Wind energy and solar energy are in most cases unpredictable and unstable and this is not good for the stability of the grid systems. These challenges are solved by smart grids through the use of better forecasting and control systems in relation to variability of renewable resources. For instance, smart grids can manage the dynamic of conventional power stations or provide energy storage to level the supply-demand interaction in response to variation in the renewable production (Gijsenbergh et al., 2021). There is also something called as demand response management in smart grid. The smart grid can encourage the consumers to change their consumption pattern through the demand response programs. For example, when demand is high and/or supply low the consumers can be given incentives to either conserve energy or use it during off-peak periods. This DSM contributes to alignment of demand with supply and in minimizing the pressure on the grid hence the need for more power generating plants (Fang et al., 2012). Smart grids also help in enhancing the energy efficiency as well as its conservation. Smart grids also by using smart meters and other feedback tools give consumers the necessary information about their energy consumption, so they can make rational decisions. This increased visibility a has positive impacts, such that there will be conservation of energy and hence low energy costs (Zhang et al., 2020).

5.2 Energy Storage Solutions

Energy storage systems are critical components of the contemporary energy systems, which solve the key issues associated with renewable energy sources and help to increase the stability and versatility of power systems. Different types of energy storage technologies are beneficial and used in diverse ways; all plays a significant role in making energy distribution stable and efficient. Energy storage is a key component that ensures the smooth operation of renewable energy system and one of the most frequently used methods for energy storage include battery storage. These include Lithium-ion batteries, lead-acid, and flow batteries which convert electrical energy to chemical energy and this is released when required. Lithium-ion batteries are preferred more than other batteries since they provide high energy density, longer cycle life and reduced cost. These have found wide application including portable electronics, electric vehicles and large scale storage systems (energy storage) as pointed out by Tarascon and Armand in 2001. In contrast, flow batteries possess such properties as scalability and superior cycle life that allows their usage in large-scale applications (Li et al., 2018). The second storage method involves pumped hydro storage which works with the help of gravitational potential energy to store electricity. In stored pumped hydro energy, water is transferred to an upper reservoir during periods that demand is low or generation is in excess. And when power is required, gate open up releasing the stored water to flow through turbines thereby producing electricity. Pumped hydro storage is perhaps one of the oldest technology of storage and the most used for large scale storage with long durations (Schmidt et al., 2019). Another method include the use of compressed air energy storage(CAES) which aims at storing energy in compressed air form. In the low energy consumption periods, electricity is employed to compress air then it is stored in large cavities such as the underground cavities or containers. Whenever there is need for energy the compressed air is released and expanded in order to turn the turbines and thus generate power. CAES systems have the possibility of large storage capacities and can maintain balance and contribute to the integration of renewable sources (Veerendra et al., 2024). The last one is the thermal energy storage which is the process of storing thermal energy. Main classifications of TES consist of sensible heat storage, phase change heat storage and thermochemical storage. Sensible heat storage takes a material, be it water or molten salt, and raising its temperature to later use the heat energy. Latent heat storage makes use of phase change materials (PCMs), materials that gain or release heat when changing phase – solid to liquid or vice versa, thermochemical storage on the other hand involves storing energy through chemical reactions (Zalba et al., 2003). As it will be noted, each energy storage solution comes with its benefits and is therefore ideal for specific applications. Battery storage is suitable for applications that require a short cycling duration since the technology offers quick response times. The pumped hydro system and CAES provide large scale with longer duration storage solutions while Thermal energy storage provides solutions for heat integration with electricity generation. The latter clearly depends on the scale, duration, cost, and the application of the energy storage technology in question.

5.3 Optimization Techniques in Grids

Integration of renewable resources in the grid requires that high levels of performance be obtained in order to reduce the overall costs of generation and delivery. To this end, several optimization strategies are used as a way of realizing such goals which include employing sophisticated modeling, control approaches, and computational methods. Network optimization is one of the most utilized techniques employed in the grid management. This entails improvement of the efficiency and reliability of the electrical network through the enhancement of the design and operational parameters of the electrical network. Network optimization techniques include load flow analysis which identifies the flow of electrical power to and from other networks and the reconfiguration of the network which entails a change in the structure of the network in order to minimize losses and improve reliability. Advanced calculation and analysis methods are used to solve specific network optimization issues to guarantee a proper functioning of the grid. One of the other critical optimization methods is the demand response programme. Demand response optimization deals with the control of consumers energy consumption to reflect those of the grid for purposes of reducing the peak load. Some of the methods include dynamic pricing, incentives, programs and real time feedback which tend to influence the consumers to change their usage pattern as dictated by the grid. Decentralized users utilize OPF to recognize the best Demand Response strategies that would serve both the consumers and the grid operators in the most optimal way (Fang et al., 2012).

Energy storage optimization is also another favorable strategic management in grid leadership. Strategies for managing storage systems include the way to charge and discharge the storage to get maximum benefit from it. This encompasses reduction of costs in raw material and other inputs used, optimization of the use of renewable energy, and guaranteeing that storage systems are adequately capable of providing backup power in cases of need. Realization of energy storage for effective operation is achieved through optimization control strategies, predictive algorithms, and other approaches (Veerendra et al., 2024).

Another optimization difficulty is encountered when using renewable energy sources in the framework of the electricity network. Methods like renewable energy forecasting, load balancing and integrated system management are used to address issues of variability and intermittency of generation from renewable sources. Forecasting methods rely on historical records and mathematical formulas to predict future generation of renewable energy on the other hand grid balancing employs mathematical methodologies of conventional generation and storages to balance with renewable production. In this context, hybrid system optimization deals with the management of integrated renewable electricity supply system and energy storage systems for the best results in terms of the system performance and dependability (Gijsenbergh et al., 2021).

6. ADVANCED OPTIMIZATION TECHNIQUES

Optimization strategies are very noteworthy when it comes to the improvement of diversity and performance of numerous systems especially when it comes to energy systems and or grid systems. These techniques include a class of tools comprised of a variety of approaches formulated to find solutions to diverse difficulties aimed at maximizing certain goals or objectives. This section focuses on detailed deterministic methods, stochastic and heuristic methodology combined, as well as multi-objective optimization principles, and explains each identified approach.

6.1 Deterministic Methods Overview

There is also the deterministic approach to optimization, which is derived from mathematical programming; the idea behind the deterministic approach is the solution of optimization problems with known constraints and objectives. These methods depend on mathematical models and computational procedures to achieve the best solution that exists, in the best of worlds all parameters are fixed in time and do not alter at unpredictable occasion. Linear programming (LP) is one of the most important methodologies falling under the category of deterministic methods and it revolves round the optimization of a linear criterion function under linear equality as well as inequality constraints. The LP is used in a number of different scenarios such as resource distribution, transportation and production. One of the most conventional methods for solving LP problems is the simplex algorithm, which was introduced by George Dantzig; it works by going through different feasible solutions, which are better than the previous one until an optimal solution is found (Dantzig, 1963). Concerning problems with integer or binary variables, Integer Programming (IP) and Mixed-Integer Programming (MIP), broaden the use of LP with discrete decision variables Wolsey (1999).Another deterministic method is that of quadratic programming (QP) where one has to find a solution that minimizes a quadratic function of decision variables while satisfying linear constraints. In implementing QP, the objective function contains quadratic terms since costs or risks to be optimized and controlled are quadratic in the general portfolio optimization and control problems (Nocedal & Wright, 2006). The interior-point method is a second widely used algorithm in solving QP problems due to its continual seeking of a better solution within the feasible region as proposed by Kuhn & Tucker in 1951.

Further, there is Dynamic programming (DP), which is another type of deterministic optimization approach applied in case if the problem can be divided into overlapping subproblems. Sequential decision making problems such as inventory control and scheduling therefore benefit greatly from implementation of DP. Dynamic programming is based on the Bellman equation which defines the optimal solution in terms of solutions on the sub-problems as originnal as 1957.

Linear and quadratic programming, dynamic programming belong to the set of deterministic optimization methods which are exact and suitable for solving definite problems. These methods are effective approach to solving optimization problems in almost all fields including engineering and finance.

6.2 Stochastic and Heuristic Approaches

As opposed to the deterministic methods which are intended to work out the simpler optimal problems, the stochastic and the heuristic are aimed to solve problems with optimizations when the amount of uncertainty or complexity is too large to be dealt with. They use random and approximate strategies to solve problems in an acceptable period while identifying the optimal solution is not the aim.

There are two primary types of optimization techniques: deterministic and stochastic; while deterministic methods solely use computational algorithms for optimization, stochastic optimization methods also included randomness in the optimization accord to variability/uncertainty in problem parameters. A major method under this category is stochastic programming whereby one aims at achieving optimize an objective function affected by randomness constraints. Stochastic programming employs either scenarios or distribution to describe the uncertainties and solutions are gotten from the expected results (Birge & Louveaux, 2011). For instance, in energy planning scenario, stochastic programming can be used to factor variability in sources of renewable energy and demand forecast.

The second stochastic method is Monte Carlo simulation where the technique involves producing a large number of random samples to estimate the distribution of the results. For measuring the performance of the optimization solutions and to determine the effect of uncertainty on the decisions it is important to apply Monte Carlo simulation (Metropolis & Ulam, 1949). This technique is often implemented in financial and risk assessments as well as reliability calculations. Heuristic optimization techniques are the other type of optimization technique since they rely on iteration and exploration to arrive at good solutions. These methods do not guarantee an optimal solution as in the exact approaches but is intended to find good, practical and accurate solutions in acceptable amount of time. Heuristic algorithms are of three types, namely greedy algorithms, local search and simulated annealing. Greedy algorithms, after constructing a solution, make the locally optimum choice at every step of the solution. For instance in the traveling salesman problem, there exists a greedy approach in which the shortest edge that is available is chosen at each step to form the tour. Restrictions of greedy algorithms include the fact that they are easy to implement, but can only find local optimum solutions as opposed to the global optimum solutions as noted by Korte and Vygen (2008). Neighborhood based local search algorithms refine a previous solution within the vicinity of solutions of the solution space. Local search strategies including for example the hill climbing and the tabu search which aims at moving from one solution to another in iterative manner while rejecting the ones that were chosen before (Glover & Laguna, 1997). Local search is useful when the problem has an extremely large number of feasible solutions so as to make the use of an exact approach impractical.

This is optimization algorithm known as simulated annealing which is a heuristic optimization technique that mimics the annealing process in metallurgy. It entails iterative and adaptive search such that with each move, one accepts both improving as well as non–improving solutions but with reducing probabilities. Inherently, simulated annealing is attractive for the purposes of escaping local optima and visiting various parts of solution space, for the task of optimizing complex problems (Kirkpatrick et al., 1983). Stochastic & heuristic approaches give very useful resources for the problems that it is difficult to solve with the exact solution.

6.3 Multi-objective Optimization Basics

Multi-objective optimization is the process of optimizing many objectives at the same time and in this case the objectives conflict with each other. In decision making problems many a times decision makers have to work with multiple objectives, like cost, benefit, time, etc., and this can be minimized using this technique. The techniques that are used for multi-objective optimization try to find a solution that is a compromise among the objectives which is obtainable according to the principles of Pareto. What comes to multi-objective optimization, Pareto optimality is the key point. According to the definition of Pareto optimal solution it is a solution for which one cannot find a better solution where at least one of the objectives is better off without being worse off according to the criteria used. Because the goal is to maximize at least one criterion while minimizing all other criteria the set of all Pareto optimal solutions is known as the Pareto front (Deb, 2001). It is possible to map the Pareto front and thereby facilitate awareness of the potential trade-offs and the ability to choose based on the decision maker's preferences. Among all the approaches to solve the multi-objective optimization problems, one of the most frequently utilized methods is the weighted sum method. This approach is one of the most typical as it merges several goals into one by assigning them coefficients. The most frequently used approach to deal with them is the weighted sum method which reduces the multiple objective problem into a single objective problem so as to be able to solve it by the traditional optimization techniques. However, there is a problem in selecting weights, which may affect the performance greatly and does not necessarily include all characteristics of the Pareto frontier (Miettinen, 1999).

The other popular method is the ε-constraint method whereby one objective function is optimized while the rest are regarded as constraints with certain allowed levels. This approach is usefully to set up the model and allowing the explore of different tradeoffs by changing the bounds of the constraints. The ε-constraint method is always good when you are looking for many, different Pareto optimal solutions and/or when you want to investigate various parts of the solution space (Chankong & Haimes, 1983). Multi-objective optimization is also solved using evolutionary algorithms. These algorithms include the Non-dominated Sorting Genetic Algorithm II (NSGA-II) and the Strength Pareto Evolutionary Algorithm (SPEA2), which apply the principles of the evolution to search for solution space and obtain the Pareto optimal solution. Genetic algorithms as part of the evolutionary algorithms are appropriate for the cases, where the size of the solution space is large and there are several objectives for the optimization. Multiple criterion optimization methods clearly offer various methods in case where there are two and more objectives and therefore different decisions are required. Through the comparisons of such objectives, these techniques help decision makers to find out the proper solutions on the basis of certain needs and preferences.

7. CASE STUDIES

In order to illustrate the practical concepts and approaches to optimization of energy systems, case studies are used. Through observation of case studies it will be easier to explain how optimization solutions improve performance, minimize costs and increase the use of renewable resources. In this section, solar energy optimization case, wind energy optimization case, and good hybrid system case are discussed focusing on some research results and methodologies of each example.

7.1 Solar Energy Optimization Example

In order to illustrate the practical concepts and approaches to optimization of energy systems, case studies are used. Through observation of case studies it will be easier to explain how optimization solutions improve performance, minimize costs and increase the use of renewable resources. In this section, solar energy optimization case, wind energy optimization case, and good hybrid system case are discussed focusing on some research results and methodologies of each example.

Solar energy optimization entails increasing the rate of return on investments in terms of efficiency of the solar panels, energy output and cost of operation. Future research efforts should more thoroughly consider the practical aspects of integration issues inlarge scale renewable energy plants, as demonstrated by a case of the solar photovoltaic (PV) optimization system in the Desert Sunlight Solar Farm in California, US. It's one of the biggest solar farms in the world and has been able to incorporate complex performance optimization strategies to improve on the efficiency and solidity of the system.

First Solar's Desert Sunlight Solar Farm, which was mentioned above, also uses several forms of optimization to increase output of energy. There is one area that has seen a major optimization drive; the use of tracking systems in orientation of the panels' angles to match those of the sun at various points of the day. This technique referred to as the solar tracking makes sure that the maximum field of the solar panels is exposed to light thereby producing more energy. It has single-axis as well as bi-axis tracking so as to position the panels to capture as much power as possible (First Solar, 2018). Besides tracking systems, the solar farm is equipped with sophisticated monitoring and controlling systems for the efficient running of the solar farm. These systems collect and respond to real time data of the weather condition, the temperatures of the panels and the energy they are generating.

Another optimization strategy used at the facility of Desert Sunlight depend on the use of predictive maintenance methods. Consequently, the farm can plan the maintenance activities with reference to equipment failure history and then use predictive modeling to work out possible future failures. It also means that the approach also plays a major role in minimizing situations when the system experiences unplanned downtimes and in keeping operational reliability at an optimal level. Machine learning algorithms are used for predictive maintenance where performance data is analyzed in order to forecast possible problems (Kusiak et al., 2010). The recommendation of the use of advanced optimization techniques in the Desert Sunlight Solar Farm is justified from the success achieved. The approaches such as the use of solar tracking, real-time monitoring, and predictive maintenance has helped the farm improved dramatically in energy production and operation. The case has revealed the possibility of harnessing optimization to improve operational performances of large-scale integrated solar PV systems and the lessons learnt from this work can be of immense value to other large-scale systems.

7.2 Wind Energy Optimization Example

The other optimization strategy that is being practiced at Desert Sunlight is use of predictive maintenance. This farm movements concern for carrying out maintenance activities, such that through data analysis and simulation likelihood of equipment failures the farm anticipate on them. This approach that is taken reduces the probability of the system running into a lot of problems hence increasing the operational reliability of the system. Predictive maintenance is based on the machine learning that involves analysis of performance data to predict the occurrence of failures (Kusiak et al., 2010).

The success story of the Desert Sunlight Solar Farm provides a good evidence for application of optimal techniques in solar power systems. Thus, by applying solar tracking, real-time monitoring, and predective maintenance the farm has received a substantial increase in energy yield and operational effectiveness. This paper gives one possible example of how the performance of large scale SPV systems can be improved and thus has useful lessons for comparable projects.

The wind farm also uses enhanced meteorological condition that is beneficial for the operation of turbines. The farm also effectively utilizes current and previous weather forecast models in order to quite accurately anticipate wind speed and direction. This information is then utilized to vary settings of turbine together with timing of the maintenance. Forecasting is used in the wind farm to increase the dependability of the wind farm by reducing the downtime and increasing the production of power by increasing the accuracy of the forecasts (Hodge et al., 2012).

There is another optimization going on at Horns Rev 2 and it is wake management. Interference occurs where there are several wind turbines and is characterized by the wake of one turbine impacting that of the others that are nearby. Because wake effects can be detrimental to the efficiency of the wind farm, the wind farm uses wake modeling together with wake optimization to deal with interaction between turbines. Wake turbulence is something that a wind farm needs to address so there should be efforts to increase the efficiency in the positioning and management of the turbines in order to enhance power production (Hansen et al., 2007).

The benefits of the advanced optimization strategies for the Horns Rev 2 Offshore Wind Farm are well explained in the case analysis. Some of the solutions that have been implemented at the farm include; accurate turbine control systems, meteorological forecasting, and wake management which led to efficiency in energy generation. From this example, we can learn more about the use of optimization in the future offshore wind energy constructions and possible usage of the similar approaches to other wind constructions.

7.3 Hybrid System Success Stories

That is why, the use of combined hybrid energy systems, using several energy sources with the highest efficiency and reliability, can be considered as one of the most prospective solutions for the sustainable energy production. One huge success that has emerged in this field is the Sumbawa Island project in Indonesian implemented by the hybrid energy system. In this project, solar PVs, wind energy and diesel generation are used in an integrated manner so as to produce the power needed for the island.

This study is about the use of hybrid energy system for electrification for Sumbawa Island where the power distribution is a challenge due to the unavailability of the power grid in the region. The system entails the use of solar photovoltaic power and wind power systems together with a diesel power generator to supply the energy demand of the island. The use of the hybrid energy system also promotes diversification in the supply of renewable energy that ensures energy dependence on diesel fuel is reduced thus minimizing its effects on the environment (Nolan et al., 2016). Another of the most important techniques applied at the Sumbawa Island project involves the utilization of an energy management system in order control the interaction of the several energy sources. The EMS oversee the data that is produced and consumed in the energy systems as well as the storage systems and the decision-making on which energy to use is based on the collected data. For instance, if, during periods of high generation from the solar and wind facilities, the EMS is able to reduce the operation of the diesel generator in order to reduce the consumption of fuel and the emission of greenhouse gases (Giebel et al., 2019).

The hybrid system also has an energy storage feature to boost the reliability and make levels of energy generation more balanced. Battery storage systems for the excess energy during fiber optic generation from the solar PV and wind turbines developed elsewhere can be used during low fiber optics generation or high demand. This way it achieves market equilibrium and the need to have a constant and secure power supply for the island (Chatzivasileiadis et al., 2013). The success of the Sumbawa Island hybrid energy shows that the integration of multiple energy sources for the same system can effectively maximize the performance and reliability of such system. The several techniques of generation implemented by the project with details of solar PV, wind energy, and diesel generation together with the energy management and storage technologies provides a sustainable energy supply to the concerned remote community. In this case, the authors explain the advantages of such hybrid systems and share useful experience that may be used to develop similar projects in other countries.

8. CHALLENGES IN SUSTAINABLE ENERGY OPTIMIZATION

Sustainable energy optimization strive to improve the productivity of energy systems with consideration to the negative impacts and expenses it will incur to the environment. But these goals are strategies that come with several difficulties in technical, economic, policy, and environmental spectrum. This section covers the challenges faced in the sustainable energy optimization; technical challenges, economic/political challenges and environmental challenges and briefly discusses each challenge.

8.1 Technical Challenges

The improvement of efficient renewable energy systems is met with several technical barriers which may hinder the process and efficiency. These challenges include the ability to connect renewable energy sources, the fact that storage capacity in energy sectors have risen to new heights and the use of enhanced control and monitoring systems.

The first technical issue that merits a mention is the decision as to how renewable energy sources are distributed into the existing energy grids. For instance, some sources of renewable energy like solar or wind energy is described by variability and intermittency, which poses a problem regarding the constant energy availability. To accommodate these sources, there should be more complex grid management process that has to meet the supply-demand balance and manage variations in energy production (Jenkins et al ., 2018). These challenges require application of Accurate forecasting techniques, online measurements, and demand side management techniques to maintain the stability of the electricity distribution grid. Another of the major technical issues in sustainable energy optimization that concerns researchers is the problem of energy storage. Batteries, like pumped hydro storage play an important role in the balancing of renewable power sources and ensuring that the same levels of energy is produced. However, Call-centre technologies have some issues, with regards to its performance, cost, and durability. For example, the lithium-ion batteries are commonly used as energy storage and they are costly and their energy density and cycles lives are restricted (Nagaura & Tozawa, 1990). Efforts are currently being carried out to explore new forms of storage technologies including the solid state batteries and flow batteries (Tarascon & Armand, 2001). The control and monitoring systems which are involved in the development and implementation of advanced control and monitoring systems also present technical risks. Such systems are useful for enhancing the productivity of renewable energy systems and thereafter the management of electricity energy. However, the design and development of these systems need professionals with analytical capabilities, machine learning, and system integration. Control strategies therefore have to allow for flexibility because renewable energy sources can fluctuate depending on certain facets of performance and may require the application of certain control mechanisms in real time (Hodge & Milligan, 2011). In addition, the integration of these systems to the existing framework might be challenging and needs huge capital investments in equipment and human resource.

8.2 Economic and Policy Barriers

Barriers stem from economic and policy perspectives have been shown to exert large influence over the topology of sustainable energy management. These barriers affect the applicability and growth of renewable energy technologies as well as determine investments. The major economic factor is the initial cost that is incurred when it comes to the adoption of renewable energy technologies and installation of renewable energy systems. The advantages of the green power are visible in the long-run while the cost of harnessing renewable energy such as through solar power or wind power, energy storage systems and other technologies can be expensive at the onset. This high capital cost could be a discouragement factor towards investors and developers especially in restricted regions of financial capability or restricted finance sources. This can only happen if subsidies, tax incentives, and financial support policies to the development of renewable energy projects are developed (Sawin et al., 2015). In the same regard, the policy barriers are also responsible for determining the use of renewable energy technologies. This is due to the fact that sometimes policies when not well developed or tend to be irregular could cause a number of concerns that can hamper investment cases in sustainable energy projects. For instance, the fact that specific renewable energy targets, policies and incentives, as well as associated regulatory frameworks are not well-defined might deter stakeholders from investing in renewable energy (Barbose et al., 2016).

8.3 Environmental Considerations

Environmental concerns play a very crucial role in the management of energy as the purpose of the optimization is to enhance energy use while at the same time having minimal effects on the environment. Some of the environmental concerns that are important in energy technologies are lifecycle footprints of the technologies, disruption of land and habitats, and consumption of resources.

The environmental cost of technologies in the energy sector looks at the effects on the environment throughout the lifecycle of the technologies beginning from the processes used in developing them, installation and operations of the technologies and up to the stage where they are removed or dismantled. For instance, manufacturing of the solar panels involves some raw materials that have to be mined through processes that may lead to negative impacts on the environment including exhaustion of natural resources and pollution. Further, the Solar panels has an environmental impact at their disposal and recycling stage when once they are overused or get to their end of useful life (Choi et al., 2019). Likewise, usage of wind power entails costs in manufacturing and installing wind turbines and these affect local environments (Meyer et al., 2014). A fundamental prerequisite for real and effective life cycle optimization of energy technologies is still the evaluation and minimization of their overall impacts during their entire life cycle. It also affects land use and habitat which are two crucial regulations to environment concerned issues. Thus, large-scale facilities for the production of electric energy from renewable sources, including solar and wind stations, entail considerable area of land usage, thus impacting on ecosystems. For instance, solar farms change the patterns of land utilization and affect plant life and animals; wind power in form of wind turbines is lethal to birds and bats (Klein et al ., 2019). These effects call for proper assessments of the sites for the renewable energy projects, Nigeria cannot afford to have renewable energy projects that have negative impacts on the environment. Another environmental factor is resource utilization or use here in a relative to the scarcity or availability of elements incorporated into energy technologies. For instance, development of advanced and efficient photovoltaic systems needs rare earth elements and metals which may come with the supply chain vulnerabilities and the effects of processing and mining (Jha et al., 2015).

9. FUTURE DIRECTIONS IN ENERGY OPTIMIZATION

It is therefore important to manage energy systems towards sustainability especially as the world sees unprecedented demand for energy. Modern trends, technological developments, and the combined capacity of artificial intelligence (AI) and machine learning bring a new direction to the energy optimization. This section covers these future directions, thereby developing insights into the current trends and giving the future possibilities which can contribute toward energy optimization. The industry related to energy optimization is dynamic because of the opportunities stimulated by information technology and the demand for environment friendly solutions. There are several trends which are emerging as well as technologies that are in the process of developing that will be very instrumental in the future of energy systems. These are; the enhancing of renewable energy sources, increasing on flexibility through the emergence of smart grid systems and coming up with improved energy storage systems. Another major trend of the energy optimization is the progressive development of the renewable sources of energy. Renewable energy sources such as solar photovoltaic (PV) and wind energy have registered significant enhancement in conversion efficiency and affordability in the last decade. As the intertidal and ocean depths open for renewable energy production, innovations such as bifacial solar panels, which collect light on the front and back of a solar panel, and floating wind turbines to utilised fairly vast areas (Kang et al., 2020; Ponta et al., 2017). Also, the developments of the new solar cell material like perovskite solar cell is expected to contribute to increase efficiency and decrease cost of solar energy (Green et al., 2018). Another emerging technology is smart grids that is a combination of digital technologies applied to improve the distribution of electric power. Smart grids are based on sensors and technologies for communicating and analyzing the current and future state of the electrical network (Podder et al., 2025). This also ensures greater incorporation of renewable power supplies, optimal demand side management and enhanced network stability (Zhou et al ., 2018).Energy storage systems, on the other hand, are also developing fast with an aim of trying to overcome the intermittency problem associated with the renewable energy sources. New types of battery technologies like solid-state batteries and flow batteries provide higher energy density, longer life durability and enhanced safety as compared to the standard lithium-ion batteries (Tarascon & Armand, 2001). Furthermore, various studies are conducted on emerging storage types, including compressed air energy storage and pumped hydro storage that present new prospects for enhancing energy systems Veerendra et al., 2024). Due to these qualities, adoption these emerging technologies in energy system has a potential of raising efficiency levels while at the same time lowering cost and increase sustainable production and consumption of energy. Therefore, further research and development in these areas will be paramount in order to enhance the effectiveness of these technologies as well as in handling the other issues related to efficient energy usage. This is important as Artificial intelligence (AI) and machine learning can be used in the field of energy optimization to help analyze data, predict performance and have enhanced methods for analysis. These technologies have the potential to bring development on the various elements of the energy systems, and thus bring improvements on the efficiency of various aspects in energy management. Data mining and knowledge discovery technologies have potential benefits in energy systems because of their ability to make use of massive datasets in order to generate precise forecasting on and promote better decision making. For example, machine learning algorithms can help in predicting the demand of energy, generation plan and also in identifying a faulty equipment. These types of models are therefore useful as they allow historical data to be reviewed and patterns to be identified that help to improve the functioning of a system (Goodfellow et al., 2016).AI is also used in determining efficiency in energy usage with an example being the coming up with a predictive maintenance plan. Predictive maintenance allows improving maintenance operations based on the data collected by sensors and equipment using such machine learning algorithms as deep learning and others. This aspect make it possible to have proper scheduled preventive and predictive actions that helps in minimizing periods of equipment stand still and increases their service life (Goh et al., 2019).AI also helps to control the efficiency of renewable energy systems as well as assists in their management. In solar PV systems, AI algorithms can utilise weather conditions, or characteristic and performance of the panel to position and regulate the positioning of the solar panels. Likewise in wind energy system AI can be used for fine tuning the control settings of the wind turbine for capturing better energies through adjusting blade angle according to the seen wind conditions. AI and machine learning improves energy optimization drastically in many ways including efficiency, cost of service, and reliability. Yet, this process also imposes challenges that are linked to issues of data protection, the lack of open algorithm disclosure, and high demands for professionals who work in this field. Solving these issues will be critical for the development of AI and ML as approaches to energy management. Sustainability in energy models for the future has been seen to involve higher levels of integration, robustness and versatility. This change includes the integration of the diverse instruments of technology, commercial structures, and policies for a smarter and environmentally friendly power system. Nowadays there are some important trends and changes that influence the further development of sustainable energy systems. One trend is toward decentralization of energy systems which have become very important in today's world. In traditional energy patterns, generation and distribution of energy such as electricity are mainly controlled by power stations that provide the power needs of a large region. On the other hand, the decentralized energy systems aim at generation and utilization of energy required at the place or location where it is needed most. Such strategies involve creation of distributed energy resources (DERs) like the roof-top solar systems, micro-wind power plants, and miniature electricity storage networks. Another emerging effective model of the future sustainable energy is also visible with the focus on the energy-as-a-service model. Energy as a service actually means selling energy services as opposed to energy itself. The new business model for these utilities came from transforming the energy provider's traditional focus on owning and operating critical pieces of energy infrastructure to delivering personalized energy services. For instance, the opportunities may include provision of energy management services, performance commitment, and outcome-based agreements to deliver energy services that do not require initial infrastructure investment according to Singh et al. (2022). This model is in line with the emerging needs of a more customer-oriented and versatile electricity supply as well as the development of renewable energy technologies. Another crucial trend is connected with the circular economy concepts in energy systems. The concepts of circular economy include making the least use of new resources in as many loops as possible and reusing and recycling them. In the case of energy systems it entails creation of goods, devices and processes that are energy efficient, durable and those that ease the process of recycling or reusing them. For instance, the reutilization of components from end-of-life solar photovoltaic panels and wind turbine blades can assist in attaining a circular and sound energy system and minimise the pollution of energy technologies (Bauwens et al., 2016). Energy system interactions with other sectors, which include transportation and buildings systems are also other important aspects of future sustainable energy models. Integration of the energy, transportation and building management systems mean that they can be effectively optimized simultaneously. For instance, EVs can connect within the energy network as both load and storage devices for the realization of V2G_scheme for leveling out supply and demand (Kempton & Tomić, 2005). Likewise, buildings integrated with smart technology such as energy management system, are capable of managing energy usage, as well as engage with the grid for demand response and energy efficiency (Hong et al., 2009). Sustainable energy models, presented in this paper, will further develop based on different aspects such as technology and related business models, and policy frameworks. Therefore, it can be concluded that cooperation with governments and other industries and organizations, including academic institutions, will be crucial in the development of various technologies and the achievement of the set targets for sustainable energy. Thus, following new tendencies, utilizing modern technologies, and using the integrated strategies, the energy sector can develop toward the better future, which is safer, more sustainable, and reasonable.

10. CONCLUSION

Energy optimization over the recent past has been subjected to tremendous changes in growth influencing key aspects such as need for sustainability, technological factor and integration of methodologies. In so doing, there is no doubt that finding ways of improving efficient sustainable energy models is well within reach through the consideration and blend of these factors with the evolving trends and technologies that underpin them. Given the rising importance of the sustainable energy systems, the need to comprehend the basic theories and the trends of the future is crucial.

The advancement in Bio energy technologies including solar, wind, biomass, geothermal and others are some of the indications of improved energy efficiency and sustainability. Advancements in these fields are continuously pushing the envelope of how renewable energy sources can be used to cater for the increasing global energy consumption needs as well as grappling with issues such as variability of the resource base. In addition, the continuity of growth and application of hybrid as well as integrated energy systems emphasizes the integration of several forms of energy and variety of technologies to get better performance in addition to dependability. This means by applying the complementary characteristics of various resources as well as organizing them into unified energy systems, the improved cohesiveness can be provided. Real-life examples prove the effectiveness of hybrid models which illustrates that such approaches could provide efficient and stable sources of energy. Smart grid and energy storage systems are new step forward improvement in energy systems which are more optimal. Smart grids are the advances in the provision of electric energy, by using modern control and monitoring techniques. Energy storage solutions on the other hand help to mitigate the issue of intermittency that is characteristic of some of the renewable energy sources and are instrumental in grid stability and energy management. The further advancement of these technologies will have a central part in determining the direction of energy optimization. Standard optimization methods, random and heuristic solutions, and the innovative methods of the multi-objective optimization form prominent tools for managing energy systems.Advanced techniques are crucial for optimizing energy production, distribution, and consumption, ensuring optimal performance and sustainability. However, challenges in sustainable energy optimization include technical issues like system integration and performance optimization, economic and policy barriers like investment costs and regulatory frameworks, and environmental considerations like the lifecycle impact of energy technologies. In the future, emerging trends and technologies will shape energy optimization landscapes, such as AI and machine learning integration. These technologies offer potential for enhancing efficiency, accuracy, and decision-making, enabling advanced data analysis, predictive maintenance, and optimization strategies. The future of sustainable energy models will be characterized by greater integration, decentralization, and adaptability, with a shift towards decentralized energy systems, energy-as-a-service business models, and circular economy approaches. The convergence of energy systems with other sectors like transportation and buildings will further enhance energy use and management optimization. A holistic approach to energy optimization, considering technical, economic, and environmental factors, is essential for a more sustainable and efficient energy future. By embracing emerging technologies, addressing challenges, and pursuing innovative solutions, we can contribute to a greener and more sustainable world.

REFERENCES

Aghaei, J., & Alizadeh, M.-I. (2013). Demand response in smart electricity grids equipped with renewable energy sources: A review. *Renewable & Sustainable Energy Reviews, 18*, 64–72. DOI: 10.1016/j.rser.2012.09.019

Anastasovski, A. (2021). Improvement of energy efficiency in ethanol production supported with solar thermal energy – A case study. *Journal of Cleaner Production, 278*, 123476. DOI: 10.1016/j.jclepro.2020.123476

Bauwens, T., Huybrechts, B., & Dufays, F. (2016). Understanding the diverse scaling strategies of social enterprises as hybrid organizations: The case of renewable energy cooperatives. *Organization & Environment, 29*(3), 263–283.

Bazaraa, M. S., Sherali, H. D., & Shetty, C. M. (2005). *Nonlinear Programming.* https://doi.org/DOI: 10.1002/0471787779

Birge, J. R., & Louveaux, F. (2011). *Introduction to Stochastic Programming.* Springer New York., DOI: 10.1007/978-1-4614-0237-4

Boyd, S., & Vandenberghe, L. (2004). *Convex Optimization.* https://doi.org/DOI: 10.1017/CBO9780511804441

Fang, Y., & Zhang, W. (2019). Enhancing traffic management efficiency using predictive modeling and AI. *Transportation Research Part C, Emerging Technologies, 101*, 108–122. DOI: 10.1016/j.trc.2019.02.008

Financial Ratio Analysis in Launchpad Learning Labs Pvt Ltd. (2019). *International Journal of Recent Technology and Engineering, 8(2S8)*, 30–35. https://doi.org/DOI: 10.35940/ijrte.B1308.0882S819

Fleten, S.-E., & Pettersen, E. (2005). Constructing Bidding Curves for a Price-Taking Retailer in the Norwegian Electricity Market. *IEEE Transactions on Power Systems, 20*(2), 701–708. DOI: 10.1109/TPWRS.2005.846082

Goh, J., Tjahjono, B., & Brawn, D. (2019). Predictive maintenance: Towards the next generation of maintenance management. *Journal of Industrial Information Integration, 15*, 1–10.

Gong, Q., & Jia, H. (2019). Frameworks for healthcare delivery optimization in smart cities. *Smart Health (Amsterdam, Netherlands), 10*, 100131. DOI: 10.1016/j.smhl.2019.100131

Goodfellow, I., Bengio, Y., & Courville, A. (2016). *Deep Learning.* MIT Press.

Green, M. A., Ho-Baillie, A., & Snaith, H. J. (2018). The emergence of perovskite solar cells. *Nature Photonics*, *8*(7), 506–514. DOI: 10.1038/nphoton.2014.134

Hong, T., Chou, S. K., & Bong, T. Y. (2009). Building simulation: An overview of developments and information sources. *Building and Environment*, *44*(4), 848–858.

Kang, J., Kim, H., Kim, J., & Cho, H. (2020). Recent advances in floating solar photovoltaic systems: A review. *Renewable & Sustainable Energy Reviews*, *110*, 332–346.

Kempton, W., & Tomić, J. (2005). Vehicle-to-grid power implementation: From stabilizing the grid to supporting large-scale renewable energy. *Journal of Power Sources*, *144*(1), 280–294. DOI: 10.1016/j.jpowsour.2004.12.022

Kruger, W., & Eberhard, A. (2018). Renewable energy auctions in sub-Saharan Africa: Comparing the South African, Ugandan, and Zambian Programs. *WIREs Energy and Environment, 7(4)*. Portico. https://doi.org/DOI: 10.1002/wene.295

Loulou, R., Waaub, J.-P., & Zaccour, G. (Eds.). (2005). *Energy and Environment*. Springer US., DOI: 10.1007/b107228

Lund, H., & Mathiesen, B. V. (2009). Energy system analysis of 100% renewable energy systems—The case of Denmark in years 2030 and 2050. *Energy*, *34*(5), 524–531. DOI: 10.1016/j.energy.2008.04.003

Mallick, C. B., Aditya, A. R. N., & Bhattacharya, C. (2020). Adaptive load balancing strategy for a distributed computing environment. *Journal of Cloud Computing: Advances, Systems and Applications*, *9*(1), 1–19. DOI: 10.1186/s13677-020-00197-8

Nature-Inspired Optimization Algorithms. (2014). *Nature-Inspired Optimization Algorithms*. https://doi.org/DOI: 10.1016/B978-0-12-416743-8.00016-6

Podder, A., Roy, P., Ghosh, S., Sarkar, S. K., & Choudhury, R. R. (2025). Citizen Engagement and Smart Governance. In Ahmad, M., Mumtaz, R., & Khan, M. (Eds.), *Leveraging IoT and Machine Learning for Smart Urban Planning* (pp. 35–68). IGI Global Scientific Publishing., DOI: 10.4018/979-8-3693-9030-6.ch002

Pokharel, S. (2007). Kyoto protocol and Nepal's energy sector. *Energy Policy*, *35*(4), 2514–2525. DOI: 10.1016/j.enpol.2006.09.015

Ponta, F. L., Amador, J., & Ponta, O. (2017). *Harvesting ocean energy: Mechanical conversion systems*. Springer.

Pouryousefi, M., & Ashuri, T. (2019). Optimizing multi-objective problems with Pareto Frontier in Renewable Energy Systems. *Energy Reports*, *5*, 198–206. DOI: 10.1016/j.egyr.2019.08.043

Prasad, S., Venkatramanan, V., & Singh, A. (2020). Renewable Energy for a Low-Carbon Future: Policy Perspectives. *Sustainable Bioeconomy*, 267–284. https://doi.org/DOI: 10.1007/978-981-15-7321-7_12

Rad, M. M., & Khosravi, F. (2021). Machine learning for optimal waste management in smart cities. *Renewable & Sustainable Energy Reviews, 137*, 110622. DOI: 10.1016/j.rser.2020.110622

Ru, Y., Kleissl, J., & Martinez, S. (2012). Battery sizing for grid-connected PV systems with fixed minimum charging/discharging time. *2012 American Control Conference (ACC)*, 270–275. https://doi.org/DOI: 10.1109/ACC.2012.6314783

Singh, M., Jiao, J., Klobasa, M., & Frietsch, R. (2022). Servitization of energy sector: Emerging service business models and startup's participation. *Energies, 15*(7), 2705. DOI: 10.3390/en15072705

Soltani, A., & Mohsenian-Rad, H. (2020). Game theory for energy distribution in smart cities. *Energy, 202*, 117784. DOI: 10.1016/j.energy.2020.117784

Sonar, D. (2021). Renewable energy-based trigeneration systems—technologies, challenges, and opportunities. *Renewable-Energy-Driven Future*, 125–168. https://doi.org/DOI: 10.1016/B978-0-12-820539-6.00004-2

Tarascon, J. M., & Armand, M. (2001). Issues and challenges facing rechargeable lithium batteries. *Nature, 414*(6861), 359–367. DOI: 10.1038/35104644 PMID: 11713543

2015. *Transforming governance for the 2030 agenda for sustainable development*. https://doi.org/DOI: 10.18356/e5a72957-en

Veerendra, A. S., Mohamed, M. R. B., & García Márquez, F. P. (2024). Energy management control strategies for energy storage systems of hybrid electric vehicle: A review. *Energy Storage, 6*(1), e573. DOI: 10.1002/est2.573

Verma, O. P., Manik, G., & Sethi, S. K. (2019). A comprehensive review of renewable energy source on energy optimization of black liquor in MSE using steady and dynamic state modeling, simulation and control. *Renewable & Sustainable Energy Reviews, 100*, 90–109. DOI: 10.1016/j.rser.2018.10.002

Wang, Q., & Zhao, L. (2023). Data-driven stochastic robust optimization of sustainable utility system. *Renewable & Sustainable Energy Reviews, 188*, 113841. DOI: 10.1016/j.rser.2023.113841

2022. *World Energy Outlook2022*. World Energy Outlook. https://doi.org/DOI: 10.1787/3a469970-en

Wu, D., & Ehsani, M. (2010). The role of energy storage in high-power hybrid electric vehicles. *Proceedings of the IEEE*, *95*(4), 678–689. DOI: 10.1109/jproc.2007.914016

Zhang, X., & Chien, S. (2018). Machine learning techniques for integrating renewable energy in urban grids. *Renewable Energy*, *119*, 666–675. DOI: 10.1016/j.renene.2017.12.012

Zhang, Y., Gu, C., Yan, X., & Li, F. (2020). Cournot oligopoly game-based local energy trading considering renewable energy uncertainty costs. *Renewable Energy*, *159*, 1117–1127. DOI: 10.1016/j.renene.2020.06.066

Zhou, K., Fu, C., & Yang, S. (2018). Big data driven smart energy management: From big data to big insights. *Renewable & Sustainable Energy Reviews*, *56*, 215–225. DOI: 10.1016/j.rser.2015.11.050

Chapter 3
Introduction to Advanced Nuclear Systems

M. D. Mathew
https://orcid.org/0000-0003-0101-2671
Saintgits College of Engineering, India

Bhanu Sankara Rao
Indian Institute of Technology, Hyderabad, India

ABSTRACT

Nuclear energy is clean energy. It does not produce any greenhouse gases. Hence, it is a game-changing technology in addressing global energy demands, climate change mitigation, and sustainable development. This chapter describes the principles of nuclear fission, and demonstrates that nuclear fission is a mature and efficient method of large-scale electricity production. The different types of commercial thermal reactors such as Pressurized Water Reactors, Boiling Water Reactors, and Pressurized Heavy-Water Reactors, as well as Advanced Gas-Cooled Reactors are discussed and evaluated based on their moderator type, fuel enrichment levels, coolant mechanisms, and safety features. Development of advanced nuclear technologies like Small Modular Reactors and Generation IV reactors is also discussed. Fast reactors and emerging fusion reactor technologies hold great promise to supplement thermal reactors and renewable energy sources to replace fossil fuels to address the challenges of global warming.

DOI: 10.4018/979-8-3373-0685-8.ch003

1. INTRODUCTION

The economic prosperity of a nation and the quality of life of its people are directly related to the amount of energy consumed (Mathew, 2022). The per capita consumption of energy is considered an indicator of the standard of living and industrial development (González & García, 2016). Global electricity generation has been increasing at a rate of about 2.5% per annum (Global energy & CO_2 status report 2019, 2019). By 2023, the total electricity output worldwide had reached about 28,500 terawatt-hours (TWh), with renewable sources accounting for 30% (World Energy & Climate Statistics – Yearbook, 2024). The Asia-Pacific region, particularly China and India, led in electricity demand growth during this decade (World Nuclear Association, 2025). To face the challenges of global warming, energy security, and ecological degradation, nuclear energy plays an important role in a sustainable future (Edenhofer et al., 2014). With advancements in technology and supportive policy frameworks (United Nations Framework Convention on Climate Change, 2015), nuclear energy has emerged as a proven, practically implementable, and environment-friendly alternative in reducing the usage of fossil fuels (Lenzen, 2008; World Nuclear Association, 2025)

2. OVERVIEW OF NUCLEAR ENERGY

Nuclear energy is one of the most efficient sources of energy. It is primarily generated through nuclear fission, a process in which heavy atomic nuclei like Uranium-235 or Plutonium-239 are split when bombarded with neutrons, releasing substantial amounts of energy. For more than sixty years, nuclear technology has played a crucial role in global electricity generation. As of 2025, 10% of the world's electricity is provided by nuclear power, supported by over 400 active reactors operating across over 30 countries with a combined installed capacity of 390 GWe (World Nuclear Association, 2025). Despite concerns surrounding safety and radioactive waste disposal, nuclear power continues to be a significant contributor to efforts aimed at lowering carbon footprint and addressing climate change.

A key advantage of nuclear energy is its ability to produce electric power with virtually no direct greenhouse gas emissions. Unlike traditional fossil fuels such as coal, oil, and natural gas, nuclear power plants do not release carbon dioxide during their operation. Nuclear energy, therefore, is a valuable low-carbon alternative (United Nations Framework Convention on Climate Change, 2015). Moreover, nuclear fuel possesses exceptionally high energy density, enabling small quantities of material to produce vast amounts of power, far greater than the efficiency of other energy sources.

In spite of the advantages, nuclear power faces several challenges. Two nuclear disasters, Chernobyl in 1986 (International Atomic Energy Agency, 1992) and Fukushima in 2011 (International Atomic Energy Agency, 2015) have heightened public fears, and this has led to increased regulatory scrutiny. Moreover, the issue of handling radioactive waste, which has the potential to stay dangerous for thousands of years, presents significant environmental and political hurdles. The large initial investment and lengthy construction periods required for nuclear plants further limit their adoption, especially in less developed nations.

Emerging technologies like Small Modular Reactors (SMRs) and advancements in nuclear fusion research offer promising pathways for the evolution of nuclear energy. These innovations aim to boost safety, cut costs, and solve persistent challenges related to waste and operational efficiency. As global energy demands continue to rise and the challenge to combat climate change grows, nuclear energy is the reliable pathway for transitioning towards a cleaner, and sustainable energy.

2.1 Introduction to Nuclear Fission

Nuclear fission is the fundamental process used to generate energy in nuclear power plants. It involves the splitting of the nucleus of a heavy atom, typically isotopes like uranium-235 or plutonium-239, into two or more lighter nuclei, on bombarding with neutrons (Glasstone, 1955).

Figure 1. The nuclear fission process in which a heavy nucleus absorbs a neutron, becomes unstable, and undergoes fission, releasing a significant amount of energy. This process can initiate a self-sustaining chain reaction if the emitted neutrons induce further fission events. (retrieved from International Atomic Energy Agency, 2022, CC BY 4.0).

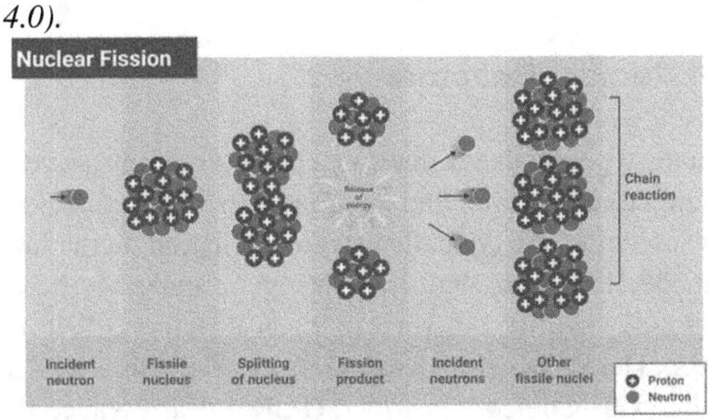

Fig. 1 is a schematic of the nuclear fission process. (International Atomic Energy Agency, 2022). The fission reaction releases a large amount of energy. The released energy primarily manifests as kinetic energy of the fission products and thermal energy (heat). The energy released is harnessed in a nuclear reactor to convert water into steam. The steam then drives turbines connected to generators, and produces electricity. A critical feature of nuclear fission is its ability to sustain a chain reaction. The neutrons produced in each fission event trigger more fission events in nearby atoms, leading to a continuous release of energy. In a nuclear reactor, this chain reaction is precisely regulated using control rods made of materials called nuclear poisons, which can absorb neutrons and thus stabilize a controlled rate of fission to generate a steady amount of electricity.

Nuclear energy offers several advantages. It provides a highly concentrated source of energy; a large amount of electricity can be generated from a small amount of electricity. Nuclear fuel is ~100,000× more energy-dense than coal. For example, nuclear fuel required for a 500 MWe power plant is about 15 tons of 3–5% enriched uranium. A coal power plant of similar capacity will require about two million tons of coal. On the other hand, nuclear fuel does not emit carbon dioxide or other greenhouse gases during nuclear fission, thus making it a green alternative for power generation.

However, nuclear fission presents unique challenges. The reaction generates radioactive byproducts that remain hazardous for thousands of years. This requires secure handling, storage, and disposal. Additionally, the fission chain reaction can lead to overheating and, in extreme cases, reactor meltdowns as seen in accidents at Chernobyl and Fukushima. Despite these risks, ongoing advancements in nuclear technology and safety systems continue to improve the reliability and environmental sustainability of fission-based power generation. Nuclear power has become an imperative to combat global warming.

2.2 Types of Nuclear Reactors

Various reactor types utilize different technological approaches and structural designs to ensure effective, safe, and environmentally sustainable power production. The nuclear reactors are typically categorized based on the fuel type, cooling systems, and moderators used, with each design optimized to balance efficiency, safety, and economic viability. Pressurized Water Reactors (PWRs), Boiling Water Reactors (BWRs), Heavy Water Reactors (HWRs), Fast Breeder Reactors (FBRs), High Temperature Gas Cooled Reactors (HTGRs) and Advanced Modular Reactors (AMRs) are already in the commercial domain or have achieved technological success (World Nuclear Association, 2025; Glasstone, & Sesonske, 1994).

2.3 Thermal Reactors

Thermal reactors are nuclear reactors that sustain a chain of fission reactions using slow-moving "thermal" neutrons. Most neutrons produced in nuclear fission are having an average energy of around 2 MeV (Million electron volts). In a thermal reactor, these neutrons are slowed down (moderated) when they collide with atoms of light elements such as hydrogen in water or carbon in graphite, until they reach thermal energies (~0.025 eV), for which the probability to cause fission in fuel nuclei is high. On the other hand, fast reactors operate with higher-energy neutrons without slowing down and hence with no moderator. Thermal reactor technology has been the cornerstone of nuclear power since the 1950s. Thermal reactors typically use uranium fuel with or without enrichment of the fissile isotope U-235 and rely on materials like water or graphite to moderate neutrons and sustain a stable chain reaction. The decades of proven performance, and improved designs over decades have made them the workhorses of commercial nuclear energy. Thermal reactors are classified primarily by the type of moderator and coolant used, which determine their core design and operating principles. The major categories of thermal reactors are described below.

2.3.1 Pressurized Water Reactors

In PWRs, the fuel is typically uranium dioxide (UO_2) enriched to ~3–5% U-235 (balance U-238), formed into fuel rods and contained in a robust steel reactor pressure vessel. Control rods inserted from the top regulate the fission reaction by absorbing neutrons. PWRs use ordinary light water (H_2O) as the coolant as well as the neutron moderator. Water flowing through the reactor core is kept at very high pressure (≈155 bar) so that it remains liquid at about 300 °C and does not boil.

In a steam generator, the superheated water is circulated, and during this process, the water exchanges its heat with water in a secondary loop. The water in the secondary loop boils and produces steam. The steam runs a turbine, and in that process, it generates electricity, while the primary coolant water is pumped back to the reactor core. Because the water in the core is kept liquid, the reactor and turbine systems are isolated (no radioactive steam in the turbine). Fig. 2 shows a schematic of a typical PWR design (International Atomic Energy Agency, 2022).

Figure 2. Cross-sectional schematic diagram of a Pressurized Water Reactor (PWR). The reactor core (fuel assemblies and control rods) is contained inside a steel reactor pressure vessel filled with high-pressure water. (retrieved from International Atomic Energy Agency, 2022, CC BY 4.0).

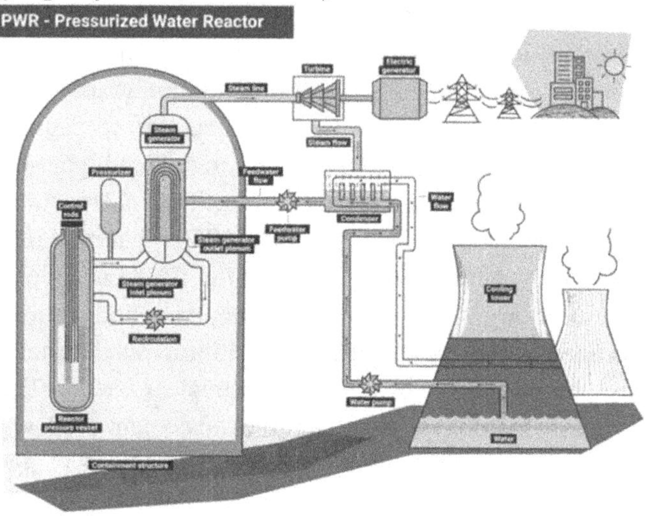

2.3.2 Boiling Water Reactor

Boiling Water Reactors (BWRs) also use slightly enriched uranium fuel (3–5% U-235). BWRs also use light water as a coolant and moderator. In a BWR, the reactor core water is at a slightly lower pressure (~75 bar) and is allowed to boil inside the reactor vessel. Thus, steam is generated directly in the reactor core and sent to the turbine, eliminating the need for steam generators. The steam, after expanding through the turbine, is condensed and pumped back to the reactor as feedwater. One adverse consequence of the BWR design is that the steam and water flowing to the turbine are radioactive (since they come from the core); therefore, shielding and safety systems requirements are more severe. BWRs typically have internal recirculation pumps or jet pumps in the reactor vessel to control water flow and power level. The simpler boiling design allows a more direct energy conversion. Both PWR and BWR designs are collectively referred to as light-water reactors (LWRs) and forms the majority of operating reactors. Fig. 3 shows a schematic of a typical BWR design (BWR – Boiling water reactor, 2014).

Figure 3. Schematic diagram of a boiling water reactor (BWR). The key components are the reactor pressure vessel, nuclear fuel assemblies, control rods, recirculation pumps, and turbine. (retrieved from: BWR – Boiling water reactor, 2014, CC BY 4.0).

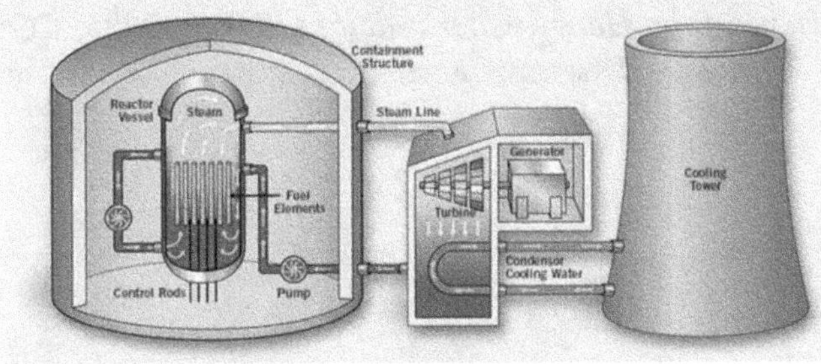

2.3.3 Pressurized Heavy-Water Reactors

Pressurized Heavy-Water Reactors (PHWRs), like the Canadian CANDU design, use heavy water (D_2O) both as the moderator and as the coolant. Heavy water is an excellent moderator (due to its low neutron absorption), which allows PHWRs to use natural uranium fuel (0.7% U-235) without enrichment. In a PHWR, the heavy water, i.e., the moderator, at near atmospheric pressure, is contained in a large cylindrical tank called a calandria. The fuel is contained in hundreds of horizontal pressure tubes that pass through this moderator tank. A separate heavy-water coolant under high pressure (~100 atm) is pumped through the pressure tubes to extract heat from the fuel, reaching around 290 °C, and then transfers that heat to boil light water in steam generators (similar to a PWR). The pressure-tube design enables on-load refueling. Fuel bundles can be replaced in individual channels by fueling machines while the reactor is at power, without the need for shutting down the operations. PHWRs typically produce more energy per kilogram of mined uranium than LWRs, but their use of natural uranium (and lower burn-up) means they generate a larger volume of spent fuel per unit of electricity. Notable examples of PHWRs are the CANDU reactors in Canada and similar heavy-water units in India. A typical PHWR design (schematic) is shown in Fig. 4 (Heavy water reactor (PHWR), 2025).

Figure 4. Schematic of a CANDU-type pressurized heavy-water reactor (PHWR). Heavy water moderator fills the calandria (horizontal tank), slowing neutrons in the core. The fuel is contained in horizontal pressure tubes running through the calandria. (retrieved from Heavy water reactor (PHWR, 2025, CC BY 4.0).

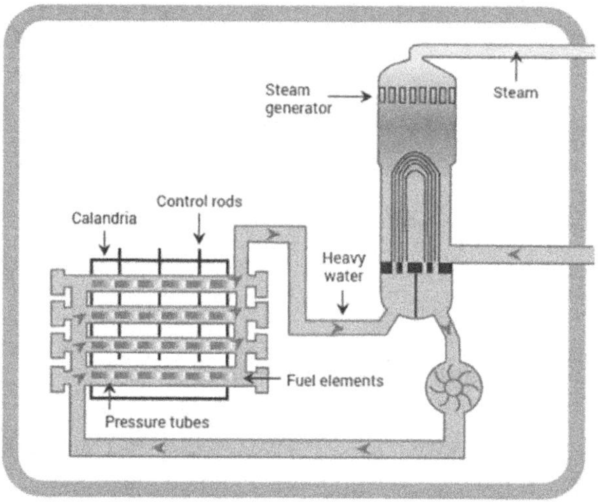

2.3.4 Gas-Cooled Reactor

Gas-Cooled Reactors (GCRs) use gas coolant and a solid graphite moderator. GCRs use carbon dioxide (CO_2) gas as the coolant, circulating through a graphite-moderated core. Fuel is uranium dioxide enriched to about 2.5–3.5% U-235, clad in stainless steel tubes. The CO_2 coolant exits the core at a high temperature (around 650 °C). This higher temperature gives AGRs a thermal efficiency of ~41%, which is greater than that of water-cooled reactors. Like PHWRs, AGRs are designed for online refueling. All existing gas-cooled power reactors are in the UK, and other countries have transitioned to light-water designs. Fig. 5 shows a schematic of a typical GCR design (Gas-cooled reactor and advanced gas-cooled reactor (AGR), 2025).

2.3.5 Graphite-Moderated Boiling Water Reactor

Graphite-Moderated Boiling Water Reactor (RBMK/LWGR) is a Soviet-designed reactor type, known as a light-water graphite reactor (LWGR). It features a large graphite moderator block, and the coolant used is light water in numerous pressure channels running through the graphite block. The water boils inside these channels (tubes), producing steam that drives the turbines (a direct cycle like a BWR). Fuel is slightly enriched uranium dioxide (typically ~2% U-235). The RBMK can be refueled online by extracting and replacing fuel assemblies from the pressure channels using specialized refueling machines. This was the of Chernobyl nuclear power plant. The RBMK design had inherent safety issues such as a positive void coefficient and lack of a full containment structure that were tragically demonstrated by the 1986 Chernobyl accident. Fig. 6 shows a schematic of a typical graphite moderated reactor design (World Nuclear Association, 2025).

Figure 5. Schematic Diagram of a Gas-Cooled Reactor (GCR). The reactor core consists of fuel assemblies containing uranium, surrounded by a graphite moderator. (retrieved from Gas-cooled reactor and advanced gas-cooled reactor (AGR), 2025, CC BY 4.0).

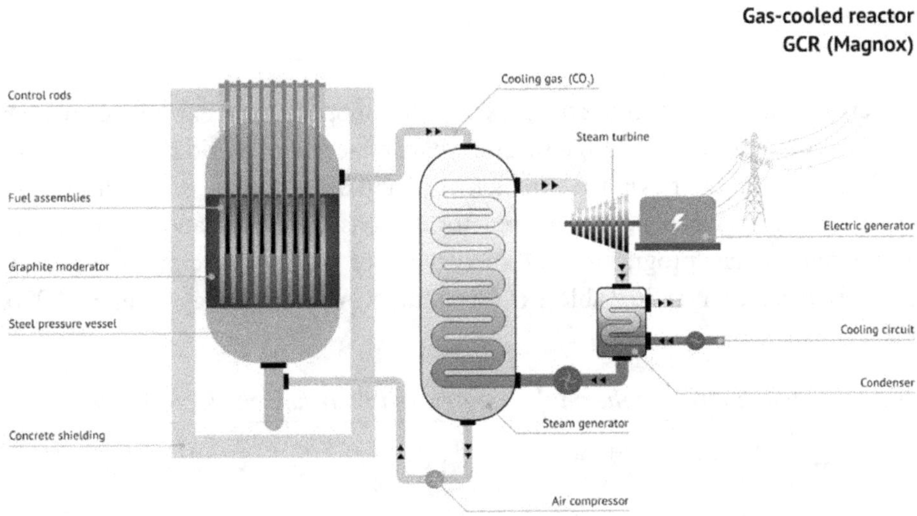

Figure 6. Schematic Diagram of a Graphite-Moderated Light Water Reactor. (retrieved from World Nuclear Association. (2025), CC BY 4.0).

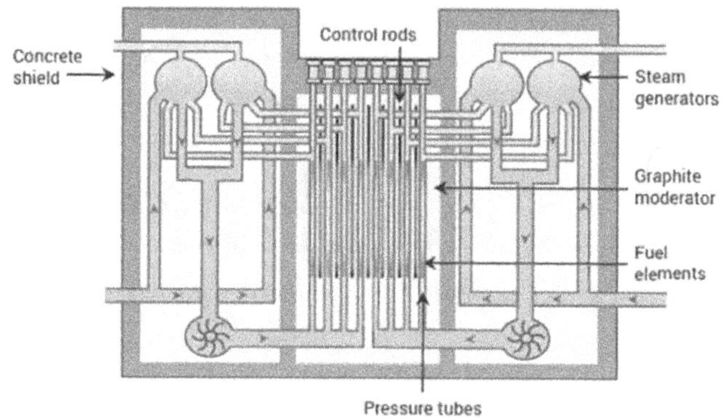

Thermal reactors have more than six decades of operational experience, which is equivalent to more than 18,000+ reactor-years of service (World Nuclear Association, 2025). Given their extensive track record and the broad base of manufacturing and regulatory experience, thermal reactor designs have been adopted by most of the national nuclear programs. Common types of thermal reactors and their key characteristics are given in Table 1 (International Atomic Energy Agency, 2020a).

Table 1. Common types of thermal reactors and their key characteristics.

Reactor Type	Neutron Moderator	Primary Coolant	Typical Fuel & Enrichment	Notable Features
Pressurized Water Reactor (PWR)	Light water (H_2O)	Light water (liquid, high pressure)	UO_2 fuel, ~3–5% U-235 (enriched)	Most widespread type (U.S., France, China). Primary loop kept liquid under ~155 bar; requires steam generator for secondary loop. Robust steel pressure vessel.
Boiling Water Reactor (BWR)	Light water (H_2O)	Light water (boiling in core)	UO_2 fuel, ~3–5% U-235 (enriched)	Boiling in core sends steam directly to turbine (single-loop). Simplified design (no steam generator) but turbine handling radioactive steam. Mostly in U.S., Japan.
Pressurized Heavy-Water Reactor (PHWR)	Heavy water (D_2O)	Heavy water (pressurized)	UO_2 fuel, natural or ~0.7% U-235	Uses *unenriched* uranium fuel (saves enrichment cost). Heavy water moderator in separate calandria; pressure-tube design (no large vessel) allows on-power refueling. CANDU (Canada) and derivatives (India).

continued on following page

Table 1. Continued

Reactor Type	Neutron Moderator	Primary Coolant	Typical Fuel & Enrichment	Notable Features
Advanced Gas-Cooled Reactor (AGR)	Graphite	Carbon dioxide (gas)	UO_2 fuel, ~2–3% U-235 (enriched)	Graphite-moderated, CO_2-cooled. High outlet gas temperature (~650 °C) yields high thermal efficiency (~40%). Stainless steel pressure vessel and fuel cladding. Online refueling. Only in UK (AGR fleet).
RBMK (LWGR) Reactor	Graphite	Light water (boiling in channels)	UO_2 fuel, ~2% U-235 (enriched)	Large core with graphite moderator and many pressure tube channels. Direct steam cycle like BWR. Capable of online refueling. Historical safety issues (Chernobyl). Remaining units only in Russia.

2.4 Challenges and Limitations of Thermal Reactors

Despite their operational success and technological maturity, thermal reactors face several scientific and engineering challenges. These are fuel utilization efficiency, management of radioactive waste, and reactor safety enhancements. The efficiency of fuel utilization in thermal reactors is low. In the once-through fuel cycle employed by most LWRs, only a small fraction of the uranium's energy content undergoes fission and is converted to electricity. The U-235 isotope (which constitutes ~3–5% of fresh enriched fuel) is primarily consumed, along with a portion of U-238 that is converted to plutonium and then fissioned. However, the vast majority of the U-238 (over 90% of the initial uranium) remains unutilized in the spent fuel. Existing thermal reactors extract a few percent of the energy potential of the natural uranium fed into them. By contrast, fast neutron reactors can fission U-238 and thus utilize almost all of the uranium, potentially achieving ~60–70 times more energy yield for the same amount of uranium. The limited fuel utilization in thermal reactors necessitates continual mining and enrichment of new uranium fuel. Strategies to improve fuel burn-up in thermal reactors (such as using higher enrichment, advanced fuel designs, or multi-cycle reuse of plutonium in MOX fuel) are areas of ongoing research. Some countries recycle plutonium from spent fuel into mixed-oxide fuel that can be burned in thermal reactors, and there are proposals for thorium fuel cycles that breed uranium-233 in situ for better utilization. These challenges provide motivation for advanced fuel cycle research and the eventual deployment of more fast reactors.

2.5 Fast Reactors

Fast reactors, also known as fast neutron reactors or fast-spectrum reactors, are a class of nuclear reactors that differ fundamentally from thermal reactors. They utilize high-energy (fast) neutrons to sustain the fission chain reaction. Unlike thermal reactors, which rely on moderators to slow down neutrons, fast reactors operate without a moderator, allowing neutrons to maintain their high kinetic energy. This design (Fig. 7) enables greater fuel efficiency, waste reduction, and resource sustainability (Mathew, 2022, Wikipedia. (2025).

Figure 7. Schematic diagram of a sodium-cooled fast breeder reactor (FBR). FBRs generate more fissile material than it consume, utilizing fast neutrons to sustain the fission chain reaction (retrieved from Wikipedia. (2025), CC BY 4.0).

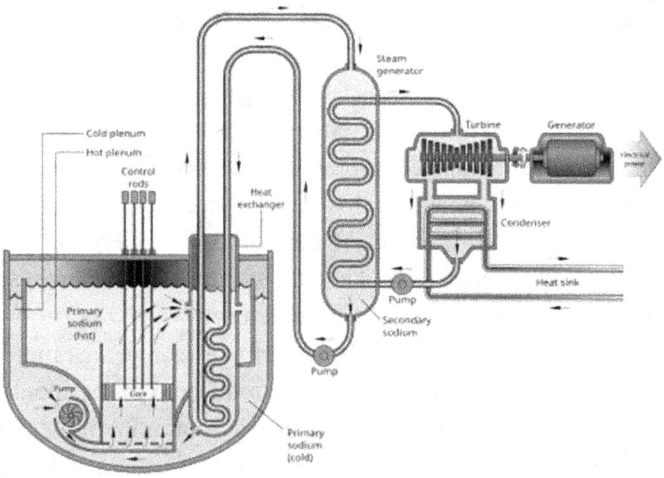

One of the most prominent features of fast reactors is their ability to "breed" fuel and are hence called fast breeder reactors. By converting fertile isotopes such as uranium-238 or thorium-232 into fissile isotopes like plutonium-239 or uranium-233, respectively, fast breeder reactors (FBRs) can generate more fissile material than they consume. This not only extends the availability of nuclear fuel but also contributes to a more closed and sustainable fuel cycle. Fig. 8 illustrates the nuclear fuel cycle, comparing the open and closed fuel cycles (Nuclear energy: Fuel cycle. (2025). An open fuel cycle (also called a once-through fuel cycle) is a system where nuclear fuel is used once in a reactor and then treated as waste after irradiation, without reprocessing or recycling. A closed fuel cycle in nuclear energy refers to a system in which spent nuclear fuel is reprocessed and reused, rather than being treated as waste after a single use in a reactor. The primary goal is to maximize fuel utilization and minimize nuclear waste. Fast reactors are particularly well-suited for using mixed oxide

Figure 8. Illustration of the nuclear fuel cycle, comparing the open and closed fuel cycles (retrieved from Nuclear energy: Fuel cycle. (2025), CC BY 4.0).

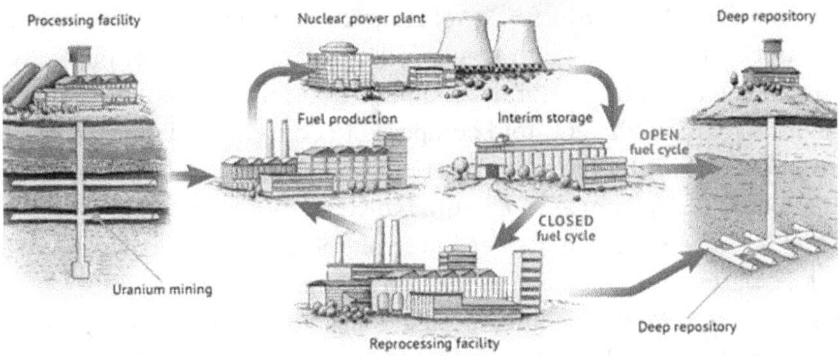

(MOX) fuel and for recycling actinides and plutonium from spent nuclear fuel, thus reducing the long-lived radiotoxicity and volume of nuclear waste (World Nuclear Association, 2025).

Fast reactors typically use liquid metal coolants such as sodium, lead, or a lead-bismuth eutectic, as these substances have excellent thermal conductivity and do not moderate neutrons. Among these, fast reactors which use liquid sodium as the coolant (SFRs) have proven to be the most developed and have been successfully demonstrated in several countries, including the U.S. (EBR-II), France (Phénix and Superphénix), Russia (BN-600 and BN-800), and India (FBTR). These reactors operate at lower pressures than water-cooled reactors, enhancing certain safety aspects, although the chemical reactivity of coolants like sodium introduces unique operational risks. It may be mentioned that the operation of EBR-II and Superphénix have been stopped while PFBR is in an advanced stage of commissioning (International Atomic Energy Agency, 2021).

Nevertheless, fast reactors are increasingly seen as the cornerstone of advanced nuclear energy strategies. They offer a viable pathway towards improved fuel utilization, long-term sustainability, and significant waste minimization. As part of next-generation nuclear programs, including Generation IV and advanced modular reactor initiatives, fast reactors will have an important role in shaping a cleaner, more efficient, and resilient nuclear energy future.

2.5.1 Advantages of Fast Reactors

The advantages of fast reactors are substantial and align with many of the strategic goals of modern nuclear energy systems as well as those of sustainable low-carbon energy resource (World Nuclear Association, 2025).

1. ***Efficient Fuel Utilization***: Fast reactors, during their operation, can convert non-fissile uranium-238 into fissile plutonium-239, and this makes them far more efficient is using natural uranium. This can potentially extend uranium resources by more than 60-fold compared to conventional thermal reactors.
2. ***Waste Reduction***: Fast reactors can burn long-lived transuranic elements (e.g., neptunium, americium, and curium), which are typically found in spent nuclear fuel. This significantly reduces the volume, heat load, and radiotoxicity of nuclear waste, thereby easing long-term storage concerns.
3. ***Closed Fuel Cycle***: By enabling fuel recycling through reprocessing, fast reactors support a closed nuclear fuel cycle, enhancing the availability of nuclear fuel for several hundred years in future, without much dependence of fresh uranium mining.
4. ***Energy Security***: Breeder reactors reduce dependency on enriched uranium, offering strategic advantages for countries with limited enrichment infrastructure or uranium resources.

2.5.2 Limitations of Fast Reactors

Despite their advantages, fast reactors face a range of technical, economic, and policy-related challenges that have hindered their widespread deployment. The use of liquid metal coolants demands sophisticated materials and maintenance protocols, particularly to manage corrosion, reactivity, and thermal stresses (International Atomic Energy Agency,2020b).

1. ***Complex Engineering***: Fast reactors require advanced materials and components capable of withstanding high radiation fields and temperatures. The clad and wrapper materials must withstand high neutron fluence and shall possess high resistance to void swelling and irradiation embrittlement. The use of liquid metal coolants also introduces challenges such as corrosion, coolant leakage, and chemical reactivity.
2. ***High Capital Costs:*** The development and construction of fast reactors involve higher initial costs compared to conventional thermal reactors, partly due to the complexity of the systems and the need for specialized infrastructure.

3. ***Safety Concerns*:** While fast reactors can be designed with passive safety features, their operation with fast neutrons and liquid metal coolants necessitates rigorous safety standards, particularly in handling coolant leaks or core overheating scenarios.
4. ***Proliferation Risks:*** The breeding and separation of plutonium in fast reactors can raise concerns about nuclear proliferation if not properly safeguarded and monitored.

3. ROLE OF NUCLEAR POWER IN A CARBON-FREE FUTURE

Climate change has become the defining global challenge of our times. Greenhouse gas emissions are driving global warming, with severe consequences for weather extremes and sea levels. The international Paris Agreement in 2015 pledges to keep global temperature rise "well below 2°C" and ideally limits warming to 1.5°C above pre-industrial levels as the reference temperature.

Figure 9. Global CO_2 emissions by sector. Energy (electricity, heat, transport, and industry) accounts for the bulk of emissions. (retrieved from U.S. Environmental Protection Agency, 2016). CC BY 4.0.

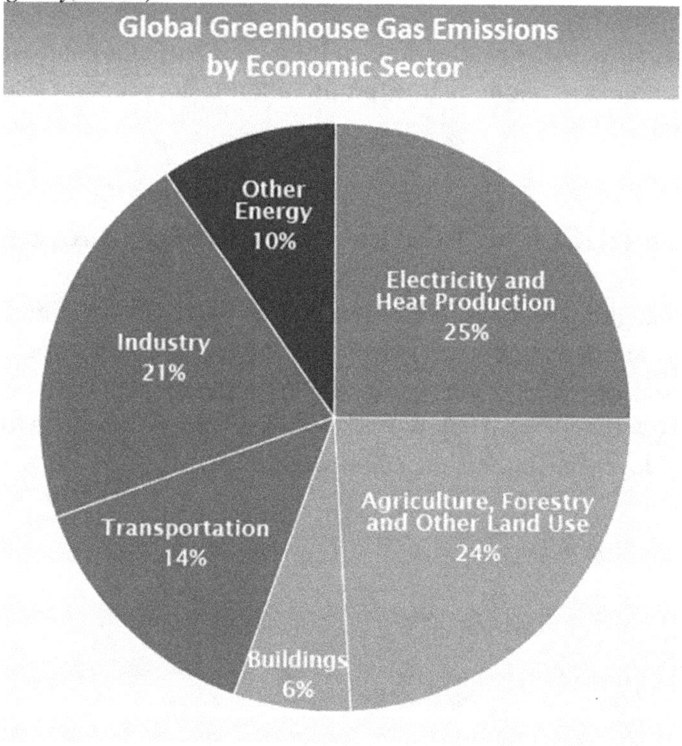

Despite growing awareness and rapid growth of renewables like wind and solar, fossil fuel-related carbon dioxide (CO_2) emissions hit a record 33.1 billion tons in 2018 (International Energy Agency, 2019). The Intergovernmental Panel on Climate Change (IPCC) warns that at the current pace, the world could exceed 1.5°C of warming as early as the 2030s (Edenhofer et al., 2014). This situation demands deep decarbonization of the global energy system urgently, leveraging all available low-carbon technologies. Energy (electricity, heat, transport, and industry) accounts for the bulk of emissions, underscoring the need to decarbonize power generation (U.S. Environmental Protection Agency, 2016) (Fig. 9). Transforming the electricity sector is especially critical because electricity production not only contributes about 40% of energy-related CO_2 emissions but it also has a role for decarbonizing other sectors through electrification (e.g. electric vehicles, heat pumps, industrial electrification). Alongside renewables, nuclear power has emerged as a key contender in the portfolio of carbon-free energy sources needed to meet climate goals.

3.1 Comparison of Nuclear Power to Other Low-Carbon Sources

A nuclear plant operates on a similar thermal cycle as a coal or gas power plant, i.e. using heat to produce steam which in turn drives a turbine. But the heat source is the atomic nucleus rather than the combustion of fossil fuels. During operation, nuclear reactors emit no CO_2 or other greenhouse gases, making them a truly carbon-free power source at the point of generation. Over the full life-cycle (including mining, construction, and decommissioning), nuclear's carbon footprint is on the same order as wind power, and only a fraction of that of solar PV. A comprehensive IPCC analysis found that nuclear power has a median life-cycle emission of about 12 g CO_2-equivalent per kWh, comparable to onshore wind (11 g) and dramatically lower than coal (~800 g) or natural gas (~500 g) per kWh. Fig. 10 illustrates the range of life-cycle emissions for various electricity sources. (World Nuclear Association, 2025; Lenzen, 2008).

Figure 10. Life-cycle emissions of electricity options. On a life-cycle basis, nuclear power emits just a few grams of CO_2 equivalent per kWh of electricity produced. (retrieved from World Nuclear Association. (2025) CC BY 4.0.

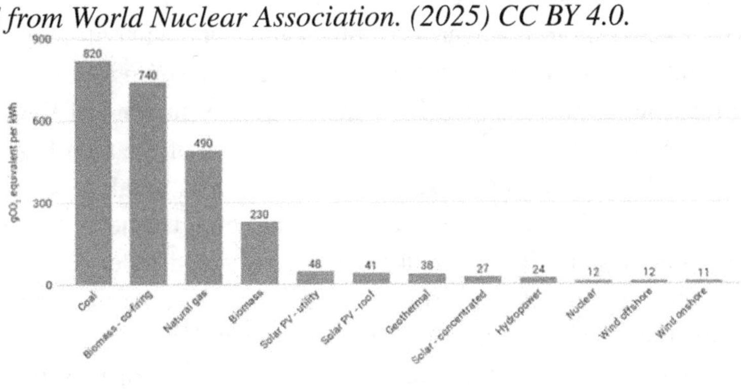

One remarkable attribute of nuclear energy is its fuel energy density. Uranium fuel contains orders of magnitude more usable energy per unit mass than chemical fuels. For example, a single uranium fuel pellet (roughly the size of a fingertip) can yield as much energy as one ton of coal or 149 gallons of oil. This extreme energy density means nuclear plants require far less fuel (and generate far less solid waste by volume) to produce a given amount of electricity, compared to fossil plants. Hence, nuclear has a relatively small land footprint. On a life-cycle basis, considering mining, plant site, and associated facilities, nuclear is the most land-efficient major energy source. Per unit of electricity generated, nuclear needs about 50 times less land than coal and around 18–27 times less land than solar.

Another key characteristic of nuclear power is reliability. Nuclear power plants operate at very high capacity factors, i.e., they produce near their maximum output most of the time. In the United States, for instance, nuclear plant capacity factors have averaged around 90% or more, which is higher than any other energy source.

While renewable energy sources such as wind, solar, and hydropower are also vital low-carbon resources, they have different strengths and limitations. Wind and solar are intermittent; their output fluctuates with weather and time of day, and so they often require backup generation or energy storage to meet demand at all times . Hydropower is dispatchable and can provide baseload or peak power, but is geographically limited and can be impacted by droughts or seasonal water availability. But nuclear plants typically generate power 24/7 except during periodic refueling or maintenance (usually only every 1.5–2 years). This around-the-clock availability of nuclear power is far higher as compared to wind (20–40%) or solar PV (10–30%). Modern nuclear reactors can also perform load-following to some extent (adjusting their output up or down) to help balance variable renewables.

3.2 Technological Innovations in Nuclear Energy

Recent advancements in nuclear energy technology are reshaping the future of clean power generation. Nuclear technology has evolved continuously through several "generations" of reactor designs aimed at improving safety, efficiency, and economics. Generation III and III+ reactors represent state-of-the-art nuclear reactors being built today. These are advanced versions of the Gen II reactors that dominate the existing power reactors. Innovations such as Small Modular Reactors (SMRs) offer enhanced safety, scalability, and cost-effectiveness, making nuclear power more accessible for diverse regions and energy grids. Generation IV reactor designs, including molten salt, gas-cooled, and fast breeder reactors, promise greater fuel efficiency, reduced waste, and improved sustainability. In parallel, closed fuel cycle technologies and advanced reprocessing methods aim to minimize long-lived radioactive waste and maximize the use of existing uranium resources. These breakthroughs position nuclear energy as a vital component of the global strategy to achieve deep decarbonization and long-term energy security.

3.2.1 Small Modular Reactors

Small Modular Reactors (SMRs) represent emerging designs in thermal reactors which are compact, and scalable systems that often employ advanced passive safety features and simplified cooling systems. Many SMR designs are based on PWR or BWR technology but are intended for deployment in remote regions, smaller grids, or specialized applications like desalination or hydrogen production. SMRs have a capacity below 300 MWe . The "modular" aspect implies that key components can be prefabricated and shipped to site, allowing serial production and much shorter construction time. The motivation for SMRs is to overcome the high upfront cost and inflexibility of large plants by offering smaller, incremental capacity additions that are easier to finance and deploy. Many SMR designs also embrace simpler, inherent safety feature and rely on passive cooling. They could also potentially replace retiring fossil fuel plants. As of 2024, there are dozens of SMR designs under development globally. If these efforts succeed, the late 2020s and 2030s could see the commercial rollout of SMRs, potentially lowering the cost barrier and expanding nuclear's use cases (including providing heat for industry or desalination, not just electricity). However, further research is needed to establish the safe isolation zone between the reactor site and the surrounding villages. Fig. 11 is a cut-away section of NuScale SMR design showing the key components (Neutron Bytes, 2020).

3.2.2 Next Generation Reactors

Next Generation Reactors Gen III/III+ designs incorporate enhanced safety features such as passive cooling systems, which can operate without human intervention or external power, and longer grace periods (often 72 hours or more) during which the plant remains stable after shutdown without active cooling. These reactors also feature more standardized, modular designs to streamline construction, and improved fuel technology allowing higher burn-up (extracting more energy per unit of fuel). Notable Gen III+ models include the AP1000 (with passive emergency cooling tanks), the EPR (European Pressurized Reactor with double containment and core-catcher), and the APR1400 and Hualong One, among others, which are being deployed in countries like China, South Korea, France, and the UAE. These reactors typically have design lifetimes of 60+ years and tougher containment against external events (e.g., aircraft impact).

Figure 11. Cut-away section of NuScale SMR design showing the key components (retrieved from Neutron Bytes, 2020), CC BY 4.0.

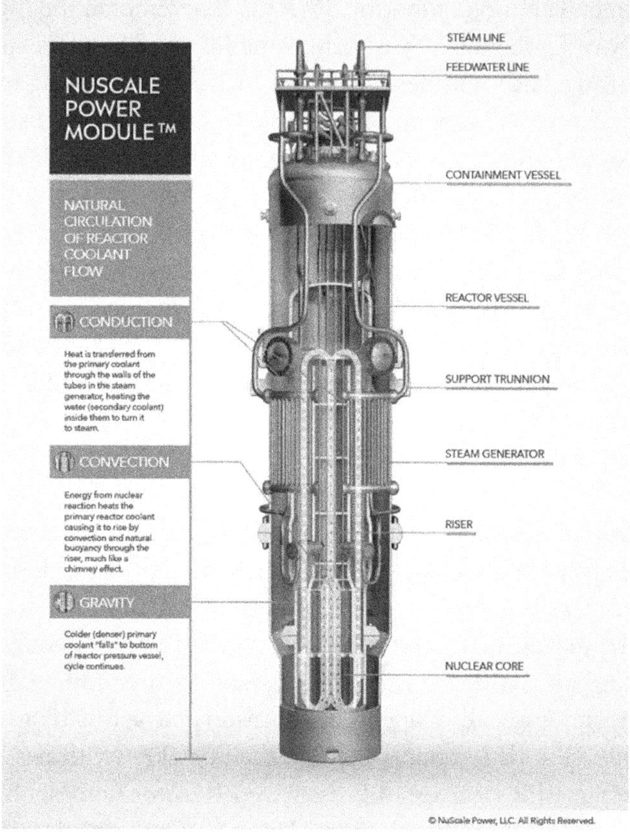

Generation IV (Gen IV) nuclear reactors represent a class of advanced fission reactor designs that aim to address the limitations of current-generation reactors in terms of sustainability, safety, efficiency, and proliferation resistance. Unlike conventional thermal reactors, many Gen IV systems utilize fast neutrons to sustain the fission chain reaction, eliminating the need for a moderator and enabling the efficient breeding of fissile material from fertile isotopes such as U-238 or Th-232. This fast-spectrum operation, particularly in designs like the Sodium-cooled Fast Reactor (SFR) and the Lead-cooled Fast Reactor (LFR), allows for a closed fuel cycle that minimizes long-lived radiotoxic waste and extends fuel resources significantly by recycling actinides.

3.2.3 Key Features of Next-Generation Reactors

i. Passive Safety Systems

Next-generation reactors incorporate passive safety mechanisms that rely on natural physical laws such as gravity, convection, and heat dissipation, rather than active mechanical or human intervention. These systems greatly reduce the risk of core damage during accidents.

ii. Closed Fuel Cycles and Waste Reduction

Several advanced reactor types are capable of using spent nuclear fuel, depleted uranium, or thorium as the fuel. Through breeding and recycling, these reactors minimize the volume and toxicity of nuclear waste, and extend fuel resources by a factor of 50–100.

iii. Advanced Materials and Coolants

Several Gen IV designs employ innovative coolants and structural materials to achieve higher operating temperatures, often exceeding 500°C and in some cases approaching 1000°C, compared to the ~300°C typical of current Light Water Reactors (LWRs). For instance, the Gas-cooled Fast Reactor (GFR) and Very High Temperature Reactor (VHTR) use helium gas as a coolant to allow for higher thermodynamic efficiency and potential integration with hydrogen production and other industrial processes. Molten Salt Reactors (MSRs), another Gen IV concept, operate with fuel dissolved directly in molten fluoride or chloride salts, offering inherent safety benefits through passive heat removal, low operating pressure, and strong negative temperature coefficients that self-regulate reactivity.

Despite their promise, most Gen IV reactors remain in the research, development, or early prototype stages due to the significant technical challenges involved. These include material degradation at high temperatures, corrosion from novel coolants (e.g., molten salts or liquid metals), complex fuel cycle management, and the need for advanced control systems to ensure safe and autonomous operation. International collaboration through initiatives like the Generation IV International Forum (GIF) continues to guide the development of these reactors, with demonstration plants expected in the coming decades as enabling technologies mature and regulatory pathways evolve (International Atomic Energy Agency, 2012; OECD Nuclear Energy Agency, 2018; World Nuclear Association, 2025)

3.3 Fusion reactors

Nuclear fusion, the process powering the sun, involves the fusion of light atomic nuclei such as deuterium and tritium to form heavier nuclei, releasing substantial energy due to mass-to-energy conversion described by Einstein's $E=mc^2$. Unlike nuclear fission, fusion produces much less long-lived radioactive waste and inherently avoids risks of runaway reactions. The primary technical challenge lies in achieving the conditions required for fusion—extreme temperatures (~150 million °C), sufficient confinement time, and adequate plasma density in order to surpass the breakeven point where energy output exceeds energy input. Magnetic confinement using devices like tokamaks and stellarators, as well as inertial confinement with laser or ion beams, represent the two pathways for fusion technology. (Wesson, 2011).

The International Thermonuclear Experimental Reactor (ITER) is the world's largest fusion experiment and a major international collaboration between USA, Europe, Russia, Japan, China, South Korea, and India. It is under construction in Cadarache, France, The ITER is based on the tokamak design and aims to demonstrate a fusion power output of 500 MWe from an input energy of 50 Mwe. ITER's key innovations include superconducting magnets for magnetic confinement, advanced divertor systems for heat removal, and remote maintenance technologies. As of 2025, major components like the cryostat, toroidal field coils, and vacuum vessel sectors are in place, with integrated commissioning expected to continue into the late 2020s. ITER is a critical milestone towards a demonstration of fusion power plant (DEMO) by the 2040s (ITER (n.d.)). Fig. 12 shows the interiors of ITER, the Tokamak is the central component of the ITER project, designed to replicate the fusion process of the sun. ITER is 24 m high, 30 m wide, weighs 23,000 tons, has 1,000,000 components and 830 m^3 plasma volume.

Figure 12. Inside ITER, the Tokamak is the central component of the ITER project, designed to replicate the fusion process of the sun. ITER is 24 m high, 30 m wide, weighs 23,000 tons, has 1,000,000Number of components and 830 m³plasma volume (retrieved fromITER (n.d.)). CC BY 4.0.

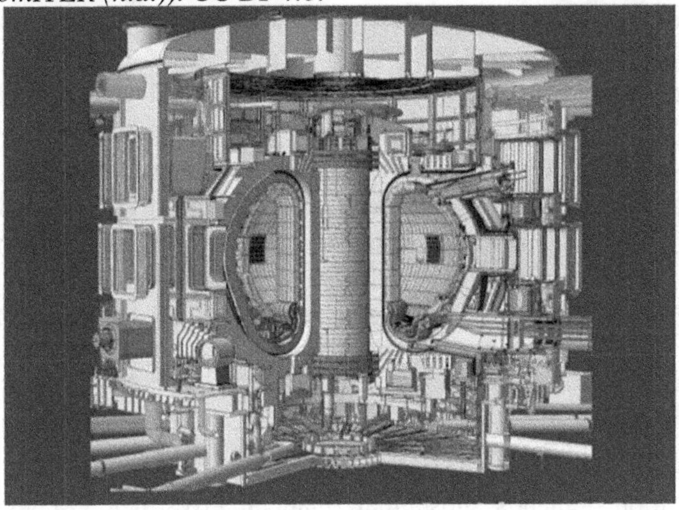

3.3.1 Key Concerns of Nuclear Fission and Nuclear Fusion Technologies

While nuclear power offers a low-carbon, baseload energy source critical to global decarbonization goals, both fission and fusion technologies face complex challenges. For fission, issues include long-term waste management, high capital costs, public acceptance, and regulatory hurdles. For fusion, despite significant advances, sustained net energy production and material endurance under neutron bombardment remain unresolved. However, breakthroughs in high-temperature superconductors and advanced plasma control have raised optimism. Nuclear fission continues to evolve through Small Modular Reactors (SMRs) and Generation IV systems, while fusion holds long-term promise for abundant, clean energy with minimal environmental impact (World Nuclear Association. (n.d.)a; González & García, 2016).

4. CURRENT GLOBAL ENERGY CHALLENGES

Faced with global warming, there is a compelling reason to move from fossil fuels to low-carbon energy mix. Transitioning to renewable energy sources and nuclear power is a critical strategy to combating climate change. The transition away from fossil fuel dependence raises multifaceted challenges in integrating renewable energy sources and navigating the socio-economic complexities associated with this transition.

4.1 Technological and Infrastructure Challenges

Integrating renewable energy sources like solar and wind with fossil-fueled power grids requires significant technological advancements since solar and wind sources of power are not steady and are subject to the availability of sunlight and wind, respectively. Hence, the electricity generated has to be stored for later use. Therefore, it is required to develop efficient energy storage solutions and smart grid technologies to ensure a stable and reliable energy supply. Additionally, upgrading and expanding the current infrastructure to accommodate decentralized energy production is both complex and costly.

4.2 Economic and Financial Hurdles

The transition demands substantial financial investment. Renewable energy projects often face high initial costs, and the economic viability can be influenced by fluctuating market conditions. For instance, Finnish utility Fortum concluded that constructing new nuclear power capacity is currently not commercially viable due to low prices in the Nordic power market. Consequently, they are focusing on extending the lifespan of existing plants and investing in renewable energy. Furthermore, renewable energy companies may encounter financial losses due to environmental factors. Zen Energy, for example, experienced significant financial setbacks during a "wind drought," highlighting the risks associated with renewable energy ventures.

4.3 Policy and Regulatory Uncertainties

Inconsistent policies and regulatory frameworks can impede the energy transition. In the United Kingdom, debates over net-zero policies have revealed challenges in policymaking, affecting business investment confidence. Critics argue that ambitious green promises lack transparency regarding actual costs and economic implications, complicating the transition process

4.4 Public Perception and Social Acceptance

Transitioning to new sources of energy and associated technologies successfully requires strong consumer support. Surveys indicate that a significant portion of the population anticipates potential problems if energy production shifts significantly from fossil fuels to renewables. Concerns include possible increases in consumer prices and the reliability of energy supply during the transition period.

4.5 Phasing Out Fossil Fuel Power Plants

Reducing reliance on fossil fuels is critical for achieving global climate objectives. However, the transition away from coal, oil, and natural gas poses economic and social challenges, especially for regions dependent on these industries.

5. SUMMARY

This chapter presents a comprehensive introduction to advanced nuclear systems. The role of nuclear energy in addressing global energy demands, climate change, and sustainable development is emphasized. Nuclear fission is a mature and efficient technology for large-scale electricity generation. Key thermal reactor types such as Pressurized Water Reactors (PWRs), Boiling Water Reactors (BWRs), Pressurized Heavy-Water Reactors (PHWRs), and Advanced Gas-Cooled Reactors (AGRs) are described. Each reactor design is evaluated based on moderator type, fuel enrichment, coolant medium, safety features, and global adoption. The chapter outlines the limitations of thermal reactors, including poor fuel utilization and radioactive waste concerns.

The article highlights the promise of Small Modular Reactors (SMRs) and Generation III/III+ and IV designs, including fast reactors which prioritize safety, scalability, and sustainability. SMRs are seen as cost-effective and flexible solutions for diverse energy needs, while Gen IV reactors, including molten salt and gas-cooled designs, aim to extend fuel resources and minimize nuclear waste generation. A prominent feature of fast reactors is their ability to breed fuel, and as a result, it extends the availability of nuclear fuel. Fast reactors offer higher efficiency, closed fuel cycle, and less long-lived waste through the use of fast neutrons and advanced coolants like liquid sodium. Fusion reactor technology which is based on nuclear fusion of light atomic nuclei, the source of endless energy of the sun and the stars, promises to provide an unlimited source of clean energy once the International Thermonuclear Experimental Reactor is successfully commissioned, operationalized and the technological lessons learnt are used to design commercial fusion reactors. Nuclear energy, along with solar, wind, and geothermal sources, offers the pathway to combat global warming.

ACKNOWLEDGEMENT

The authors thank Asst. Prof. Tonny Binoy, Saintgits College of Engineering (Autonomous), for a critical reading and editing of the manuscript.

REFERENCES

BWR – Boiling water reactor. (2014). Nuclear Power. https://www.nuclear-power.com/bwr-boiling-water-reactor/

Edenhofer, O., Pichs-Madruga, R., Sokona, Y., Farahani, E., Kadner, S., Seyboth, K., Adler, A., Baum, I., Brunner, S., Eickemeier, P., Kriemann, B., Savolainen, J., Schlömer, S., von Stechow, C., Zwickel, T., & Minx, J. C. (Eds.). (2014). *Climate change 2014: Mitigation of climate change. Contribution of Working Group III to the Fifth Assessment Report*. Cambridge University Press.

Gas-cooled reactor and advanced gas-cooled reactor (AGR). (2025). Energy Encyclopedia. https://www.energyencyclopedia.com/en/nuclear-energy/the-nuclear-reactors/gas-cooled-reactor-gcr-and-advanced-gas-cooled-reactor-agr

Glasstone, S. (1955). *Principles of nuclear reactor engineering*. D. Van Nostrand Company.

Glasstone, S., & Sesonske, A. (1994). *Nuclear reactor engineering: Reactor systems engineering* (4th ed.). Springer International Publishing AG.

Global energy & CO_2 status report 2019 (2019). IEA. https://www.iea.org/reports/global-energy-co2-status-report-2019

González, E., & García, J. (2016). Nuclear fission and fusion: Comparative analysis and perspectives. *Energy Policy*, (94), 51–60.

Heavy water reactor (PHWR). (2025). Energy Encyclopedia. https://www.energyencyclopedia.com/en/nuclear-energy/the-nuclear-reactors/heavy-water-reactor-phwr

International Atomic Energy Agency. (1992). *The Chernobyl accident: Updating of INSAG-1* (Safety Series No. 75-INSAG-7). IAEA.

International Atomic Energy Agency. (2012). Advances in nuclear power process heat applications. IAEA., IAEA-TECDOC-1682.

International Atomic Energy Agency. (2015). The Fukushima Daiichi accident: Report by the Director General. Technical (Vol. 5). IAEA., STI/PUB/1710.

International Atomic Energy Agency. (2020a). *Nuclear power reactors in the world* (Reference Data Series No. 2). IAEA.

International Atomic Energy Agency. (2020b). *Status of fast reactor research and technology development*. IAEA.

International Atomic Energy Agency. (2021). *Liquid metal coolants for fast reactors cooled by sodium, lead, and lead-bismuth eutectic*. IAEA.

International Atomic Energy Agency. (2022). *What is nuclear energy? The science of nuclear power*. https://www.iaea.org/newscenter/news/what-is-nuclear-energy-the-science-of-nuclear-power

ITER. (n.d.). *The Machine*. Retrieved April 15, 2025 from https://www.iter.org/machine

Lenzen, M. (2008). Renewable and nuclear electricity: Comparison of environmental impacts. *Energy Policy*, *37*(12), 5058–5068.

Mathew, M. D. (2022). Nuclear energy: A pathway towards mitigation of global warming. *Progress in Nuclear Energy*, *143*, 104080. DOI: 10.1016/j.pnucene.2021.104080

Neutron Bytes. (2020). *NuScale completes key milestone with NRC*.https://neutronbytes.com/2020/08/29/nuscale-completes-key-milestone-with-nrc

Nuclear energy: Fuel cycle. (2025). Energy Encyclopedia. https://www.energyencyclopedia.com/en/nuclear-energy/nuclear-fuel/fuel-cycle

OECD Nuclear Energy Agency. (2018). *GIF R&D outlook for Generation IV nuclear energy systems*. Gen IV International Forum.

United Nations Framework Convention on Climate Change. (2015). *The Paris Agreement*. https://unfccc.int/process-and-meetings/the-paris-agreement/the-paris-agreement

U.S. Environmental Protection Agency. (2016). *Global greenhouse gas emissions data*. https://climatechange.chicago.gov/ghgemissions/global-greenhouse-gas-emissions-data

Wesson, J. (2011). *Tokamaks* (4th ed.). Oxford University Press.

Wikipedia. (2025). *Sodium-cooled fast reactor*. Retrieved from https://en.wikipedia.org/wiki/Sodium-cooled_fast_reactor

World Energy & Climate Statistics – Yearbook (2024). Enerdata. https://yearbook.enerdata.net/

World Nuclear Association. (2025). *World Nuclear Symposium 2025*. https://world-nuclear.org/events/world-nuclear-symposium-2025

World Nuclear Association. (2025) *World energy needs and nuclear power*. Retrieved January 6, 2025, from https://world-nuclear.org/information-library/current-and-future-generation/world-energy-needs-and-nuclear-power

World Nuclear Association. (2025) *Carbon dioxide emissions from electricity*. Retrieved September 3, 2024, from https://world-nuclear.org/information-library/energy-and-the-environment/carbon-dioxide-emissions-from-electricity

World Nuclear Association. (2025) *Nuclear power reactors*. Retrieved March 21, 2025, from https://world-nuclear.org/information-library/nuclear-fuel-cycle/nuclear-power-reactors/nuclear-power-reactors

World Nuclear Association. (2025) *RBMK reactors*. Retrieved February 15, 2022, from https://world-nuclear.org/information-library/appendices/rbmk-reactors

World Nuclear Association. (2025) *Generation IV nuclear reactors*. Retrieved April 30, 2024, from https://www.world-nuclear.org/information-library/current-and-future-generation/generation-iv-nuclear-reactors.aspx

ADDITIONAL READING

MIT Energy Initiative. (2018). *The future of nuclear energy in a carbon-constrained world: An interdisciplinary MIT study*. https://energy.mit.edu/research/future-nuclear-energy-carbon-constrained-world/

Murray, R. L., & Holbert, K. E. (2019). *Nuclear energy: An introduction to the concepts, systems, and applications of nuclear processes* (8th ed.). Butterworth-Heinemann.

National Academies of Sciences, Engineering, and Medicine. (2021). *Bringing fusion to the U.S. grid*. The National Academies Press. https://doi.org/DOI: 10.17226/25991

Neilson, G. H. (Ed.). (2016). *Magnetic fusion energy: From experiments to power plants*. Elsevier B.V.

World Nuclear Association. (n.d.). *Small nuclear power reactors*. Retrieved February 16, 2024, from https://world-nuclear.org/information-library/nuclear-fuel-cycle/nuclear-power-reactors/small-nuclear-power-reactors

World Nuclear Association. (n.d.). *Nuclear fusion power*. Retrieved February 3, 2025, from https://world-nuclear.org/information-library/current-and-future-generation/nuclear-fusion-power

Wu, J., Chen, J., Cai, X., Zou, C., Yu, C., Cui, Y., Zhang, A., & Zhao, H. (2022). A review of molten salt reactor multi-physics coupling models and development prospects. *Energies*, *15*(21), 8296. DOI: 10.3390/en15218296

KEY TERMS AND DEFINITIONS

Nuclear Fission: Nuclear fission is a nuclear reaction in which the nucleus of a heavy atom, such as uranium-235 or plutonium-239, splits into two or more lighter nuclei when bombarded with a neutron. This process releases a large amount of energy, primarily as heat, along with additional neutrons that can initiate further fission reactions in a chain reaction. It is the fundamental principle behind the operation of nuclear power plants.

Nuclear Fusion: Nuclear fusion is a process where two light atomic nuclei, such as deuterium and tritium, combine to form a heavier nucleus, releasing a tremendous amount of energy. This reaction powers the sun and stars. Nuclear fusion holds great promise for providing a virtually limitless, clean source of energy on Earth.

Nuclear Reactor: A nuclear reactor is a system designed to sustain and control a chain reaction of nuclear fission to produce a large quantity of energy, typically in the form of heat. This heat is used to generate steam, which drives turbines to produce electricity in a nuclear reactor. Nuclear reactors are built with multiple safety systems to regulate the fission process and manage radioactive materials safely.

Thermal reactor: Thermal reactor is a type of nuclear reactor that uses slow-moving, or "thermal," neutrons to sustain the fission chain reaction. They rely on a moderator, such as water or graphite, to slow down fast neutrons produced during fission, increasing the likelihood of further fission events. Thermal reactors are currently the most common type of reactors used in commercial nuclear power generation worldwide.

Fast reactor: Fast reactor is a nuclear reactor that use high-energy, or "fast," neutrons to sustain the fission chain reaction without the need for a moderator to slow them down. They are designed to efficiently utilize fuel, and can also breed more fissile material, making them highly fuel-efficient. Fast reactors offer advantages like better fuel utilization and reduced long-lived radioactive waste compared to thermal reactors.

Fusion reactor: Fusion reactor is a reactor designed to replicate the process of nuclear fusion which occurs in the Sun, where light atomic nuclei combine to form heavier nuclei, under extreme temperatures and pressures, releasing vast amounts of energy. They aim to produce clean, virtually limitless power with minimal long-lived radioactive waste. Fusion reactors, such as tokamaks are still in experimental stages but hold promise for the future of sustainable energy.

Global Warming: Global warming refers to the long-term increase in Earth's average surface temperature due to the buildup of greenhouse gases like carbon dioxide in the atmosphere. It results primarily from human activities such as burning fossil fuels, deforestation, and industrial processes. Global warming leads to climate change, including more extreme weather events, rising sea levels, and ecosystem disruptions.

Greenhouse gases: Greenhouse gases are naturally occurring and human-produced atmospheric gases that trap heat in the Earth's atmosphere, leading to the greenhouse effect. These gases, like carbon dioxide, methane, and nitrous oxide, absorb infrared radiation emitted by the Earth's surface, preventing it from escaping into space and contributing to global warming.

Chapter 4
Advanced Nuclear Systems and Climate Resilience

Ranjit K. Parmar
https://orcid.org/0009-0005-7403-3141
Government Polytechnic Junagadh, India

ABSTRACT

Due to global warming and other factors, the frequency of climate induced calamities like flood, tsunami, draught are increasing. They damage the energy infrastructure a lot. To combat these challenges, existing infrastructure must be strengthened for energy resilience. Renewable energy sources are gaining momentum as they are non-polluting unlike fossil fuels, but their energy output is intermittent, unpredictable and depends on weather conditions. In this context, advanced nuclear reactors may play an important role to supply stable base load power. Small modular reactors with passive safety features can help in meeting the energy, heat and drinking water demand even in remote area where traditional rectors are unpractical. Integrating renewable energy sources with advanced nuclear reactors though smart grid can show the path for a carbon free future. This chapter discusses the role of SMRs in climatic adaptation along with related opportunities and challenges.

INTRODUCTION

(*Vasudhaiva Kutumbakam)*, a Sanskrit text mentioned in the Indian philosophy book Maha Upanishad chapter 6, Verse 71-73, states that "**The world is a family**". It tells us to live peacefully with all creatures of the world while judiciously using (and not exploiting) the precious natural resources. However, there is no end

DOI: 10.4018/979-8-3373-0685-8.ch004

to the greed of humans. We have almost overexploited all our resources, leading to climate change and its catastrophic results. The International Atomic Energy Agency, in its report, states that "to achieve carbon neutrality and limit global warming to 1.5°C, energy sector investment must be scaled up and directed towards cleaner and more sustainable technologies that support climate change mitigation and adaptation". (Fragkos, 2020). This global challenge needs to be addressed urgently with innovations and international collaboration to ensure a sustainable future.

Climate-induced disasters are escalating day by day, forcing mankind to explore additional options to meet reliable energy demands. It's no longer a theoretical concept but a real challenge we are facing today. Among the available various solutions, renewable energy sources and nuclear energy systems are solutions to meet the two ends of a thread. Advanced nuclear systems emerge as a robust and reliable option. This chapter explores the potential of advanced nuclear systems in building robust and reliable energy infrastructure, ensuring disaster recovery, and maintaining energy reliability in an increasingly unpredictable world (Wu, 2019).

The data collection of rainfall, temperature, snow, and wind patterns helps define the climate of any area over a considerable period. Climate resilience is a term given to withstand, respond and prepare for climate change. It's the capacity of an individual or a system to recognize and recover from the ill effects of climate change. It also includes environmental shocks, adaptability, and transformation to withstand the changing, challenging climate. It also integrates the ever-changing climate considerations into planning and policy framing to reduce the impact of climate-prone calamities.

However, achieving climate resilience poses many challenges. Lack of financial resources, varying government policies across the globe among different countries, lack of dedicated efforts from poor countries due to financial crises, and lack of willingness from certain developed countries, who are primarily responsible for the major pollution share, are a few noteworthy significant challenges. Lack of technologies & infrastructure, particularly in developing countries, is also a critical point as a significant challenge. All these factors are ultimately leading to irreversible damage to our ecosystem, threatening biodiversity, and the existence of human beings (Jalles, 2020).

Amid these challenges, the latest technologies are proving a new dawn to relieve the stress and burden of these challenges. Opportunities are growing in newer areas of energy, which can mitigate the effects of climate change, promoting sustainable development. Advancements in Green Energy, Advanced Nuclear Reactors, and promoting sustainable urban planning can strengthen resilience towards the ever-harsher climate change. International cooperation through frameworks like the Paris Agreement can provide a platform for sharing knowledge, funds, and technology to cop with the calamity (Canny et al, 2020). Man must understand the coexistence and role of all species, judicious use of natural resources, and interconnections of ecological, social, and economic systems to unlock the pathways for sustainable development, for safeguarding the planet for future generations.

SELF-SUSTAINED SOCIETIES

We have been taught since our childhood that defining energy is the ability to do work, but its scope extends much beyond that! In the modern era, almost every machine runs with the help of different forms of energy. Energy sustainability is essential to maintain the reliability of any system. Energy systems have become critical for building resilient societies, as they undergo economic and social growth. It ensures a reliable, sustainable energy even in the catastrophic consequences of climate-induced incidents. However, modern societies are facing an increasing number of energy outages due to ever-growing demand for power, extreme weather conditions, and the global transitions towards low-carbon emitting systems (Kabeyi & Olanrewaju, 2022).

Strengthening energy resilience involves diversification of various energy sources, decentralization of power production, adopting innovative roust technology, and integrating renewable energy sources like solar, wind, and hydro power along with energy storage possibilities while keeping an eye on enhancing system stability as well as reducing greenhouse gas emissions. Smart grid may play a vital role in reducing power outages through advanced monitoring, automation, and predictive technologies, through rapid responses (Amin, 2011). Microgrids may become helpful to meet local energy demands in remote areas, thus reducing dependence on centralized systems. The role of energy resilience also includes framing robust policies and participation of the public and governments to prioritize sustainable energy investments and innovations. Thus, more reliable energy systems are needed for an hour, demand of a time to address the future challenges to craft a sustainable global system.

CLIMATE ADAPTATION AND NUCLEAR SYSTEMS

As the world is facing an urgent challenge to reduce GHG emissions, we have no option other than relinquishing the traditional coal and petroleum-based energy sources. Renewable energy sources and nuclear systems are the ray of hope for mankind to come out of this crisis. In this chapter, we shall focus on the adaptability of nuclear systems to mitigate the effects of climate change and their climate adaptation. Advanced nuclear systems employ innovative technologies that can assure reliable energy production along with low-carbon energy generation while withstanding resilience in the event of adverse climatic conditions (Proskuryakova, L., 2023).

We have witnessed nuclear disasters in the past, forcing man to rethink the energy generation through nuclear systems, but much water has passed since then. Technology has evolved a lot to get rid of such adverse situations. Unlike traditional nuclear reactors, nowadays SMRs (Small Modular Reactors) and Generation IV reactors are much safer and more efficient (Vujić et al,2012). They have a lower environmental footprint, scalable designs with modular structures. Nuclear reactors may help in reducing the emission of GHGs in many industrial sectors. They can generate high heat to be utilized in industrial processes or may supply Hydrogen in the transportation sector. They can also be operated in parallel with wind and solar power plants to increase the grid stability and diversity. Safety is the prime key to establishing a nuclear system for power generation. Many advanced reactors utilize safety mechanisms to prevent accidents, while some others can recycle nuclear waste, thus reducing environmental impact (Forsberg et al, 1991). However, high initial cost, public perception, location, safety concerns, and regulatory framework remain the challenges for nuclear systems. Investment, Government support, international collaboration, and public engagement can lead nuclear systems as an important role in combating the climate change crisis and mitigating its ill effects.

CLIMATE RESILIENT ENERGY SYSTEMS

Climate resilient systems offer sustainable energy while combating threats of climate change. These are equipped with a set of pre-planned features to cop with such unpredictable events, which help them function even in weather affecting conditions.

1. Robust infrastructure: These systems are built with climate-resilient resilient strong materials with robust infrastructure to withstand the severe shocks of climate change in adverse weather such as floods, intense scorching heat waves, storms, etc. Such systems may have additional protective layers like elevated substations, protected transmission systems, and reliable power sources installed at safe locations.
2. Decentralized Energy sources and their diversification: Centralized energy sources are always vulnerable in calamities. Microgrids protect energy systems from power failures, relieving dependency on the main grid. Further amalgamation of diverse energy sources like wind, solar, geothermal, nuclear, along with thermal power, aids a layer of reliability.

3. Predictive Technologies: These systems use advanced tools that are smart enough to predict calamities in advance and prepare mitigation plans based on previously available data.
4. Energy storage: These systems are equipped with additional facilities like battery backup, hydrogen storage, and pumped hydro to be utilized in emergencies of high demand or low generation periods. Energy storage adds an extra layer of resilience to enhance the system's reliability.
5. Community Awareness and Policy: Social awareness among the community residing in the peripheral area about clean energy and sustainable energy sources always plays a vital role in the long run and peaceful coexistence of any power plant. Such systems always initiate steps for mass mobilization. Further, clean energy subsidies and investment-provoking policy framework enable resilient systems.
6. Low carbon footprint: Such systems adopt low carbon-emitting technologies and take steps to improve sustainability through material recycling, reducing, and reusing wherever possible.

Features of Advanced Nuclear Reactors

Advanced Nuclear Reactors are designed with the latest technology to withstand adverse conditions and emergencies, which increases their reliability, safety, and adaptability, making them a reliable partner to combat climate change in the field of energy systems for a sustainable energy future.

1. Passive Safety Features: These types of safety features allow the auto to shut down in case of an emergency. This system works in auto mode in the background without any human intervention and does not require any power to operate. They use natural powers like gravity, buoyancy, etc. (Nayak et al, 2007).
2. Smaller Modular Reactors: These smaller and modular structures have a robust structure. This leads to their easy deployment in areas where traditional reactors are not suitable (Hidayatullah et al, 2015).
3. Low waste: Modern reactors are more efficient and thus use a lesser amount of fuel. Few reactors even recycle the waste, thus further minimizing the environmental impact.
4. Integration with renewable energy sources: Renewable energy sources have unpredictable output, whereas a nuclear reactor can be designed for any variable output. This flexibility makes them ideal to integrate with renewable energy sources.

5. Climate resilience: Modern nuclear reactors are designed with a robust design that can withstand extreme climatic conditions, which permits them to be utilized in natural disasters and emergencies.
6. Carbon-free energy: Since nuclear reactors do not use any coal or petroleum-based fuel, no greenhouse gases (GHG) are generated, thus the carbon footprint of power generated is almost zero, meeting the global targets of sustainability.
7. Economic viability: Nuclear reactors are costlier to establish but are always cheaper in the long term. They also require minimum maintenance.

Thus, we may say, Advanced Nuclear Reactors not just play an important role in energy sustainability but can also help in climate climate-resilient energy future with a lower carbon footprint.

CLIMATE ADAPTATION IN REACTOR DESIGNS

Modern reactor projects plan a rigorous site evaluation to minimize the climate-induced risks like floods, earthquakes, and rising sea levels. Sufficient measures and protective barriers are employed in flood-prone areas for the safety of critical infrastructure. Advanced nuclear reactors are designed with highly efficient cooling systems to use water judiciously and function even during heat waves and water shortages. Hybrid systems may be designed to ensure operation in draught conditions. Their modular designs make them suitable for deployment in remote areas with limited infrastructure, along with flexible placement and easier relocation in case of any unavoidable environmental conditions. Reactor building and containment structures are always designed to withstand extreme situations due to the risk of radioactive leakage, which also helps here to withstand climatic calamities, ensuring durability and minimum civil maintenance (Eidemüller, 2021). These features make nuclear reactors ideal for supplying stable base load power and provide an alternative option in case of blackouts in climatic conditions.

SMALL MODULAR REACTORS

Small Modular Reactors, as the name itself suggests, are smaller in size. They are compact, robust, and scalable in design. They are emerging as reliable energy sources these days to effectively mitigate the impact of climate change. They are built in modules, thus allowing phased construction along with an expansion option for the future, which allows the possibility of production growth as per demand without major upfront costs as in traditional large reactors. Unlike traditional large reactors, Small Modular Reactors can be deployed in remote areas to provide local energy needs, relinquishing dependency on the main grid. This decentralization enhances energy security. They can even be installed underground owing to their small size, to provide safety and environmental protection. Modular construction and factory-based manufacturing reduce its cost and timing to be made available on an urgent basis in emergencies during natural disasters. They can easily integrate with intermittent renewable energy sources as a consistent base load to meet the demand fluctuations. They are also equipped with passive safety designs along with efficient fuel utilization, thus minimizing risks and wastes. In a nutshell, SMRs present a flexible, decentralized, and sustainable energy future (Locatelli et al, 2013).

GENERATION IV REACTORS

At present, old age reactors have been almost wiped out from operation, and these reactors are being replaced by the present reactors due to their better fuel efficiency and higher safety standards (Locatelli et al., 2014). Some noteworthy features of Generation IV reactors are:

1. Higher Efficiency: These reactors can produce power with an efficiency as good as 50%. Major targeted sectors for this achievement are higher operating temperatures (500°C – 1000°C), which improve the thermodynamic cycle efficiency. Advanced coolants like Helium, Sodium, Lead, or Supercritical CO_2 reduce heat losses and permit higher temperature operation, reprocessing, and recycling of fuel.
2. Lower Waste Production: Efficient use of Thorium & Uranium with the improved designs leads to better fuel management and lower radioactive waste. These reactors can burn actinides and other isotopes from used fuel, thus reducing long-lived radioactive waste (Lindley et al, 2015). The impact on the environment is greatly reduced due to their capability to convert higher half-life radioactive isotopes to lower half-life isotopes.
3. Designs: At present, six major designs are in operation across the globe.
    a. Gas-Cooled Fast Reactors (GFRs)
    b. Supercritical-Water-Cooled Reactors (SCWRs)
    c. Very High-Temperature Reactors (VHTRs)
    d. Molten Salt Reactors (MSRs)
    e. Sodium-Cooled Fast Reactors (SFRs)
    f. Lead-Cooled Fast Reactors (LFRs)
4. Advantages of Generation IV Reactors

Table 1. Advantages of Generation IV reactors with respect. Current reactors.

Feature	Current reactors	Generation IV Reactors
Efficiency	~33%	Up to 50%
Waste Longevity	High (thousands of years)	Reduced (by transmutation and recycling)
Proliferation Resistance	Moderate	High
Resource Utilization	~1% of uranium	Up to 80% with closed cycles

5. Challenges: Despite many advantages, Generation IV reactors have few challenges, like high initial costs for research & development and construction, regulatory approvals, and durability of materials that must withstand extreme temperatures. As research progresses, they may become an option for sustainable energy sources.

PASSIVE SAFETY FEATURES IN ADVANCED REACTORS

Those safety features that use natural forces like heat convection, gravity, pressure difference, buoyancy, etc, as an energy source for their operation are known as Passive Safety Features. They do not use any artificial power source such as electricity. These features are used in an emergency for auto shutdown in case of an emergency. Traditionally, they were in use to a limited extent, and electrical or diesel motors were heavily used. Modern designs prefer passive safety features as they can operate naturally and without requiring any power source or human intervention. They offer reliable safety features to tackle the climate-induced emergencies like floods, storms, and heatwaves, or in case of emergencies. For example, in case of flood or power failure, traditional reactors might face coolant loss, which leads to overheating; however, in case of advanced reactors with passive safety design, like gravity-fed cooling, it will work to cool the core without pumps. Natural air circulation due to temperature gradient will naturally cool even in the case that the external power fails. Advanced materials such as molten salt and liquid are used for better heat dissipation. SMRs also play a vital role due to their underground installation, thus adding a layer of safety and climate resilience in case of natural disasters. Fail-safe shutdown mechanisms further add a layer of safety and reliability. The Pebble Bed Reactor is an example of a reactor utilizing a passive safety process for all operational modes. In case of temperature rise of the fuel rod, Doppler broadening increases the neutron capturing by U_{238} rather than U_{235}, thus reducing the reactor's power output and fission process itself, leading to an upper limit on the temperature (Parsons, 2017).

ADVANCED MATERIALS

Present nuclear reactors are expected to withstand the extreme climate challenges, along with proving them ideal even in the case of the worst weather conditions for a longer period. In this regard, durable, efficient, safe, and corrosion-resistant material is required to be utilized. Certain nickel-based super alloys can withstand temperatures exceeding 1,000°C without degrading the structural deformities, making them suitable for nuclear cores and heat exchangers. Ceramics, in particular Silicon Carbide composites, are also found suitable due to their radiation resistance and thermal stability. These materials can function well for a longer period in prolonged radiation-prone areas at super-high temperatures. At present, various alloys are being tested and tried. Prominent among them are FeCrAl and Zr Alloys. Another method is to use advanced coatings and claddings. Iron alloys generally face higher corrosion in case of humid environments, thus Zirconium alloys with protective coatings and stainless steel are used to prevent corrosion, increasing the life of critical components. Modern materials like molten salts and liquid Sodium or liquid Bismuth may be used for better heat transfer and thus improving overall efficiency. These metals can also withstand extreme temperatures and pressure inside the reactor. (Guérin et al., 2009)

STABLE ENERGY SUPPLY AMIDST UNSTABLE ENVIRONMENTS

Recently, in January 2025, California, USA destroyed nearly 12,000 structures by burning more than 40,000 acres of land, along with killing 24 people. Power grids have failed in an entire area. One cannot ensure a continuous and reliable power supply in such situations. More robust power infrastructure will be required to cop with these challenges.

Adaptation Strategies for Climate Change

- Floods: More robust infrastructure is required to be designed to withstand floods and their consequences. Restoring wetlands, planting mangroves act as flood buffers. Proper urban planning to prevent construction in flood-prone zones, early warning system implementation, and community preparedness help in mitigating the effects of floods.
- Droughts: Proper water management is required to be implemented through modern irrigation techniques like drip irrigation to conserve water, along with proper storage of rainwater to fight droughts. Less water-consuming

crops may be promoted. Groundwater level may be increased through re-charge systems.

- Rising Temperatures: To reduce the heat absorption, heat-reflecting material may be used in construction. Increase green coverage by planting more trees to increase cooling. Promote crops suited for a warmer climate and take steps to improve soil quality to retain more moisture for a longer duration of time. Educate people about preparedness for heat strokes and their impacts.

INTEGRATION OF RENEWABLE ENERGY

Renewable energy sources are always considered a boon, but their intermittency always poses a big challenge. Its use with nuclear power as a base load may, however, revolutionize the power sector. Combining these two sources increases the reliability and stability of a system, along with a reduction of carbon generation. (Suman, 2018). Nuclear power may act as a base load, which provides a consistent, reliable generation balance for the variable generation of solar and wind. Advanced nuclear reactors also offer flexibility in terms of power generation, which can increase or decrease the power output by the power output of renewable sources. To further optimize, we may pair energy storage technologies like hydrogen production and batteries. During high availability of renewable power and low demand, excess power generated can be stored in batteries or as hydrogen through electrolysis. This stored energy can then be reproduced during low renewable generation and high-power demand. This type of hybrid energy system also helps efficient use of transmission lines, thus improving energy efficiency distribution and reducing overall cost. Advanced control systems and smart grids are required to enable real-time monitoring and load balancing of nuclear and renewable energy sources. The role of policy framing and market incentives like carbon pricing, investment in research, and promotion of hybrid systems plays a pivotal role in the integration of nuclear and renewable energy sources. By combining the reliability of nuclear sources and sustainable renewable energy sources, we can effectively bridge the intermittency challenges and find a way for low low-carbon energy future. The following figure shows how renewable sources can complement nuclear energy.

Figure 1. A block diagram to show the integration of renewable and nuclear energy sources.

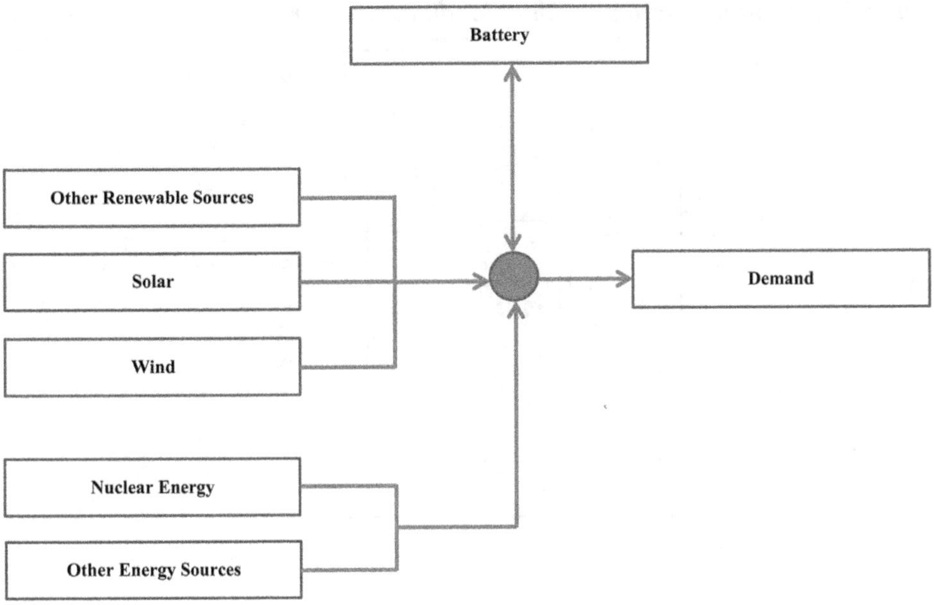

GRID INFRASTRUCTURE AND NUCLEAR POWER

Nuclear power plants generate a continuous load without emitting greenhouse gases, thus making them ideal for base load. Modern grid requires a well-designed and developed grid infrastructure to accommodate nuclear power. For this, advanced grids inculcating real-time monitoring & data analysis are essential, which are robust and smart enough to handle fluctuations of intermittent renewable energy sources along with nuclear power to ensure uninterrupted power delivery. Generally, nuclear reactors are in remote areas from the main load centres, like urban areas. So, transmission lines are required to transmit the power from the generating point to the consuming point. However, SMRs can provide flexible nuclear power and are designed to support a decentralized grid, and can complement renewable energy sources very easily (Bragg-Sitton et al, 2020).

Figure 2. Schematic diagram of Small Modular Reactor

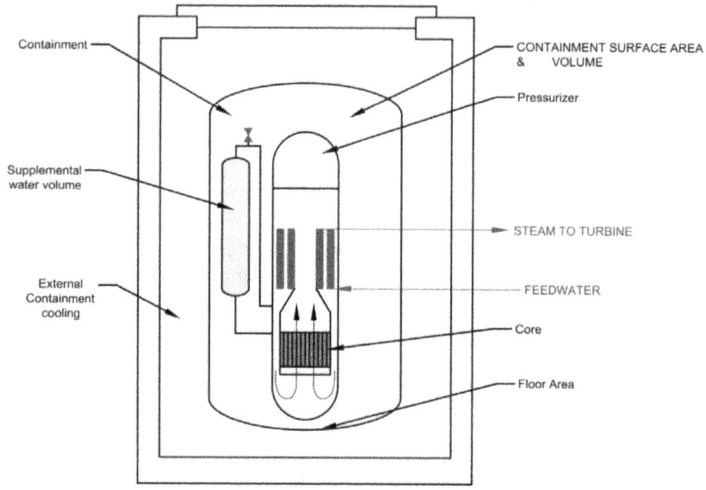

1. SMRs for Remote Areas

Smaller Modular Reactors of SMRs, as discussed earlier, are emerging as a promising alternative to traditional nuclear reactors for providing reliable, clean, and carbon-free energy. They are smaller in size and more flexible to suit the requirements of remote areas. Remote areas mostly use more polluting power sources like diesel generators, which are costly to operate. SMRs with capacities from 10 MW to 250 MW can fulfil the demand of a small community, a village, a mining operation, or a remotely located industry. This also helps reduce the transmission losses. One of the main advantages of Small Modular Reactors is their modular and scalable structure, thus significantly reducing the construction time (Luo et al., 2021).

Additional modules can be very easily added to the existing capacity as and when the need increases. They may not need to be refuelled for many years, as long as 20 years. Thus, it reduces the logistical requirements of fuel transportation. They also incorporate modern passive safety features to ensure quick, safe shutdown in emergencies. They may also be used for other purposes, like desalination plants and heating, to be more useful in remote areas. Many countries are exploring the options to deploy SMRs in the Arctic region and isolated islands.

2. Fighting Water Scarcity

Water is essential for all animals and plants. But safe drinking water has become a big challenge due to its poor availability. Billions of people lack this necessity. At present, desalination plants are used to remove the salts from saline or sea water to get drinking water in many parts of the world, but this process greatly depends on the energy retrieved from fossil fuels, which further contributes to greenhouse gas emissions (Buzzetti et al, 2024). However, nuclear power plants offer a sustainable solution to tackle water scarcity without increasing the environmental impact. Nuclear power provides a continuous base supply for the power requirement for reverse osmosis (RO) as well as continuous heat for the desalination process, such as multistage flash (MSF) or multi-effect distillation (MED). Many countries, including India, Japan, and Russia, are exploring such desalination plants due to their cost-effectiveness. BARC (Bhabha Atomic Research Centre) has already commissioned a few plants in India to get fresh water from seawater, for instance, one in Kalpakkam (a hybrid MSF-RO plant) and another in Bhuj (Gujarat) along the Creek border to provide fresh water to Border Security Forces. Desalination plants have many advantages, like utilizing the excess heat of a nuclear power plant, providing fresh drinking water in remote or water-scarce areas, providing water to make up for a power plant, and overall increasing the efficiency of a plant while reducing the carbon footprint (Buzzetti et al,2024). Thus, proving them more suitable as an ideal power source in remote water-scarce areas while combating global climate challenges. Figure No. 2 shows a block diagram of a nuclear power plant also supplying drinking water through two methods, viz., distillation and RO. The following figure represents the extraction of water using nuclear energy.

Figure 3. Schematic diagram to extract water using nuclear energy.

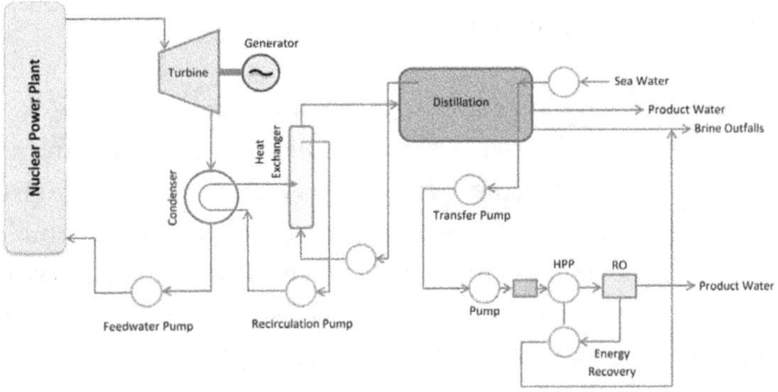

CASE STUDIES

Fukushima: Lessons in Climate and Disaster Preparedness

It was an earthquake and then a tsunami in March 2011, which devastated the Fukushima Daiichi power plant, leading to a meltdown in three reactors, paralysing communities nearby (Lukner & Sakaki, 2013). Authorities and administration failed to implement proper safety mechanisms. They had underestimated the power of natural disasters. They were not prepared for such an unprecedented event. It exposed the need for robust plans to meet such challenges. It highlighted the need for risk assessments & improved communication while dealing with hazardous infrastructures like nuclear power plants. Following this mishap, Japan has established a new regulatory mechanism. It may be termed as a near-miss catastrophe. Luckily wind blew in the direction of the sea rather than densely populated areas in the south, like Tokyo, otherwise it might have caused greater damage and contamination, dislocation, and loss of life than it did. It was later discovered in the investigation that it was a result of organizational inertia and human error. Every year, Japan experiences thousands of minor earthquakes and about 150 earthquakes with a magnitude of more than 5. Even the public had faced minor and major tsunamis in the past. Yet administration and nuclear plant authority were unprepared for the plausible destruction due to the "absolutely safe" myth as it was portrayed about a nuclear village in residents around the reactor. Contingency planning was missing. For instance, they had established an off-site centre for emergency response after a previous accident, which had exposed many workers and residents to higher radiation levels, but this centre too had become inoperative due to the earthquake and tsunami. Blackouts in the area worsened the situation. Protective measures were not enough to tackle this type of incident.

Key lessons that we may learn from this incident are:

1. Never underestimate the potential severity of natural powers like tsunamis and earthquakes. The need for risk assessment plans in case of extreme scenarios
2. Design Flaws & Safety Measures: The Design was not strong enough to withstand a large tsunami. Critical safety systems were located at vulnerable locations, leading to power loss abs hence core meltdown. Design should be strong enough.
3. Poor communication among authorities and, lack of emergency plans hampered effective mitigation. Proper escape and disaster plans should be available.
4. Timely and transparent information was not shared with the public, leading to erosion of public trust and anxiety among communities.
5. Cascading effects due to interconnectedness led to power outages in cooling systems, exposing the design flaw from a safety perspective.
    - Canada's SMR Deployment for Harsh Environments

Canada is leading the path of Small Modular Reactors to meet the energy requirements in its harsh and remote areas. These small reactors are tailored to suit the needs of isolated communities or industrial areas, where our traditional systems fail. Conventional grids are not suitable and economical for Canada's vast and sparsely populated northern region. Hence, diesel generators are generally used to meet the energy needs, but they are costly, polluting, and depend on supply chain systems. SMRs are best suited to these types of requirements. They are specifically designed to work in such a harsh environment. Canada plans to achieve a target of net-zero carbon emissions by 2050. SMRs significantly reduce carbon footprints (Sam-Aggrey, 2015). Canada is playing a transformative role in the deployment of SMRs in collaboration with government, industry, and local communities to meet the energy demands of remote areas sustainably.

- France's Climate-Resilient Energy Grid

France is a global leader in opting for nuclear energy and targets its ambitious decarbonization goals. Nearly 70% of its energy demand is generated from nuclear power plants. These nuclear establishments are providing baseload in all weather conditions. Adaptive technology has evolved to meet adverse situations, for example, during heat waves, alternative cooling technologies are used to conserve water. Advanced smart grid technology has been employed for real-time monitoring and quick response. Energy storage systems and demand response mechanisms have been strengthened. Renewable energy sources like solar and wind have been widely integrated into the main grid to provide diversity in the generation (Bonjean Stanton, 2019).

POLICY AND GOVERNANCE FOR NUCLEAR-DRIVEN CLIMATE RESILIENCE

Reducing greenhouse gases (GHG) can be easily achieved through the integration of advanced nuclear power sources while enhancing resilience against climate calamities. This is an emerging strategy to combat climate change as well as to meet the energy need, as it assures reliable, low-carbon energy, decentralized energy, co-generation of heat and water desalination, along with resilience to natural disasters. Nuclear systems can provide a continuous electricity supply for critical loads like hospitals, communication systems, and emergency systems. Microgrids powered by small modular reactors also reduce the possibility and risk of cascading blackouts. Policies and a regulatory framework are required to be drafted to incorporate nuclear energy into long-term national adaptation goals. Educational campaigns are required to be launched to spread awareness among the public to build trust to get rid of misinformation about nuclear energy. Flexible financial tools are required to be developed and promoted to motivate investment in advanced nuclear energy sources, like carbon credits and carbon trading. Special feasibility studies should be planned to identify the locations best suited for nuclear systems. Train the manpower for the required latest skills.

International Cooperation and Climate Resilience Agreements

The role of international cooperation is also very important. It fosters technology sharing for the transfer of advanced inventions and skills. It enhances the safety standards and operational standards to promote quick deployment. It can encourage joint investments, public-private partnerships, and access to international funding sources and subsidies. It promotes research and development (R&D) to improve reactor design and safe disposal of waste.

Existing Frameworks:

- Paris Agreement: It is an international treaty between various countries. It promotes all countries to reduce GHGs to mitigate climate change. It was formally launched in 2015 at the UN Climate Change Conference (COP 21) in Paris, France. Its goals are to limit the global average temperature increase to below 1.5°C above pre-industrial levels and reduce GHG emissions. It supports technology transfer to developing countries, encourages all nations to increase their share of nuclear energy in their power sector to reduce greenhouse gas emissions (Allen et al, 2019).
- International Atomic Energy Agency: It promotes the peaceful use of nuclear energy for the benefit of mankind, such as electricity generation. For that, it monitors and inspects the nuclear facilities, it works to prevent the spread of nuclear weapons, and it develops the standards for nuclear safety and security. It supports countries in supporting sustainable development to address environmental issues. It also promotes international cooperation between countries on nuclear research, deployment, and safety standards. The International Atomic Energy Agency (IAEA) works under the United Nations.
- Nuclear Energy Agency (NEA): All members of this federation receive technical and legal bases for using nuclear energy safely and economically. It is an intergovernmental agency that is organized under the Organization for Economic Co-operation and Development (OECD). It was formed in February 1958. A total of 33 countries are its members, including Australia, Canada, France, Germany, Japan, South Korea, the United Kingdom, and the United States. Together they account for approximately. 85% of the world's

installed nuclear capacity. Its prominent work areas are nuclear safety and regulation, nuclear energy development, radioactive waste management, nuclear laws, radiation protection, and public health.

Resilient Nuclear Energy Frameworks: Best Practices from Leading Nations

Based on long experience, leading nations in the field of nuclear energy program have developed some best practices worth following for others, which ensure the safety of fuel, powerplant, and waste in case of natural disasters. A few of them are mentioned below:

a. Passive safety features: The USA employs a passive cooling system, which functions without power in case of a natural disaster or power failure. Post-Fukishima, Japan has adopted boiling water reactors with gravity-fed cooling systems, allowing safe shutdown of a nuclear power plant in case of emergency.
b. Advanced Reactor Technology: Russia has focused on reducing nuclear waste and resource sustainability through its BN-800 reactors. Canada has emphasized waste fuel recycling and introduced CANDU reactors, which use Thorium as fuel, thus bringing fuel diversity (Panov et al, 2021).
c. Modular Construction: The United Kingdom's Rolls-Royce SMRs are designed for modular construction, allowing deployment in remote areas very quickly.
d. In France, ASN (Autorité de Sûreté Nucléaire) looks after all safety standards across the whole country stringently. South Korea has the NSSC (Nuclear Safety and Security Commission) to adopt a transparent licensing framework incorporating best international practices.
e. Learning from past accidents, Sweden has made it mandatory to conduct regular stress tests for reactors to assess their ability to withstand extreme weather.
f. Post-Fukishima, Japan has invested in tsunami walls, elevated cooling systems. The United States has established a mandatory practice to keep portable generators, water pumps to respond in emergencies.
g. Canada introduced decentralized microgrids through SMRs for vulnerable regions like the Arctic. Russia deployed floating nuclear power plants (e.g., Akademik Lomonosov) to provide energy in isolated areas where traditional grids are impractical.
h. H.H. KAERI (Korea Atomic Energy Research Institute) of South Korea regularly organizes training programs for domestic and international professionals to share knowledge and expertise in the nuclear field. The United States, though DoE funds university programs to research nuclear technology and safety. France & India have collaborated on EPR (European Pressurized Reactor) to share innova-

tive practices and co-development of advanced systems. Finland and the United Kingdom regularly conduct consultations with the public to bring transparency, inclusivity, and trust through publicly accessible repositories like Onkalo. Many countries have initiated public campaigns to aware common citizens about the benefits of nuclear energy and the role of nuclear energy in fighting climate change. A leading example is the Canadian Nuclear Safety Commission (CNSC). Sweden, Finland, and the United States aim to increase long-term geological storage facilities that safely isolate nuclear waste. La Hague, France, reprocesses the nuclear waste, whereas Russia uses a closed fuel cycle to ensure minimum nuclear waste and efficient use of uranium (Prosvirova, 2018).

i. The European Union and ASEAN have signed MoUs for the peaceful use of nuclear energy and cross-border cooperation, and nuclear safety. Knowledge and technology are shared for smaller nations through programs like the IAEA projects.

Points discussed above have led to low incidents of operational failures, accidents, and environmental harm. Transparent communication has led to higher public trust and public acceptability, and cost-effective projects have led to the development of robust SMRs along with a reduction in greenhouse gas emissions by replacing fossil fuels.

ADDRESSING SOCIOECONOMIC DIMENSIONS OF CLIMATE RESILIENCE

The socioeconomic dimension of climate resilience nuclear power plant perspective includes job creation, sustainable development, careful management of social feelings and perceptions, and community engagement. Let us see how nuclear energy contributes to socioeconomic resilience.

Employment Opportunities in Advanced Nuclear Development

a. Contribution to socio-economic resilience: Energy security through stable and reliable, consistent electricity even during extreme weather. Decentralized energy generation reduces the vulnerabilities of centralized grids.
b. Economic development: Economic development is possible through job creation in the fields of construction, maintenance, safety, and others. Skilled and semi-skilled manpower is generally sourced from the local area. Local economy is also stimulated through supporting industries like transportation, waste man-

agement, & engineering. The stay of employees boosts the local market, right from groceries to the clothing market.

c. Rural development: Generally, nuclear reactors are deployed in rural, remote areas. This not only provides electricity access to this remote area but also reduces urban migration. It also bridges the energy gap at an affordable price to marginalized communities dwelling in these remote areas.
d. It also helps solve water scarcity issues, a major problem in vulnerable areas.
e. It also helps in providing energy for heating in colder countries as a side product, relinquishing dependency on fossil fuels.
f. Overall quality of life in the area is elevated by revenue from nuclear power plants, though fund divergence in schools, local infrastructures, hospitals, etc. (Kenley et al, 2009)

Reducing Vulnerabilities in Energy-Dependent Economies

For a nation to progress, energy security is a must to have. Reducing energy vulnerability is essential for energy-dependent economies for sustainable development. These vulnerabilities often arise from over-reliance on fossil fuels and climate change. Nuclear energy provides a robust solution, along with energy diversity, against these vulnerabilities.

- Key Vulnerabilities: Over-reliance on petroleum products always threatens economies through price volatility, supply chain disruptions, and geopolitical relations between nations affected through world geopolitics. High greenhouse gas emissions are an unwanted additional environmental vulnerability. Energy insecurity is posed in case a country is heavily dependent on imports. Energy shortages may occur during extreme weather conditions, leading to disruptions of economic activities. Economic Vulnerability may occur as price fluctuations in the petroleum market directly affect government revenues. Further, subsidies on fossil fuels may drain the nation's budget away from investment in clean energy. Pollution due to fossil fuels and petroleum products affects public health.
- Role of Nuclear Energy in Reducing Vulnerabilities: Diversification of energy sources is possible through the integration of nuclear and renewable energy sources, reducing dependency on polluting fossil fuels. Nuclear power plants offer consistent, reliable power in almost all conditions, making it ideal as a base load, ensuring power stability. Long-term operational cost of nuclear power plants is stable, thus shielding them from the volatile fossil-fuel price market. New revenue streams are created through sharing expertise in nuclear power-based electricity and heat generation. South Korea and France

are leading examples to prove it. Climate resilience is achieved as nuclear power helps to reduce greenhouse gas emissions, supporting sustainable development. It also offers a water shortage solution in drought-prone areas, reducing one more vulnerability.

- Strategy to reduce vulnerabilities: Diversification of energy sources, along with the integration of renewable sources, is essential to reduce vulnerabilities. SMRs are to be given priority rather than traditional reactors. Develop a policy to assess and manage risks in the energy market. Promote clean energy and phase out subsidies on fossil fuels gradually. Invest in smart grids that can work with renewable energy and nuclear energy sources. Promote microgrids through SMRs for electrifying disaster-prone remote areas. Use advanced reactors with passive safety features to reduce operational risks. Train workers from traditional power plants in nuclear energy and operation. Start educational programs that focus on advanced nuclear knowledge and technology. Promote advanced fuel cycles and waste-reducing technology. Increase international cooperation for nuclear technology to acquire the best safety and operational practices across the globe. Developing countries should be supported through technical and financial assistance to encourage the use of nuclear energy rather than getting energy from fossil fuels.
- South Korea has reduced its dependency on fossil fuels. Approximately 30% of its energy is produced from the nuclear sector through 24 operational reactors (Kim et al, 2021). Finland also gets 30% of its electricity from nuclear power plants. Its Olkiluoto repository has become a role model across the globe to demonstrate safe and long-term nuclear waste management. Even oil oil-rich economy like the United Arab Emirates has diversified its energy demand through the Barakah Nuclear Power Plant.

CHALLENGES AND OPPORTUNITIES FOR RESILIENT NUCLEAR SYSTEMS

To have the assurance of consistent power, we will need a reliable energy source resilient to various climatic conditions. However, key challenges associated with operational risks and natural hazards are always there.

Challenges:

a. Aging infrastructure: Nuclear systems not just require heavy investments for commissioning, but they also require heavy investments while decommissioning due to the radioactivity leakage threat. Many plants across the globe are approaching or have already exceeded their designed lifetimes. Maintaining them is a very big challenge, incurring heavy costs (Laraia, 2018).

b. Nuclear safety: Continuous improvements in safety protocol are necessary lest a nuclear accident occur. Fukushima and Chernobyl have already shown the devastating effects of nuclear accidents. Cybersecurity and cyberattacks also require a proper defence mechanism, so that this technology doesn't fall into terrorists' hands.
c. Natural hazards like extreme weather, heat strokes, tsunamis, earthquakes, rising sea levels, and climate change always pose a significant challenge. New innovative designs are required to be designed to counter these threats.
d. Long-term storage and disposal of waste always remain a critical challenge for nuclear energy.
e. Devastating effects of previous nuclear accidents have ruined the trust of public in nuclear energy in many regions. To remove such types of misconceptions & misinformation and create an amicable environment to accept a nuclear power plant in a peripheral area is always a significant challenge for authorities.

Opportunities:

a. Advancements in reactor technology, like Small Modular Reactors (SMRs), offer smaller, scalable, factory-made, easy-to-deploy reactors with enhanced safety features. Generation IV reactors are designed to have higher efficiency and lower waste with improved safety parameters.
b. Hybrid systems with the integration of renewable energy sources make nuclear energy a reliable base load to enhance the grid stability.
c. Climate change mitigation is possible through the deployment of nuclear power plants, as they do not generate greenhouse gases. Thus, contributing to global warming targets.
d. Nuclear energy establishments create many job opportunities and stimulate the local economies.
e. Global collaboration has led to international partnerships in safety enhancements, reduction of operational risks, reduction of nuclear waste, and better fuel efficiency through technology sharing.

Balancing Cost, Safety, and Climate Benefits

No doubt regarding the fact that nuclear energy provides a reliable solution that is capable of supporting decarbonization goals, but at the same time, it is also associated with high costs and safety concerns. Climate benefits are essential for sustainable development, but an equilibrium needs to be explored in depth to make it economically viable. Nuclear power plants require a very high initial investment, a long construction timeline, along high compliance with regulatory frameworks and policies. Decommissioning of old reactors and management of radioactive waste also add a heavy financial burden. Whereas renewable energy sources, through solar panels, are offering cheap electricity. Per per-watt price of solar energy is declining at a rocket speed. It can be witnessed through the Mega and Giga solar power plants. India and China are leading the race. Special surveys or studies must be conducted to find out the point of equilibrium above which the nuclear power plants become beneficial and economical in the long term. These challenges may vary for different countries or even for locations within the country.

Research and Development Needs for Future Climate Resilience

A deeper focus on Research and Development (R&D) is required to be given to harness the advantages of nuclear energy. Special target should be on reactor designs, core material, cooling material, safety protocols, and integration with renewable energy sources. The following points should be considered carefully:

- Neutron flux distribution should be optimized to enhance fuel burning and reduce waste. Integrated simulation models should be developed to study and predict the reactor behaviour under extreme conditions for fluid mechanics, heat transfer, and neutron transport. Material with higher thermal resistance is to be explored to function as control rods for reliable operation under very high temperature transients.
- Development of high-density fuel rods needs attention to reduce core volume, thus making it convenient for space exploration. More experiments are required to prevent oxidation in fuel rods under extreme temperatures in case of low cooling. Thus, high-density oxidation-proof Accident Tolerant Fuels (ATFs) are required. More R & D. and research is required to be conducted on oxide-dispersion-strengthened (ODS) steels for reactor and pipes to tolerate extreme temperatures. High entropy alloys (HEAs) should be explored. Anti-corrosion coatings need to be developed for seashore or high-humidity areas (Nie et al, 2022).
- More study is required for cooling through liquid metals like lead-bismuth or sodium for fast reactors. Compact heat exchangers should be designed and developed for higher efficiency and lower heat inertia. More investigation is required to get rid of traditional steam cycles by replacing them with thermoelectric or magnetohydrodynamic (MHD) systems to convert heat directly into electricity.

- Closed Brayton cycles with working fluid supercritical (sCO_2) should be developed for increased performance. Integrated approach needs more attention, particularly in the coproduction of hydrogen (though thermochemical cycles like Sulphur-Iodine), the coproduction of heat, production of drinking water in drought-prone areas.
- Machine learning based algorithms should be implemented to control reactor operation for response to variable load to optimize fuel usage. To work with renewable energy sources, rapid ramp-up and rapid ramp-down cycles should be planned. Through fast reactors, frequency regulation and voltage stability can be achieved with modern power electronics devices (Ahmad, 2023).
- Develop advanced thermal materials that can withstand fluctuating environmental conditions. More research is required for shock-absorbing and vibration-damping materials to cop with seismic shocks. The modular containment system for SMRs needs to be improved for flood-prone areas.
- Radiation-resistant electronics need to be developed for in-core and ex-core monitoring. Long-lived radioactive isotopes should be converted to short-lived or non-radioactive materials through special reactors. Efficient separation of actinides from spent fuel needs more exploration through pyro-processing and electrorefining.
- A system with auto-initiated shutdown in case of extreme pressure or extreme temperature, or other emergencies, should be developed, which should be capable of functioning through real-time data analysis. Fibre optics needs more exploration as they are radiation-proof.
- Use Artificial Intelligence to optimize the design to conserve fuel, reduce waste, and real-time monitoring and frequency-voltage regulation before feeding power to the grid.

FUTURE PATH IN RESILIENT NUCLEAR SYSTEMS

The future path of nuclear systems includes many critical dimensions, including opportunities and challenges. Opportunities lead to a very optimistic dawn, but at the same time, challenges are pointing towards huge issues related to nuclear proliferation, sabotage, safety, and waste management. Let us explore the role of AI, predictive analysis, and innovative reactor design in more detail, which can establish nuclear energy in the future to achieve net-zero carbon goals.

AI and Predictive Analytics for Climate-Responsive Reactor Operations

Reactor designs have been evolving since their inception. Major focus has always been on safety, radioactive leakage, better fuel management, and efficiency. When we investigate the past few decades, we can easily conclude an increased frequency of climatic calamities like floods, hurricanes, droughts, extreme temperatures, and tsunamis. Varying water availability affects the cooling of plants. Increased sea level affects the reactors located near coastal areas. Unpredictable weather patterns are affecting the operational risks associated with nuclear reactors. Present nuclear reactors should be able to withstand these climatic events. Artificial intelligence can help a lot in this area. AI-powered sensors can provide real-time monitoring data through continuous monitoring of temperature, pressure, and radiation levels, along with environmental temperature fluctuations and sea water levels. AI-based models can predict component wear and future failure and optimize maintenance schedules to reduce unintended downtime. Early warning systems and risk assessment are possible through AI. AI-driven adjustments may be helpful to optimize the cooling water usage based on real-time conditions during droughts or heat waves. Prediction analysis is possible through the integration of historical data. Simulation models can predict reactor performance under extreme climate conditions to better predict and understand the reactor design and its ability to withstand calamities. Using AI, we can forecast energy demand and integrate it with the power generation. Grid stability can also be enhanced with adjustments in reactor output. However, many challenges are also there concerning the integration of AI with nuclear reactors, like dependency and accuracy of climatic data, challenges to integrate older reactors, etc. A balance will be required to include human oversight in applying AI to ensure ethical implications, transparency in AI decisions (Huang et al, 2023).

Fusion Energy and Its Role in Long-Term Resilience

The sun is the ultimate source of energy for us and the whole Earth. The sun is using a fusion reaction to generate energy. Presently, we are using fission principles to generate energy, which divides a big nucleus of an atom into two or more smaller nuclei, whereas fusion is the process of combining two lighter atomic nuclei to form a heavy nucleus along with the production of energy. All stars, including our sun, produce energy through this principle. The simplest case of fusion reaction involves Deuterium (^2H) and Tritium (^3H). Deuterium is obtained from seawater, and Tritium is produced from Lithium (Wayne et al, 2002). Both are widely available in abundant amounts, but their fusion reaction occurs under extreme conditions. To overcome the repulsion between nuclei, a temperature greater than 100 million degrees Celsius is required. Also, high pressure or strong magnetic fields are required to maintain plasma. These conditions are very difficult to achieve outside the laboratory to sustain the reaction. Let us now compare fusion and fission energy briefly in the table below:

Table 2. Comparison of Fusion and Fission Energy

Criterion	Fusion Energy	Fission Energy
Fuel Availability	Raw material is not easily available and is costly	Raw material is comparatively cheap and easily available
Fuel Sources	Deuterium (Heavy water) Tritium (through lithium)	Uranium, Plutonium
Safety	Safe	The risk of meltdown requires strict protocols
Chain Reaction	No	Yes
Waste	Minimal radioactive waste, short-lived isotopes are generated	Long-lived radioactive isotopes are generated
Environmental effect	No greenhouse gas emissions	Low emissions of GHGs, but waste storage is a headache
Process	Light nuclei are heated to extremely high temperatures under heavy pressure to fuse.	A heavy nucleus is bombarded with a neutron to break it into smaller nuclei.

- Why Fusion energy: Global energy demand is increasing very rapidly. Fossil fuels are polluting our environment by producing greenhouse gases. Production of energy through nuclear fission poses a big threat to nuclear accidents, safety, and long-lasting radioactive waste management. Renewable energy sources are intermittent, and they are environmentally dependent. Seeing this scenario, fusion is one of the best solutions to ever ever-growing energy demand. Fusion technology can help reduce global temperature and achieve a net-zero carbon footprint by replacing fossil fuels. It reduces the dependency on petroleum products like natural gas, oil, and geopolitically sensitive fuel like Uranium and Plutonium. Unlike renewable energy sources, fusion reaction offers a continuous power supply and heat irrespective of environmental conditions or weather patterns, or availability of wind or sunlight.
- Brief history of fusion research: Since 1945, research on fusion reaction began along with fission reaction. In the 1960s, the concept of fusion reaction gained momentum after the introduction of magnetic containment through tokamaks. In 1985, a global project was launched under the umbrella of ITER (International Thermonuclear Experimental Reactor) to demonstrate the fea-

sibility of fusion reaction. However, as of today, fusion energy is still in the cradle of research and development. No commercial power plants are operating anywhere in the world. Few experimental reactors achieved fusion reaction for a considerable time, but more research and experiments are required to make it viable (El Guebaly et al, 2010).

- Current developments in fusion technology: Out of the many approaches, magnetic confinement fusion (MCF) seems most promising to achieve a sustained fusion reaction. ITER (International Thermonuclear Experimental Reactor), a multinational project comprising 35 countries based in France, is researching to demonstrate the feasibility of fusion reaction through tokamaks. Their goal is to produce 500MW of power with the input of 50MW, thus getting a net energy gain of Q>10. Major milestones achieved so far are progress in superconducting magnets and plasma-facing materials to withstand extreme temperatures and radiation. SPARC (Compact High-Field Tokamak by Commonwealth Fusion Systems), a private sector research initiative, is aiming to achieve high-temperature superconducting (HTS) magnets for net positive energy generation through fusion reaction. In 2022, EAST (Experimental Advanced Superconducting Tokamak) of China successfully achieved plasma sustainability for over 1000 seconds. India is conducting research through its tokomak facility at the Indian Plasma Research (IPR) institute located at Gandhinagar, Gujarat. To address the stability issues of tokomaks, stellarators are designed to achieve long-duration sustainability of plasma. Recently, though, stellarators, Wendelstein 7-X of Germany, achieved a remarkable performance of steady state operation of 30 minutes (Klinger et al, 2019). Inertial Confinement technology (ICF) uses powerful lasers to compress and heat the fuel pellets. National Ignition Facility (NIF), based in California, USA, in December 2022 generated net positive (Q>1) energy from lasers. They are focusing on quality fuel pallets and improved laser efficiency. Some researchers are also experimenting on hybrid concepts to combine ICF and magnetized target fusion to enhance energy yield.
- Breakthroughs in Alternative Fusion Approaches: While MCF & ICF, that is, magnetic confinement fusion and inertial confinement fusion, are the most experimented approaches, alternative concepts are also being tried by researchers to find alternative technology with reduced cost and faster commercialization. Compact fusion reactors are gaining more significance due to their compact size and easy deployment. One such company is Helion Energy, which focuses on the direct conversion of fusion energy into electricity using electromagnetic coils, using the principle of electromagnetic induction, eliminating the need for traditional steam turbines. It uses pulse magnetic compression technology to achieve fusion. Another concept is Aneutronic

Fusion. It produces minimum neutron radiation; thus, they are creating minimum radioactive waste and are safer. It uses TAT technology, which focuses on fusing Proton and Boron-11. In this case, energy will be released as charged particles rather than neutrons. Thus, heavy shielding is not required, ultimately leading to a reduction in the cost of the reactor. It also generated electricity directly through particle deceleration. Advanced machines like Norman and Copernicus are built to achieve stable high-temperature plasma. Commercialised aneutronic fusion may be achieved by 2035. Another prominent concept is Magnetized Target Fusion (MTF), which integrates the magnetic and inertial confinement to compress plasma. In this case, a liquid metal liner is used, which acts as a heat exchanger, neutron shield, and for compressing plasma. This concept simplifies the plasma compression. Field-Reversed Configurations (FRCs) use magnetic fields to confine plasma. The central core is not used in this method. It also yields high energy production. One more approach, Z-Z-Pinch fusion, uses electrical currents directly to compress plasma. It does not require complex magnetic structures (Wessel & Rostoker, 2000). It has the advantage of a simplified reactor design without complex external big magnetic coils consuming heavy power. It has potential for producing low-cost and rapid, scalable, compact reactors. The Dense Plasma Focus (DPF) method creates high-density plasma by pulsed electrical discharges. It focuses on achieving fusion through proton-boron-11 as fuel. It has the potential advantage of low-cost fusion and compact design. Alternative fuels to deuterium and tritium are proton-boron11 and Deuterium-helium-3. Helium-3 is rare, but it can be extracted from lunar resources in the future. Alternative approaches may suggest a compact reactor design rather than the present heavy, complex, and power-draining tokomak design. Present attention should be given to speed, cost effectiveness, scalability, and reduced waste.

- Innovations in Enabling Technology: Significant progress has been achieved in High Temperature Superconductors (HTS), which helps to produce stronger magnetic fields, leading to compact and efficient reactor design. Radiation-resistant materials need to be explored that can withstand high temperature, neutron flux, and plasma pressure, along with high thermal conductivity and durability.
- Towards commercialisation: In 2020, the goal of net positive energy, which is Q>1, was achieved. It is expected to deploy a pilot fusion power plant by 2030, along with the integration into the main power grid. As a long-term goal, it is expected to start large-scale commercialization by 2050 with a mix of renewable energy sources. The fusion energy concept is rapidly gain-

ing momentum to establish itself from theory to application (Vendinha et al, 2023).

- Challenges in implementing nuclear fusion energy: From the above discussion, it is very clear that nuclear fusion energy can become a main power source of the future, but before that, it must overcome certain major and minor hurdles in its way. The very big and most challenging aspect is that of initiating the nuclear fusion reaction, which requires extremely high temperature in the range of 100 to 150 million degrees Celsius, along with high plasma density confinement for a specified necessary time. Plasma, the fourth state of matter itself is unstable. Maintaining the plasma in a confined manner to extract energy is difficult. For those devices like tokomaks and stellarators, they are to be designed such that a strong magnetic field is maintained to confine the plasma. Precise control of plasma is difficult due to instabilities in the plasma caused by turbulence and disruptions. In another method of laser compression, it is technically challenging to compress and heat a very tiny mass of fuel pellets evenly. There are also challenges in material technology to prepare a strong enough material that can withstand extremely high temperatures and plasma pressure. These materials should also not be brittle on radiation exposure, having the ability to withstand thermal and mechanical stresses. Present reactors require more energy than they produce to sustain the nuclear fusion reaction (that is, $Q<1$). It is very important to get $Q >> 1$ to start economical and commercially viable reactors. Extraction of heat in an efficient manner is also a challenging task. Constructing a large fusion reactor like ITER requires a very high amount of funds. Their big size also poses challenges in scalability and affordability. A long development timeline also deters investors from R&D funding. Skilled workforce in physics, nuclear science, material technology, thermal engineering, and plasma physics will be required. Certain fuel material is also rare on Earth, for example, helium 3. It's abundantly available on the moon, but bringing it down to Earth is a daydream even today...! Tritium breeding in a nuclear reactor is still in the development stage and may take decades (Okada & Momoshima, 1993). Fusion reaction indeed produces short-lived radioactive wastes, but neutron activation may contaminate reactor material, hence, proper handling and waste disposal are necessary. Educating people will also be a big task, as people often link nuclear fusion reaction with nuclear fission reaction and related fission accidents like Chernobyl and Three Mile Island. Fusion projects require long-term term consistent funding from governments. The majority of countries do not have a policy or framework to support investment in nuclear fusion research. Concerning the climatic catastrophes and global warming, cleaner energy like nuclear fusion must be given priority for the survival of mankind at large.

Innovative Reactor Designs for Coastal and Climate-Stressed Areas

Advanced and innovative nuclear reactor designs are necessary in response to climatic calamities. The frequency and severity of climatic calamities are escalating. To tackle rising sea levels, extreme temperatures, drought, and corrosion in coastal areas, we shall have to come up with a relevant design. Small Modular Reactors (SMRs) as discussed previously may prove as best option to generate energy, heat, and water in remote vulnerable areas up to 300 MW. Micro reactors with capacities up to 10 MW, are ideal for small scale application in remote area say for a small remote and separate island community to run school, hospital, and domestic appliances. They are portable and can be airlifted to such remote areas and work autonomously with minimum maintenance for years. Like oil rigs, offshore floating reactors may be built on barges or offshore platforms. They are immune to earthquakes and tsunamis. Plenty of surrounding seawater is available for cooling. An example of a floating nuclear plant is Akademik-Lomonosov, is plant in Russia. Submersible Reactors are also in operation. They are safe from wind surges, storm surges, and earthquakes. They have the benefit of natural cooling. They also offer enhanced safety due to isolation from populated areas on land. They are more suitable for islands. Elevated reactor designs may be planned for protection from rising sea levels. At the same time, underground reactor designs may be best suited to the regions facing frequent tsunamis and hurricanes (Zhang et al,2016).

CALL TO ACTION: IT'S NOW OR NEVER

It is very clear from the previous discussion that energy demand is expected to grow in the future. Present fossil fuel resources are neither adequate nor environmentally suitable, as they are polluting our environment and are leading to climatic calamities through greenhouse gas emissions. So, it's in the interest of mankind to switch to alternative energy sources like renewable and nuclear energy sources. It is the right time and it's urgent to address this issue. Let us conclude the chapter with a discussion on policy development, international collaboration, and bridging the gap between policy, technology, and public awareness to attain global resilience through innovation.

Governments across the globe must act in the direction of framing policy and regulatory frameworks adaptable to emerging nuclear technologies. Incentives, though tax credits, grants, and subsidies, should be given for R&D to explore next-generation reactors and advanced materials capable of withstanding extreme temperatures and a radioactive environment. International cooperation is necessary through various collaborations for non-proliferation, safety, and technology sharing (Meier, 2013). The active and transparent partnership is expected to achieve success for projects like the International Thermonuclear Experimental Reactor (ITER). The International Atomic Energy Agency (IAEA) should proactively share technology with developing countries for peaceful purposes, like the generation of electricity. Auto shutdown technology must be shared among all stakeholders to avoid nuclear accidents. Investments in fusion energy, molten salt reactors, and alternative fuels should be promoted. Advanced recycling technology is required to reduce the half-life of radioactive isotopes. A proper waste management policy is required to be implemented in a stricter way to safeguard land, sea creatures, and groundwater. AI and predictive analysis may play a pivotal role in coming future. To remove the scepticism and misinformation, public education should be given priority, along with priority in the jobs to the local population, highlighting environmental benefits. Careful study of nuclear accidents and leakage should be considered, and appropriate measures should be taken to avoid such incidents in the future.

REFERENCES

Ahmad, A. (2023). Advances in machine learning for monitoring, control, and optimization of the temperature of reactors. *Journal of Chemical Engineering and Process Optimization, 15*(2), 101–115. DOI: 10.20944/preprints202309.1318.v1

Allen, M., Antwi-Agyei, P., Aragon-Durand, F., Babiker, M., Bertoldi, P., Bind, M., & Zickfeld, K. (2019). Technical summary. In *Global warming of 1.5 °C: An IPCC special report on the impacts of global warming of 1.5 °C above pre-industrial levels and related global greenhouse gas emission pathways, in the context of strengthening the global response to the threat of climate change, sustainable development, and efforts to eradicate poverty* (pp. 33–114). Intergovernmental Panel on Climate Change., https://www.ipcc.ch/sr15/

Amin, S. M. (2011). Smart grid: Overview, issues, and opportunities. Advances and challenges in sensing, modelling, simulation, optimization, and control. *European Journal of Control, 17*(5–6), 547–567. DOI: 10.3166/ejc.17.547-567

Bonjean Stanton, M. C. (2019). *Weathering the storm: Managing physical climate risks in the electricity sectors in the United Kingdom and France* (Doctoral dissertation, University of Leeds). University of Leeds. https://etheses.whiterose.ac.uk/id/eprint/25248/

Bragg-Sitton, S. M., Gorman, J., Burton, G., Moore, M., Siddiqui, A., Nagasawa, T., . . . Cox, J. (2020). *Flexible nuclear energy for clean energy systems* (NREL/TP-6A50-77088). National Renewable Energy Laboratory. https://www.nrel.gov/docs/fy20osti/77088.pdf

Buzzetti, R., Frano, R. L., & Cancemi, A. S. (2024). Feasibility study of a desalination plant powered by SMR. *Nuclear Engineering and Design, 418*, 112897. DOI: 10.1016/j.nucengdes.2023.112897

Canny, A., Kaziba, C., Mitchell, N., & Prest, E. (2020). *Effective and sustainable capacity-building for achieving the Paris Agreement: A special report on climate finance and digitalization.* Climate Finance & Digitalization Institute.

de Vendinha Ferreira, M. (2023). *Possible impacts of nuclear fusion deployment on spatial planning practices: Case studies of ITER and NIF projects.*

Eidemüller, D. (2021). Radioactive incidents and disasters. In *Nuclear power explained* (pp. 195–240). Springer., DOI: 10.1007/978-3-030-72670-6_10

El-Guebaly, L. A. (2010). Fifty years of magnetic fusion research (1958–2008): Brief historical overview and discussion of future trends. *Energies*, *3*(6), 1067–1086. DOI: 10.3390/en30601067

Forsberg, C. W., & Reich, W. J. (1991). Worldwide advanced nuclear power reactors with passive and inherent safety: What, why, how, and who. Oak Ridge National Laboratory., , ORNL/TM-11907.DOI: 10.2172/6365276

Fragkos, P. (2020). Global energy system transformations to 1.5 °C: The impact of revised Intergovernmental Panel on Climate Change carbon budgets. *Energy Technology (Weinheim)*, *8*(9), 2000395. DOI: 10.1002/ente.202000395

Guérin, Y., Was, G. S., & Zinkle, S. J. (2009). Materials challenges for advanced nuclear energy systems. *MRS Bulletin*, *34*(1), 10–19. DOI: 10.1557/mrs2009.6

Hidayatullah, H., Susyadi, S., & Subki, M. H. (2015). Design and technology development for small modular reactors–Safety expectations, prospects, and impediments of their deployment. *Progress in Nuclear Energy*, *79*, 127–135. DOI: 10.1016/j.pnucene.2014.11.010

Huang, Q., Peng, S., Deng, J., Zeng, H., Zhang, Z., Liu, Y., & Yuan, P. (2023). A review of the application of artificial intelligence to nuclear reactors: Where we are and what's next. *Heliyon*, *9*(3), e12738. DOI: 10.1016/j.heliyon.2023.e13883 PMID: 36895398

Jalles, J. T. (2020). The impact of financial crises on the environment in developing countries. *Annals of Finance*, *16*(2), 281–306. DOI: 10.1007/s10436-019-00356-x

Kabeyi, M. J. B., & Olanrewaju, O. A. (2022). Sustainable energy transition for renewable and low carbon grid electricity generation and supply. *Frontiers in Energy Research*, *9*, 743114. DOI: 10.3389/fenrg.2021.743114

Kenley, C. R., Klingler, R. D., Plowman, C. M., Soto, R., Turk, R. J., Baker, R. L., Close, S. A., McDonnell, V. L., Paul, S. W., Rabideau, L. R., Rao, S. S., & Reilly, B. P. (2009). Job creation due to the nuclear power resurgence in the United States. *Energy Policy*, *37*(11), 4894–4900. DOI: 10.1016/j.enpol.2009.06.045

Kim, J. H., & Yoo, S. H. (2021). Comparison of the economic effects of nuclear power and renewable energy deployment in South Korea. *Renewable & Sustainable Energy Reviews*, *135*, 110236. DOI: 10.1016/j.rser.2020.110236

Klinger, T., Andreeva, T., Bozhenkov, S., Brandt, C., Burhenn, R., Buttenschön, B., & Hollfeld, K. P. (2019). Overview of the first Wendelstein 7-X high-performance operation. *Nuclear Fusion*, *59*(11), 112004. DOI: 10.1088/1741-4326/ab03a7

Laraia, M. (2018). *Nuclear decommissioning*. Springer., DOI: 10.1007/978-3-319-75916-6

Lindley, B. A., Fiorina, C., Gregg, R., Franceschini, F., & Parks, G. T. (2015). The effectiveness of full actinide recycles as a nuclear waste management strategy when implemented over a limited timeframe–Part I: Uranium fuel cycle. *Progress in Nuclear Energy*, *85*, 498–510. DOI: 10.1016/j.pnucene.2015.07.020

Locatelli, G., Bingham, C., & Mancini, M. (2014). Small modular reactors: A comprehensive overview of their economics and strategic aspects. *Progress in Nuclear Energy*, *73*, 75–85. DOI: 10.1016/j.pnucene.2014.01.010

Locatelli, G., Mancini, M., & Todeschini, N. (2013). Generation IV nuclear reactors: Current status and prospects. *Energy Policy*, *61*, 1503–1520. DOI: 10.1016/j.enpol.2013.06.101

Lukner, K., & Sakaki, A. (2013). Lessons from Fukushima: An assessment of the investigations of the nuclear disaster. *Asia-Pacific Journal: Japan Focus, 2013*(19). https://apjjf.org/2013/19/19/Lukner/3959/article.html

Luo, B., Huang, G., Chen, J., Zhang, X., & Zhao, K. (2021). A chance-constrained small modular reactor siting model: A case study for the Province of Saskatchewan, Canada. *Renewable & Sustainable Energy Reviews*, *148*, 111320. DOI: 10.1016/j.rser.2021.111320

Meier, O. (2013). *Technology transfers and non-proliferation: Between control and cooperation*. Routledge. DOI: 10.4324/9780203798973

Nayak, A. K., & Sinha, R. K. (2007). Role of passive systems in advanced reactors. *Progress in Nuclear Energy*, *49*(6), 486–498. DOI: 10.1016/j.pnucene.2007.07.007

Nie, J., Jia, T., Pan, L., Zhang, X., & Zou, J. J. (2022). Development of high-energy-density liquid aerospace fuel: A perspective. *Transactions of Tianjin University*, *28*(1), 1–5. DOI: 10.1007/s12209-021-00302-x

Okada, S., & Momoshima, N. (1993). Overview of tritium: Characteristics, sources, and problems. *Health Physics*, *65*(6), 595–609. DOI: 10.1097/00004032-199312000-00001 PMID: 8244708

Panov, A. V., Kuznetsov, V. K., Isamov, N. N., Geshel', I. V., Trapeznikov, A. V., & Korzhavin, A. V. (2021). Assessment of the influence of BN-800 operation on the radioecological situation in the vicinity of Beloyarsk NPP. *Atomic Energy*, *129*(5), 297–304. DOI: 10.1007/s10512-021-00751-6

Parsons, D. K. (2017). *Some notes on neutron up-scattering and the Doppler-broadening of high-Z scattering resonances (No. LA-UR-17-28809)*. Los Alamos National Laboratory., https://www.osti.gov/servlets/purl/1371457 DOI: 10.2172/1396101

Proskuryakova, L. (2023). The contribution of low-carbon energy technologies to climate resilience. *Climate (Basel)*, *11*(12), 231. DOI: 10.3390/cli11120231

Prosvirova, Y. (2018). *International radioactive waste for long-term disposal* (Doctoral dissertation, University of Wien). Wien University.

Sam-Aggrey, H. (2015). Opportunities and challenges related to the deployment of small modular reactors (SMRs) in the Northern Territories of Canada. *Journal of Nuclear Engineering*, *44*(2), 23–34.

Shahzad, S., Abbasi, M. A., Ali, H., Iqbal, M., Munir, R., & Kilic, H. (2023). Possibilities, challenges, and future opportunities of microgrids: A review. *Sustainability (Basel)*, *15*(8), 6366. DOI: 10.3390/su15086366

Suman, S. (2018). Hybrid nuclear-renewable energy systems: A review. *Journal of Cleaner Production*, *181*, 166–177. DOI: 10.1016/j.jclepro.2018.01.262

Vujić, J., Bergmann, R. M., Škoda, R., & Miletić, M. (2012). Small modular reactors: Simpler, safer, cheaper? *Energy*, *45*(1), 288–295. DOI: 10.1016/j.energy.2012.01.078

Wayne, M., Farokh, N., John, S., & John, S. (2002). Role of fusion energy in a sustainable global energy strategy. *Energy & Environment*, *13*(4-5), 647–665. DOI: 10.1260/095830502320939606

Wessel, F. J., & Rostoker, N. (2000). *Controlled thermonuclear fusion in a staged Z-pinch, August 1, 1993–December 31, 1998*. University of California., DOI: 10.2172/765161

Wu, Y. (2019). *Neutronics of advanced nuclear systems* (Vol. 765). Springer., DOI: 10.1007/978-981-13-6520-1

Zhang, M., Huang, Y., & Bao, Y. (2016). The mechanism of shallow submarine landslides triggered by storm surge. *Natural Hazards*, *81*(2), 1373–1383. DOI: 10.1007/s11069-015-2112-0

ADDITIONAL READING

Alonso, G., Del Valle, E., & Ramirez, J. R. (2020). *Desalination in nuclear power plants*. Woodhead Publishing., DOI: 10.1016/C2017-0-04915-7

Bratt, D. (2022). SMRs in Canada: Federal-provincial cooperation in pursuing net-zero emissions. *Canadian Foreign Policy*, *28*(3), 305–320. DOI: 10.1080/11926422.2022.2116063

Hoogstad, M. J. P. (2025). Learning from the past to enhance the resilience of nuclear power plants against natural disasters. *Nuclear Safety and Engineering*, *45*(1), 57–70. DOI: 10.1016/j.nse.2024.12.012

Managi, S., & Guan, D. (2017). Multiple disasters management: Lessons from the Fukushima triple events. *Economic Analysis and Policy*, *53*, 114–122. DOI: 10.1016/j.eap.2016.12.002

Misra, B. M. (2007). Seawater desalination using nuclear heat/electricity—Prospects and challenges. *Desalination*, *205*(1-3), 269–278. DOI: 10.1016/j.desal.2006.03.555

Murty, K. L., & Charit, I. (2008). Structural materials for Gen-IV nuclear reactors: Challenges and opportunities. *Journal of Nuclear Materials*, *383*(1-2), 189–195. DOI: 10.1016/j.jnucmat.2008.08.044

Pioro, I. (2017). *Handbook of generation-IV nuclear reactors*. Woodhead Publishing. DOI: 10.1115/1.4035327

Portugal-Pereira, J., Esteban, M., & Araújo, K. (2024). Exposure of future nuclear energy infrastructure to climate change hazards: A review assessment. *Energy Strategy Reviews*, *53*, 101365. DOI: 10.1016/j.esr.2024.101365

Sabharwall, P., Bragg-Sitton, S., Boldon, L., & Blumsack, S. (2015). Nuclear renewable energy integration: An economic case study. *The Electricity Journal*, *28*(8), 85–95. DOI: 10.1016/j.tej.2015.09.003

Zinkle, S. J., Terrani, K. A., & Snead, L. L. (2016). Motivation for utilizing new high-performance advanced materials in nuclear energy systems. *Current Opinion in Solid State and Materials Science*, *20*(6), 401–410. DOI: 10.1016/j.cossms.2016.10.004

KEY TERMS AND DEFINITIONS

Resilience: It is the quality and or ability of a person or a system to quickly recover from difficulties and adjust to challenges.

Nuclear Fission: It is the process of breaking a heavy nucleus into two or more smaller nuclei, along with the generation of energy.

Weather: A day-to-day atmospheric condition of any place at any time. It includes data related to temperature, pressure, humidity, wind, etc.

Climate: It is a long-term average of weather conditions. Like weather, it also includes temperature, pressure, humidity, wind, etc.

Self-sustained Society: Also known as an eco-village. It minimizes ecological footprint, prioritizing local production and local consumption.

Climate Adaptation: To prepare or adjust to current climate change, like heat, floods, droughts, and other disasters.

GHG: Greenhouse gases, which are emitted due to the burning of fossil fuels and lead to global warming.

Passive Safety Features: This system works in auto mode in the background without any human intervention and does not require any power to operate. They use natural powers like gravity, buoyancy, etc.

Smaller Modular Reactors: These are smaller and modular nuclear reactors capable of deployment in areas where traditional reactors are not suitable.

Nuclear Waste: A radioactive waste from the spent fuel of a nuclear reactor.

Carbon-free energy: Energy generated using resources that do not produce greenhouse gases.

Carbon footprint: It refers to the generation of greenhouse gases like carbon dioxide and methane.

Neutron capturing: It refers to the collision of a nucleus to merge with an atom. Unlike electrons, neutrons do not have any charge, so they can easily enter a positively charged nucleus.

FeCrAl: It is an alloy of iron, chromium, and aluminium. It is used in high-temperature applications.

Neutron flux: It is used to characterize the distribution of neutrons in the reactor. It is a scalar quantity used in nuclear physics.

Fuel rods: They are the sealed tubes containing fuel pellets. They are grouped into fuel assemblies to make core of the nuclear reactor.

Brayton cycles: It is a thermodynamic cycle that explains the working of certain engines with constant pressure. There are two types of Bryton cycle: Closed and Open. In a Closed cycle, working gas stays inside the engine.

Auto-initiated shutdown: Here, it refers to a situation where a nuclear reactor automatically proceeds towards shutdown. It is an auto system specifically developed to safeguard the safety of a nuclear reactor in case of an emergency, like a power failure or excessive heat generation.

Chapter 5
Innovations in Nuclear Fuel Technology:
Accident Tolerance, Efficiency, and Sustainability

J. Mangaiyarkkarasi
https://orcid.org/0000-0003-1431-9584
Nadar Mahajana Sangam S. Vellaichamy Nadar College, India

J. Shanthalakshmi Revathy
https://orcid.org/0000-0003-1724-7117
Velammal College of Engineering and Technology, India

ABSTRACT

This chapter highlights the development of nuclear fuel and the current status with special reference to accident tolerable fuels and efficiency and sustainability. Low-gamma materials like chromium-coated zirconium and FeCrAl alloys enhance the safety by operation at harsh conditions and long-lived harmless materials; sophisticated fuels like HALEU and thorium-based fuels increase energy extraction efficiency, decrease radioactive waste, and help in long reactor cycles. Technologies in recycling also advance sustainability since areas of uranium usage are reduced and unsafe burial also eliminated by pyroprocessing. The chapter has devoted due attention to the economic and regulation factors with concern to the fact that the overall costs are high in compare to the perceived long term advantages. The book examines the roles of advanced fuels in new generations of reactors, including SMRs and Gen IV reactors, and how they improve safety, efficiency, and sustainability to guarantee nuclear energy's competitive and sustainable future.

DOI: 10.4018/979-8-3373-0685-8.ch005

1. INTRODUCTION

The application of nuclear fuel technology has remained the primary means for attaining sustainable and safe efficient energy from its original discovery. Worldwide increasing energy consumption and growing environmental concerns force the nuclear industry to explore intense innovation strategies. The current stage of technical nuclear fuel advancements concentrates on developing safer and more effective nuclear fuel systems. Operational safety stands as the top priority among nuclear energy goals which combine effective cost management and green energy initiatives for the industry (Rebak, 2023). Traditional nuclear fuels, primarily uranium dioxide (UO_2), have served the industry well but present limitations, particularly in the context of severe accident scenarios. The Fukushima Daiichi disaster of 2011 revealed fundamental limitations of nuclear reactor fuels which motivated the development of materials that maintain critical reactor integrity under extreme conditions. The objective behind developing ATFs focuses on increasing nuclear reactor resistance to severe heat conditions to prevent destructive consequences and radioactive leaks. It is essential to improve fuel efficiency simultaneously to achieve economic competitiveness for nuclear power systems. Scientists working in materials science have found ways to develop fuel that enhances the burnup rate capacity while extending fuel lifetime resulting in increased electricity production from the same amount of annual supply. The enhanced reactor performance through these innovations allows operators to deploy their systems longer while decreasing expenses and waste output (Yin et al., 2021).

Modern nuclear fuel technology development requires the core sustenance concept that drives current energy policy frameworks. Research developers maintain sustainable nuclear fuel systems by developing recycling operations and material reprocessing and by inventing waste-reducing materials. This section examines initial developments in nuclear fuel systems while explaining scientific approaches together with their existing obstacles for creating advanced accident-resistant systems and environmental benefits.

Objectives of this chapter

This chapter performs comprehensive research about contemporary nuclear fuel system advancements which enhance both power sustainability and safety security protocols. The global shift toward sustainable low-carbon alternatives continues to make nuclear power one of the essential power sources that combat carbon dioxide emissions. For nuclear technology to maintain its future utility it must solve essential problems about security practices and fuel efficiency as well as waste disposal and environmental sustainability. The research investigates how sustainable fuel cycle concepts unify with accident-tolerant fuels to resolve critical nuclear fuel issues. The main research goal investigates the advancement of accident-tolerant fuels because these new nuclear fuel designs represent significant improvements in safety standards. Chemical engineers have developed three improved fuel and cladding components for extreme conditions: chromium-coated zirconium along with FeCrAl alloys and uranium silicide. Research explores cell behavior under extreme events and thermal and structural attributes together with temperature resistance properties of these fuels. Added to their immediate safety benefits nuclear reactors receive enhanced operational performance from ATFs due to increased fuel performance and minimized core damage risks.

Researchers have developed nuclear fuel cycle technologies to produce maximum resource efficiency improvements combined with minimal waste generation outcomes. This section examines three potential fuel choices: HALEU uranium alongside thorium-based fuels alongside mixed oxide MOX fuels to enhance burn-up capacity as well as minimize resource usage. The section discusses recycling techniques and nuclear fuel cycle systems that retrieve spent nuclear materials to utilize them again while decreasing waste-related environmental consequences and lowering the need for nuclear waste storage facilities. This chapter establishes a complete understanding by using regulatory and environmental viewpoints with economic evaluation to present the technical developments of nuclear fuel. A complete cost analysis combined with licensing examinations and future benefits assessment of advanced fuels exists alongside testing their operational compatibility with future nuclear reactor technology such as Generation IV systems and small modular reactors (SMRs). The main objective of this chapter explain how nuclear fuel development improves both the safety and operational performance of nuclear power technology and environmental sustainability to build public trust towards nuclear energy for long-term environmental protection.

Research gap

Since technological advancements in nuclear fuel were developed for better accident tolerance operational efficiency and sustainability, there still exist fundamental research gaps that prevent the large-scale implementation of nuclear fuel innovations. The current body of information contains minimal performance records of Accident-Tolerant Fuels (ATFs) under reactor operating situations. The encouraging laboratory findings and brief irradiation trials need more investigation to determine the long-term behaviors of these kinds of fuels when they encounter regular operational stresses and unexpected conditions. Insufficient testing of fuel integrity together with uncertain fission gas release and corrosion resistance throughout an entire fuel cycle prevents regulatory approval and industry trust. There exists insufficient thorough economic analysis between the standard fuel cycles alongside the developing advanced fuel systems along with recycling technology solutions. Research on the practical financial advantages between waste reduction and improved burn-up requires more detailed investigation of total fabrication expenses alongside regulatory challenges and system maintenance expenses over time. Such situations make it challenging for both power facilities and potential investors to decide about fuel system transitions. Pyroprocessing together with advanced aqueous reprocessing technologies continue development yet persist at different stages of technological readiness. The resolution of important issues which include proliferation dangers and process scalability together with waste management precision demands increased research effort. The transition to closed fuel cycles together with thorium-based nuclear systems requires attention to supply chain development as well as reactor compatibility and international regulatory systems.

One significant barrier appears with the integration of next-generation reactor technologies which includes Small Modular Reactors (SMRs) and Generation IV reactors. Research about fuel compatibility with advanced systems along with passive safety performance and system optimization data remains insufficient despite frequent mentions of such systems with advanced fuels. The early stages of cooperation between fuel developers and reactor designers produce a gap between new fuel research and the timing of reactor implementation. The relationship between public acceptance of new nuclear technologies and policy structure needs further exploration. The development of new nuclear technology keeps advancing yet both regulatory modifications to welcome new nuclear systems and public acceptance of such technologies have not kept pace. Research efforts must unite science from multiple fields to help technological progress blend with regulatory updates and community trust-building processes. The successful advancement of nuclear fuel technology demands additional specific investigation into performance confirmation and economic assessment along with fuel cycle maturity and reactor implementation and policy synchronization.

2. LITERATURE REVIEW ON TRADITIONAL NUCLEAR FUEL TECHNOLOGIES

Traditional nuclear fuel technologies have been the backbone of the nuclear energy industry for decades, playing a crucial role in generating reliable, low-carbon electricity worldwide. The most commonly used fuel is uranium dioxide (UO_2), typically enriched to (Garud & Rebak, 2023) about 3-5% uranium-235 (U-235) for use in light-water reactors (LWRs), the most widespread reactor type. The proven effectiveness of uranium-based fuel remains under scrutiny due to its present limitations that affect both safety measures and operational efficiency and waste disposal. The properties of uranium dioxide support its effectiveness as a fuel substance. This substance functions well at high temperatures because of excellent radiation resistance despite easy prediction in reactor systems (Chandler et al., 2019). The features render this material suitable for operating within LWR conditions of elevated temperatures and pressures. Fission of uranium-235 nuclei releases major energy thus creating heat to operate turbines that generate power. Nuclear power plants successfully maintain consistent operations thanks to this fuel system because their capacity to produce electricity reaches above 90% levels without interruption. The use of UO_2 fuel comes with major operational constraints. UO_2 encounters difficulties during thermal conductivity because it poorly conducts heat energy. The temperature irregularities produce hot spots that heighten the danger of fuel degradation when reactors operate at maximum power levels. The operating effectiveness of reactors is limited by this drawback because plant operators need rigorous temperature and power maintenance to avoid temperature rise hazards and fuel system breakdowns. Additionally, UO_2 fuel requires frequent replacement due to limited "burnup," or the amount of energy that can be extracted from the fuel before it needs to be replaced (Y. Li et al., 2024).

Standard uranium-based fuel requires special attention regarding its reaction behavior when accidents occur. The UO_2 fuel rods risk severe destruction when their temperature rises due to lost cooling as experienced during the Fukushima Daiichi incident in 2011. The high temperatures that occur generate dangerous chemical reactions between water and zirconium-alloy cladding materials which produce explosive hydrogen gas. Studies on accident-tolerant fuels (ATFs) have intensified because of this vulnerability since these fuels are meant to endure severe conditions without generating harmful chemical reactions. Regular nuclear fuel processing generates environmental problems because it produces radioactive waste materials without proper management solutions (Setty et al., 2017). The fuel rods preserve large amounts of radioactive substances along with plutonium and fission products after completing their reactor duty. After its use in fuel rods the spent material needs proper management and storage in dedicated long-term facilities because it maintains its radioactivity over time. Some countries possess reprocessing technologies which separate usable material from waste but this process remains both complex and expensive and finds limited application in the current industry context. The finite supply of uranium introduces a separate limitation factor. The sufficient amount of uranium exists yet the required high-grade uranium stands scarce for efficient reactor operations (Hsiao et al., 2021). The expanding international demand for nuclear energy requires new sources or better nuclear fuel management systems which will protect the sustainability of fuel resources. UO_2 has allowed nuclear power to develop into a dependable energy source through its traditional fuel role although it encounters serious operational constraints. The list of limitations includes low thermal conductor properties, safety challenges during accidents along with tough waste management procedures and usage of finite resources. Research for advanced nuclear fuel technologies started due to recognizing safety issues as well as the need for efficient and sustainable operations. The established research into new materials together with fuel cycle technologies works to overcome these limitations and will establish nuclear energy as a safer and more efficient and environmentally friendly power generation system (DeHart et al., 2018).

3. ADVANCES IN ACCIDENT TOLERANT FUELS: ENHANCING SAFETY AND RELIABILITY

Accident Tolerant Fuels (ATFs) improve reactor safety during normal and accident conditions by offering accident tolerance capabilities. The creation of Accident Tolerant Fuels resulted from Fukushima Daiichi lessons in 2011 which exposed weaknesses of standard nuclear fuel during extreme operating conditions. Advanced recovery fuel technology has three fundamental targets, which include improved fuel performance in regular operations and emergency situations and decreased core damage chances and limited radioactive content release when accidents occur (Yun et al., 2021). Figure 1 shows an accident tolerant fuel system.

Figure 1. Accident tolerant fuel system

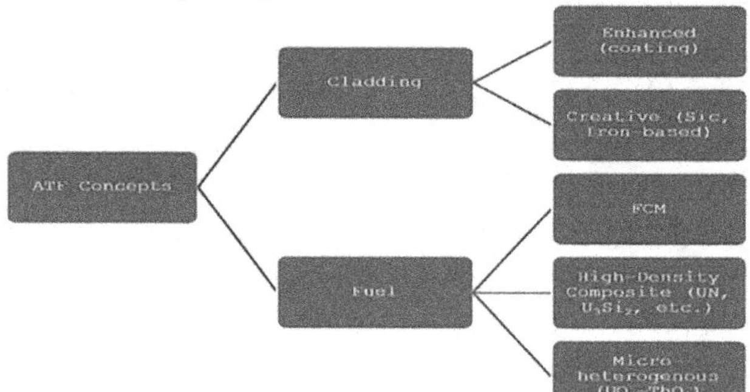

Here are ten key points about the advances in Accident Tolerant Fuels:

Enhanced Cladding Materials: The traditional use of zirconium-based alloys for cladding in nuclear fuel rods provides effectiveness yet allows heat-driven reactions between water and this material to form hydrogen that poses safety risks in accident situations. The cladding materials for ATFs use silicon carbide composites combined with coated zirconium alloys and advanced steel variants. The alternative materials display improved resistance to oxidation thus minimizing hydrogen production risks during extreme temperature conditions.

Chromium-Coated Zirconium Cladding: Research has developed a chromium coating procedure to protect traditional zirconium cladding. The application of chromium on zirconium cladding enables oxidation protection at elevated temperatures thus improving resistance to fuel damage following accidents. The coated material demonstrates substantial resistance against oxidation in hot steam-rich atmospheres thus guarding systems during coolant leak accidents (Gong et al., 2020).

Silicon Carbide (SiC) Composites: Silicon carbide (SiC) functions as an advanced ceramic material because it presents high temperature stability along with robust corrosion resistance capabilities. The high temperature resistance of SiC-based cladding ensures safety because it operates without hydrogen production at extreme conditions when searching for ATF applications. The high melting point together with material resilience works to maintain fuel structure stability during emergencies to avoid rod deformation issues.

Iron-Chromium-Aluminum (FeCrAl) Alloys: FeCrAl alloys function as a substitute for zirconium cladding since they offer exceptional high-temperature oxidation protection. These materials present stable properties which make them corrosion-resistant while remaining durable even under severe operating conditions. FeCrAl cladding survives accident conditions for extended periods according to testing results thus allowing plant operators to activate emergency procedures in a more timely manner.

Uranium Silicide (U3Si2) Fuel: U_3Si_2 is under investigation as an ATFs fuel material because its thermal conductivity exceeds that of UO_2. The higher thermal conductivity of U_3Si_2 minerals allows fuel systems to operate more efficiently by minimizing dangerous temperature concentrations. Higher burnup rates are achievable using U_3Si_2 allowing power production from the fuel to reach greater levels before replacement needs occur thus decreasing operational costs and cutting down on waste production (Parisi et al., 2020).

Chromium-Doped Fuels: Energy researchers have developed improved UO_2 fuel pellets by integrating small chromium content into standard pellets. The incorporation of chromium increases fuel durability under hot operating conditions along with decreasing the odds of pellet damage. Such an addition protects fuel tanks from structural failure which could result in leakage by enhancing operational performance and safety factors in accident situations (Rebak, 2023).

Reduced Hydrogen Production: Accidents will produce less hydrogen due to the safety advantage ATFs provide. Silicon carbide along with FeCrAl alloys have shown effective resistance to water reactions at high temperatures thus preventing hydrogen formation which resulted in Fukushima Daiichi explosions. The feature increases reactor safety through its reduction of hydrogen accumulation and explosion possibilities.

Extended Reactor Cooldown Times: Active cooling is not necessary for ATFs because their engineering design allows them to endure damage and maintain structure for lengthy periods. The increased resistance of operators to system failures provides them with additional time to fix cooling problems or execute reactor site evacuations. An extended time period for reactor cooldown delivers critical advantages to effective accident response actions.

Compatibility with Existing Reactor Designs: The design of ATF technologies allows easy compatibility with existing reactor systems thus making their adoption process straightforward. The adoption of cladding materials including chromium-coated zirconium and FeCrAl alloys requires mostly standard reactor infrastructure improvements. Widespread industry adoption of ATF depends on this compatibility because it simplifies the shift even when using existing reactor designs (Nelson, 2023).

Reduced Environmental Impact: ATFs provide additional environmental as well as safety benefits to the nuclear industry. Advanced materials combined with better designs enable the extraction of higher quantities of energy per fuel rod that contributes to reduced frequency of replacement. The efficient operation of ATFs decreases the manufacturing output of nuclear waste while reducing the requirements of storage sites and lowering their environmental footprint.

Nuclear fuel technology has taken a significant stride with the development of Accident Tolerant Fuels that specifically solve major safety and operational reliability problems that plagued the nuclear industry. The implementation of silicon carbide and FeCrAl and uranium silicide in accident tolerant fuel designs increases material resistance to high temperatures and lowers hydrogen generation as well as improves their operational safety in emergency events. The adoption of existing reactors is feasible due to ATFs and their efficient operations lead to waste reduction and sustainable environmental outcomes. ATFs have been developed to ensure safer and more dependable nuclear power facilities because they represent the basis of contemporary reactor systems.

4. HIGH-PERFORMANCE FUEL MATERIALS FOR IMPROVED EFFICIENCY

Modern nuclear power needs high-performance fuel materials to enhance the operational efficiency and industrial competitiveness of this energy technology. The specific materials enable reactors to reach higher operating temperatures while extracting more thermal energy from each piece of fuel, while requiring less reactor maintenance, which drives substantial economic gains (Aguirre et al., 2022). The improvement of efficiency stems mainly from material advancements that produce fuel substances with enhanced thermal abilities and structural properties, which result in extended fuel duration and higher burnup levels and lower waste output. This research examines various promising high-performance fuel substances alongside their function in improving nuclear energy operational efficiencies.

Uranium Silicide (U_3Si_2) for Enhanced Thermal Conductivity

U_3Si_2 stands as an emerging material which demonstrates major potential as a high-performance fuel solution. UO_2 fuel faces a significant shortcoming due to its low thermal conductivity level because it creates thermal distribution inequalities throughout the reactor core. The fuel damage risk increases when power reactors operate above their safe generation levels because of this effect. The thermal conductivity of U_3Si_2 material exceeds its counterpart UO_2 by about five times thus enabling better heat transfer throughout the reactor core. The improved thermal conductivity element bolsters safety of fuel while enabling increased operating temperatures that produce greater energy generation (Doncel et al., 2024).

Thorium-Based Fuels for Extended Fuel Cycles

The attention grows towards thorium-based fuels which include thorium-uranium and thorium-plutonium mixed fuels because of their abilities to enhance efficiency rates in nuclear systems. The natural abundance of Thorium along with various other welfare benefits makes it superior to uranium as an element. Unified irradiation of thorium-232 leads to the creation of the fissile isotope uranium-233. Testing has demonstrated that thorium-derived nuclear materials produce longer operating periods since they utilize high burnup characteristics alongside reduced radioactive isotope creation. The extended lifetime of nuclear fuel decreases regular shipments to the reactor, which leads to less downtime between operations and decreased expenses for fuel procurement. Figure 2 provides insights into Thorium-Based Fuels for Extended Fuel Cycle.

Figure 2. Thorium-Based Fuels for Extended Fuel Cycle

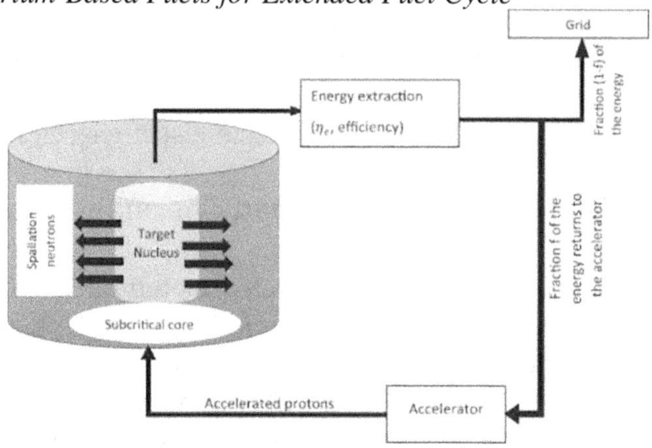

High-Assay Low-Enriched Uranium (HALEU)

The enrichment percentage in high-assay low-enriched uranium (HALEU) ranges from 5% to 20% uranium-235 placing it between conventional low-enriched uranium and weaponizable uranium. Reactors powered by HALEU can generate more power while needing fewer changes of fuel because of its extended service time inside the reactor core. Operations running on HALEU material enable longer-lasting fuel operations and improved efficiency because it allows higher temperature reactor designs including those utilized in advanced reactors. HALEU functions as a promising candidate for reactor fuel because it enables safety-critical operations at elevated burnup levels in present-day and next-generation reactor systems (Brown, 2022).

Doped Fuels for Improved Structural Integrity

The performance of nuclear fuel receives significant enhancement from using small doping amounts as additives. Laboratory tests demonstrate how chromium additions to uranium dioxide structures improve fuel pellets by strengthening their structures to prevent cracks while enhancing their operational safety. Additive substances known as dopants stabilize the fuel matrix under high temperatures while protecting against radiation damage that could otherwise make fuel degrade with time. The addition of dopants enables cells to process energy at elevated burnup levels making them more efficient in energy extraction.

Mixed Oxide (MOX) Fuel for Waste Reduction and Efficiency

The blend of plutonium with natural or depleted uranium exists as Mixed oxide (MOX) fuel which serves two essential functions in advancing power efficiency and waste management. The use of MOX fuel delivers nuclear fuel recycling opportunities for spent fuel plutonium while allowing generation of higher energy output rates than standard UO_2 fuel systems. Plutonium mixed with fuel allows reactors to reach elevated burnup levels and make the most of their nuclear material before disposal. Waste storage accumulations of plutonium decrease through recycling and the energy output increases along with improved fuel efficiency (Capps et al., 2020).

Advanced Fuel Pellet Designs for Optimized Performance

Pellet development innovations represent a method to enhance fuel efficiency. High-density pellet designs enable manufacturers to insert greater amounts of uranium into each fuel pellet, thus expanding the energy output from reactor loads. The use of annular (hollow) pellets enables better heat distribution across the fuel, ensuring both operational safety and maximized electrical output. The development of advanced fuel pellet shapes represents a fundamental requirement for generating superior performance fuel materials since designed pellets enable greater power densities and longer fuel cycles at safe operating levels (Sanchez-Torrijos et al., 2023).

Nanotechnology for Enhanced Fuel Performance

Nuclear fuel performance benefits from nanotechnology through nanostructured materials development which enhances temperature conductivity and decreases radiation-induced degradation. Nanoscale design of fuel pellets enhances properties related to heat transfer and radiation tolerance to defend fuel pellets against high temperatures and neutron bombardment during operational conditions. The implementation of nanostructured fuels brings two benefits to nuclear reactors since the materials demonstrate increased durability within these harsh operating environments.

The development of high-performance fuel materials will drive the future of nuclear energy through improved reactor functions and efficiency and operational safety as well as cost effectiveness. The advancement of nuclear efficiency depends on uranium silicide together with thorium-based fuels and HALEU and MOX fuel because these materials enable higher operational temperatures and extended fuel cycles that produce superior burnup rates. Nuclear power sustainability receives improvement through fuel reliability enhancement achieved with doped fuels combined with advanced pellet designs and nanotechnology applications. The nuclear industry will establish a future energy solution for the twenty-first century through its focus on creating highly efficient fuel materials (Petrie et al., 2023).

5. INNOVATIONS IN FUEL RECYCLING AND WASTE MINIMIZATION

Fuel recycling together with waste minimization strategies represents essential innovation areas for the nuclear industry because the world needs cleaner and sustainable energy alternatives. The traditional operation of nuclear fuel cycles creates extensive waste streams that contain radioactive elements that need secure long-term storage exceeding thousands of years. Researchers develop innovative waste minimization technologies together with advanced recycling methods to enhance fuel consumption and minimize environmental harm and nuclear waste volume and content.

Closed Fuel Cycles for Sustainable Resource Use

The closed fuel cycle system functions differently from standard open fuel cycle systems because it allows nuclear specialists to reprocess and recycle spent fuel materials. In a closed cycle, valuable fissile materials such as uranium and plutonium are recovered from spent fuel and reintroduced into reactors as mixed oxide (MOX) fuel or other advanced fuels. The recycling approach reduces uranium mining needs while saving resources, along with minimizing high-level radioactive waste amounts. The recycling approach to nuclear fuel material has been pioneered by France and Japan, who show waste reduction through reused materials (Nickel et al., 2002).

Mixed Oxide (MOX) Fuel for Plutonium Recycling

MOX fuel stands as an advanced technology to recycle spent nuclear fuel because it combines uranium with plutonium. Using plutonium extracted from spent fuel through MOX fuel production lowers storage amounts and decreases the requirement for new uranium. Each assembly of MOX fuel enables reactors to extract greater amounts of energy before replacement because of its capability to tolerate higher burnup rates. Through its use in fuel production MOX fuel promotes resource efficiency and waste reduction because it decreases uranium resource demands and extends nuclear materials' operational period.

Advanced Partitioning and Transmutation Techniques

Partitioning and transmutation represent developing technologies which demonstrate promising capabilities in diminishing the long-term toxic nature of nuclear waste. The partitioning process separates particular isotopes including americium alongside neptunium and curium which possess long half-lives from spent fuel. Particle accelerators and nuclear reactors become vital in transmutational processes through the irradiation of specific isotopes which transforms them into stable or shorter-lived equivalent forms. The combination of these two processes greatly decreases the nuclear waste's dangerous characteristics while lowering both its heat production levels and the duration it can harm. (Demkowicz et al., 2019).

Innovative Solvent Extraction Processes

Solvent extraction stands as a tested method to separate fissile materials from spent fuel and ongoing advancements have improved its efficiency as well as environmental friendliness. The PUREX (Plutonium-Uranium Redox EXtraction) process previously used for decades continues as an improvement through new extraction methods together with advanced solvents under the name UREX+ (Uranium Extraction Plus). With UREX+ uranium gets separated from the waste mixture while plutonium remains intact which minimizes the risk of nuclear proliferation and promotes efficient management of uranium-containing streams. Modern solvent extraction technology focuses on both the recovery of valuable fission products including rare earth elements that can lead to industrial usage while turning radioactive waste into practical resources.

Pyroprocessing for Metal-Based Fuel Recycling

Dry reprocessing under the name Pyroprocessing provides an advanced method to recycle metal-based nuclear fuels specifically intended for fast reactors. The processing technique delivers uranium together with plutonium and actinides through high-temperature salt solutions rather than standard water methods commonly used in reprocessing operations. This methodology provides excellent benefits for both the recycling process of spent fuel with high burnup along with the management of complicated reactor fuel compositions. Through pyroprocessing operations high-level waste gets condensed into smaller volumes while the production of advanced fuel nuclear materials ensures both better long-term efficiency and reduced radiation output therefore promoting sustainability in nuclear power generation (Di Lemma et al., 2020).

Accelerator-Driven Systems (ADS) for Waste Reduction

ADS serves as an innovative waste reduction technology through their capability to conduct transmutation of persistent radioisotopes within subcritical reactor systems. A particle accelerator within an ADS system generates neutrons to power fission reactions in materials that lack sustainability in open chain reactions. Through this technology, specific actinide elements become transmuted into less dangerous substances which decrease radioactivity and heat levels in nuclear waste. Scientists have demonstrated ADS as an emerging solution to decrease the amount of nuclear waste that future generations will need to handle. Figure 3 depicts accelerator driven system.

Figure 3. Accelerator driven system

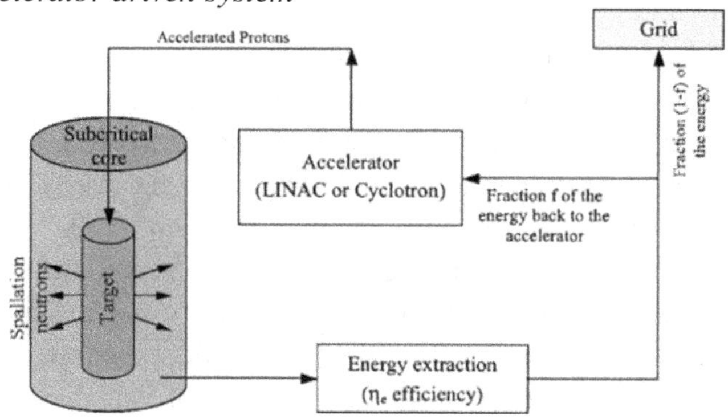

On-Site Recycling and Small Modular Reactor (SMR) Integration

SMRs create new prospects for localized on-site waste recycling which enables smaller-scale waste management capabilities. SMRs provide enhanced flexibility compared to conventional reactors which makes them perfect for hosting the adoption of compact-scale pyroprocessing recycling units. Site-based nuclear resource recycling reduces both transportation of fuel and associated safety issues and enables communities to handle resources better which supports sustainability and improves public support for nuclear power (Burns et al., 2020).

Reducing High-Level Waste Volume through Advanced Reactor Designs

Fast reactors including sodium-cooled fast reactors (SFRs) and lead-cooled fast reactors (LFRs) have a designed operation scheme utilizing the closed fuel cycle to burn waste products from advanced fuel cycles. Fast reactors become capable of breaking apart actinides through their high-energy neutron operations which would remain as long-lasting waste otherwise. The technology enables power generation from spent fuel while lowering the size of high-level waste storage materials which presents an effective answer to long-term waste storage needs (Bergeron & Crigger, 2018).

Nuclear energy progresses toward a greener directional future through new innovations that boost waste reduction and enhance fuel recycling programs. The nuclear industry can minimize its environmental influence on waste products through closed fuel cycles along with improved separation methods and transmutation techniques and new reactor design work. These innovative systems will enhance nuclear safety while improving efficiency which leads to better resource management and lower waste generation to enable future groups to partake in nuclear power's worldwide energy role. The development of these innovations will create opportunities to increase public support for using nuclear energy as a safe sustainable power generation method (Betzler et al., 2020).

6. SUSTAINABILITY IN NUCLEAR FUEL: ENVIRONMENTAL AND ECONOMIC PERSPECTIVES

Adopting sustainable nuclear fuel practices has become essential because the world demands power systems with efficient operation and reliable performance and environmental and economic sustainability. Nuclear power serves as an essential climate strategy because of its low carbon footprint but various challenges persist regarding waste disposal and economics along with resource consumption issues. The nuclear industry uses new approaches in nuclear fuel development together with recycling methods and waste reduction techniques to enhance sustainable nuclear energy. Sustainable fuel cycles move forward because they deliver environmental benefits alongside economic stability which helps extend nuclear power operations into the long-term (Hajipour et al., 2025).

Environmental Sustainability in Nuclear Fuel

Nuclear power faces its most significant environmental obstacle through proper handling of radioactive waste. Open fuel cycles that utilize spent fuel only once produce vast amounts of high-level waste which future generations will need to keep safely stored through millennia. Nuclear energy faces sustainability challenges because permanent waste storage creates future generational burdens triggering long-term sustainability concerns. New developments in fuel recycling together with closed fuel cycles have successfully tackled this issue. Through fuel reprocessing facilities spent radioactive materials are recovered for reuse in making uranium and plutonium which becomes reintroduced into nuclear power plants (Cheng et al., 2019). The waste reduction strategy decreases the volume of hazardous material since it improves resource utilization while decreasing the nuclear power sector's environmental impact.

Accident Tolerant Fuels (ATFs) along with other advanced fuel materials now exist to minimize waste generation while providing greater resistance to critical conditions. Advanced fuel burnups enable power plants to draw greater energy amounts from their fuel elements which leads to fewer necessary refueling procedures and a reduction in generated radioactive waste. The technology reduces wastage in parallel with improved reactor operational performance. The incorporation of uranium silicide (U_3Si_2) in new fuel designs and advanced materials enhances both thermal conductivity and structural strength which improve reactor withstand capability against environmental contamination incidents (Gussev et al., 2021).

Advanced nuclear reactors such as fast reactors together with molten salt reactors utilize nuclear materials ranging from depleted uranium to recycled plutonium among other components during operation. The reactor technology operates within closed cycles while processing long-lived isotopes that represent the most complicated elements of nuclear waste. The operation of fast reactors transforms difficult-to-store long-lived isotopes into either more short-lived forms or stable stable forms which reduce both their radioactivity and storage requirements. The nuclear energy system will achieve sustainability through minimal waste production along with recycling previously discarded materials for creating an advanced fuel cycle.

Economic Sustainability in Nuclear Fuel

The global energy market can sustain nuclear power when it maintains economic sustainability. The traditional uranium-based fuel cycle incurs high expenses because it demands controlled storage facilities together with waste disposal operations and the constant extraction and enrichment of new uranium fuel. The future of low-cost nuclear power becomes possible through advanced fuel technologies which deliver longer-lasting fuel as well as increased waste reduction and material reuse capabilities.

The closed loop fuel cycle system enables reactors to maintain valuable materials which lowers their requirement for uranium obtained from fresh mining operations. The processes involved in uranium mining prove costly and generate damaging effects to both habitat ecosystems and water resources (Cramer et al., 2022). Closed fuel cycles cut down the necessity of extracting fresh uranium which consequently decreases both economical costs and environmental impacts to create sustainable nuclear power. Since France has established reprocessing along with spent fuel recycling programs their nuclear fuel expenses show stability while global uranium prices rise and fall (Di Lemma et al., 2020).

The advanced Small Modular Reactor design (SMR) provides economic advantages through decentralized power generation mechanisms that decrease the expenses related to building massive nuclear facilities. On-site recycling technologies fit perfectly with Small Modular Reactors because they enhance the deployment simplicity and safety while decreasing waste transportation and storage expenses. Continual operation of Small Modular Reactors in remote areas and off-grid locations serves as an important recycling method because such areas do not have centralized storage capabilities (Z. Li et al., 2023).

The economic sustainability of nuclear power plants depends on high-burnup fuels along with Accident Tolerant Fuels (ATFs) because they provide better operational efficiency. High-burnup fuels reduce refueling operations thus minimizing operational downtime as well as associated expenses. ATFs provide both operational durability and extended service life in harsh conditions which minimizes reactor maintenance expenses and prolongs the operational lifespan of fuel assemblies (Crawford et al., 2007).

Sustainability in nuclear fuel management from both environmental and economic perspectives is critical to the future of nuclear power. By moving toward closed fuel cycles, advanced recycling techniques, and innovative reactor designs, the nuclear industry can significantly reduce its environmental impact and improve resource efficiency. The economic advantages of these advancements can decrease expenses which makes nuclear energy more affordable and market competitive. Sustainable nuclear fuel practices have become vital for turning nuclear power into the foundational component of worldwide renewable energy development as power demands increase. The nuclear industry works to establish sustainability in the 21st century by addressing waste management together with resource utilization and cost reductions.

7. FUTURE DIRECTIONS AND EMERGING TECHNOLOGIES IN NUCLEAR FUEL DEVELOPMENT

Nuclear fuel development for future use focuses on safety improvements of reactors and enhanced fuel performance and waste reduction and environmental protection. Nuclear power's critical role in cleaner energy transitions gets supported by improvements in materials science coupled with advancements in reactor design along with better waste management solutions. The current development of emerging technologies aims to enact two-fold improvements: it achieves enhanced safety and lower cost operation while solving persistent waste issues and improving sustainability. Click to view the most forward-looking approaches and developing technologies within nuclear fuel advancement.

Accident Tolerant Fuels (ATFs)

Researchers have devoted significant efforts toward developing Accident Tolerant Fuels (ATFs) since the 2011 Fukushima Daiichi disaster occurred. New nuclear fuel technologies were developed for operating under extreme conditions such as high temperatures and complete coolant loss while showing reduced potential for system breakdown. The advanced cladding materials employed in ATFs utilize silicon carbide as well as chromium-coated zirconium because they display superior resistance to oxidation and hydrogen production compared to standard zirconium alloys. The performance benefits of uranium silicide (U_3Si_2) nuclear fuels improve in high-temperature settings due to their higher thermal conductivity as well as their premium fuel properties. Modern designers consider ATFs as an immediate safety technology which existing reactor systems could utilize to deliver enhanced operational protections.

Advanced Recycling Techniques and Closed Fuel Cycles

The traditional fuel cycle produces massive waste materials which must be stored for extended periods. A distinctive characteristic of closed fuel cycles is their ability to recycle spent fuel along with its reprocessing methods. Pyroprocessing and solvent extraction methods represented by UREX+ perform advanced waste separation through their ability to extract uranium and plutonium from final products. The recycling approach decreases waste by volume while simultaneously lowering both new uranium consumption and waste radioactivity. New transmutation technologies aim to convert long-lived nuclear waste into safe or more short-lived substances which minimize future waste management needs.

High-Assay Low-Enriched Uranium (HALEU)

The fuel material High-Assay Low-Enriched Uranium (HALEU) receives recognition as a flexible fuel alternative since it includes uranium-235 concentrations ranging from 5% to 20%. Higher burnup rates are possible through HALEU that enables reactors to extract greater energy from their fuel so more time elapses before needing replacement thus reducing operational costs. Advances in nuclear power performance require HALEU due to its energy-dense potential which benefits advanced reactor designs especially Small Modular Reactors (SMRs) seeking smaller and more efficient operations. Higher burnup levels of the fuel contribute to sustainability by minimizing waste generation and optimizing resource consumption.

Molten Salt Reactors (MSRs)

MSRs mark a revolutionary design for both reactor systems and fuel compositions. The liquid fuel known as molten salts operates within both high temperatures and low pressure conditions because of which MSR technology improves safety alongside increased operational efficiency. MSRs provide a capability to reprocess fuel inside the reactor facility which enables a continuous fuel cycle operation. The designed system eliminates the requirement for external waste storage while permitting MSRs to leverage thorium together with recycled actinides from nuclear waste therefore improving sustainability.

Advanced Fuels for Fast Reactors

The fission process in fast reactors happens through high-energy (fast) neutrons that enables the use of diverse fuel materials including recycled elements along with the spent nuclear fuel produced by conventional reactors. These devices function as optimum systems for consuming long-lasting radioactive particles which reduces nuclear waste storage requirements. The advanced fuels used in fast reactors consist of metallic fuels alongside metallic uranium or uranium-plutonium alloys because they exhibit exceptional thermal conductivity while offering superior power density capability. Fast reactors demonstrate potential as the next-generation technology because they effectively reduce waste output while enhancing the efficiency of the nuclear fuel process. Nuclear fuel development will advance through the combination of improved materials applied to reactor designs together with waste management techniques that minimize cycle releases. The advancement of nuclear power technology includes ATFs combined with advanced recycling techniques and HALEU as well as MSRs together with fast reactor fuels to enhance sustainability and operational safety and efficiency. Nuclear power production will advance global energy targets by operating as a dependable low-carbon system which produces reduced waste with increased fuel efficiency because these technologies continue to evolve and regulatory systems adapt. The current developments predict a positive future where nuclear power demonstrates compatibility with both environmental sustainability and economic requirements of modern times.

CONCLUSION

Nuclear fuel technology innovations allow for a fresh business model of nuclear energy by ensuring safety performance along with efficiency enhancement and sustainability advancement. The emergence of Accident Tolerant Fuels (ATFs) represents a vital scientific breakthrough that lets reactors endure harsh operating conditions while stopping catastrophic failures, thus improving safety operations significantly. Higher temperature operation and increased energy extraction from fuel units are made possible through the use of uranium silicide fuel alongside HALEU. The innovations enhance both the economic feasibility of nuclear power together with decreased frequency of refueling while decreasing costs through reduced fuel production and transportation needs.

The combination of different recycling techniques within nuclear fuel materials cycles enhances operational sustainability in the industry. The waste reduction approaches recover valuable fissile materials from the waste stream to make them suitable for forthcoming reactors therefore creating reduced waste amounts with lower toxicity levels. Fast reactors and molten salt reactors along with other advanced designs provide optimization conditions for resources and prolonged radioactive isotope consumption thus establishing circular economy principles. The creation of new nuclear power systems demonstrates potential to achieve sustainable nuclear power generation by maintaining carbon-neutral operations. Nuclear fuel technology becomes essential through perpetual research funding and regulatory support as the world establishes environmentally friendly sustainable power systems with secure energy reserves.

REFERENCES

Aguirre, T. G., Cramer, C. L., & Mitchell, D. J. (2022). Review of additive manufacturing and densification techniques for the net- and near net-shaping of geometrically complex silicon nitride components. *Journal of the European Ceramic Society*, *42*(3), 735–743. DOI: 10.1016/j.jeurceramsoc.2021.11.001

Bergeron, A., & Crigger, J. B. (2018). Early progress on additive manufacturing of nuclear fuel materials. *Journal of Nuclear Materials*, *508*, 344–347. DOI: 10.1016/j.jnucmat.2018.05.058

Betzler, B. R., Ade, B. J., Wysocki, A. J., Jain, P. K., Chesser, P. C., Greenwood, M. S., & Terrani, K. A. (2020). Transformational Challenge Reactor preconceptual core design studies. *Nuclear Engineering and Design*, *367*, 110781. DOI: 10.1016/j.nucengdes.2020.110781

Brown, N. R. (2022). Engineering demonstration reactors: A stepping stone on the path to deployment of advanced nuclear energy in the United States. *Energy*, *238*, 121750. DOI: 10.1016/j.energy.2021.121750

Burns, J. R., Hernandez, R., Terrani, K. A., Nelson, A. T., & Brown, N. R. (2020). Reactor and fuel cycle performance of light water reactor fuel with 235U enrichments above 5%. *Annals of Nuclear Energy*, *142*, 107423. DOI: 10.1016/j.anucene.2020.107423

Capps, N., Sweet, R., Wirth, B. D., Nelson, A., & Terrani, K. (2020). Development and demonstration of a methodology to evaluate high burnup fuel susceptibility to pulverization under a loss of coolant transient. *Nuclear Engineering and Design*, *366*, 110744. DOI: 10.1016/j.nucengdes.2020.110744

Chandler, D., Betzler, B., Cook, D., Ilas, G., & Renfro, D. (2019). Neutronic and thermal-hydraulic feasibility studies for High Flux Isotope Reactor conversion to low-enriched uranium silicide dispersion fuel. *Annals of Nuclear Energy*, *130*, 277–292. DOI: 10.1016/j.anucene.2019.02.037

Cheng, L., Gao, R., Yan, B., Yang, Z., Li, B., Xu, Q., Liu, P., Chu, M., Bai, B., & Zhang, P. (2019). SiC whisker strengthened UO2–Mo composite pellets with superior thermal-mechanical properties. *Ceramics International*, *45*(17), 23377–23383. DOI: 10.1016/j.ceramint.2019.08.039

Cramer, C. L., Ionescu, E., Graczyk-Zajac, M., Nelson, A. T., Katoh, Y., Haslam, J. J., Wondraczek, L., Aguirre, T. G., LeBlanc, S., Wang, H., Masoudi, M., Tegeler, E., Riedel, R., Colombo, P., & Minary-Jolandan, M. (2022). Additive manufacturing of ceramic materials for energy applications: Road map and opportunities. *Journal of the European Ceramic Society*, *42*(7), 3049–3088. DOI: 10.1016/j.jeurceramsoc.2022.01.058

Crawford, D. C., Porter, D. L., Hayes, S. L., Meyer, M. K., Petti, D. A., & Pasamehmetoglu, K. (2007). An approach to fuel development and qualification. *Journal of Nuclear Materials*, *371*(1–3), 232–242. DOI: 10.1016/j.jnucmat.2007.05.029

DeHart, M. D., Karriem, Z., Pope, M. A., & Johnson, M. P. (2018). Fuel element design and analysis for potential LEU conversion of the Advanced Test Reactor. *Progress in Nuclear Energy*, *104*, 117–135. DOI: 10.1016/j.pnucene.2017.09.007

Demkowicz, P. A., Liu, B., & Hunn, J. D. (2019). Coated particle fuel: Historical perspectives and current progress. *Journal of Nuclear Materials*, *515*, 434–450. DOI: 10.1016/j.jnucmat.2018.09.044

Di Lemma, F. G., Liu, X., Holschuh, T. V.II, Folsom, C. P., Murray, D. J., Teng, F., & Jensen, C. B. (2020). Investigation of the microstructure evolution of alpha uranium after in pile transient. *Journal of Nuclear Materials*, *542*, 152467. DOI: 10.1016/j.jnucmat.2020.152467

Doncel, N., Martínez, L., Aragón, P., Feria, F., Herranz, L. E., Queral, C., Fernandez-Cosials, K., Ruiz-Hervias, J., Cristobal-Beneyto, M., Castro, E., Sanchez-Torrijos, J., & Soler, A. (2024). R&D in advanced technology fuels (ATFs) in Spain. *Nuclear Engineering and Design*, *424*, 113246. DOI: 10.1016/j.nucengdes.2024.113246

Garud, Y. S., & Rebak, R. B. (2023). Effect of surface oxides on tritium entrance and permeation in FeCrAl alloys for nuclear fuel cladding: A review. *Corrosion Reviews*, *41*(2), 143–169. DOI: 10.1515/corrrev-2022-0033

Gong, B., Yao, T., Lei, P., Harp, J., Nelson, A. T., & Lian, J. (2020). Spark plasma sintering (SPS) densified U3Si2 pellets: Microstructure control and enhanced mechanical and oxidation properties. *Journal of Alloys and Compounds*, *825*, 154022. DOI: 10.1016/j.jallcom.2020.154022

Gussev, M. N., Sridharan, N., Babu, S. S., & Terrani, K. A. (2021). Influence of neutron irradiation on Al-6061 alloy produced via ultrasonic additive manufacturing. *Journal of Nuclear Materials*, *550*, 152939. DOI: 10.1016/j.jnucmat.2021.152939

Hajipour, M., Ansarifar, G. R., & Yeganeh, M. H. Z. (2025). Assessment and predicting the effect of accident tolerant fuel composition and geometry on neutronic and safety parameters in small modular reactors via artificial neural network and adaptive neuro-fuzzy inference system. *Nuclear Engineering and Design*, *433*, 113837. DOI: 10.1016/j.nucengdes.2025.113837

Hsiao, C. Y., Ou, Y., Sheng, N., & Wei, X. (2021). Measuring contagion effects of nuclear energy policies and events. *International Journal of Energy Research*, *45*(8), 11510–11525. DOI: 10.1002/er.6421

Li, Y., You, E., Wang, T., Xing, D., Xu, J., & Zhang, H. (2024). Developments, challenges and prospects in thermal-hydraulic research on accident tolerant fuel. *Heliyon*, *10*(19), e38999. DOI: 10.1016/j.heliyon.2024.e38999 PMID: 39498276

Li, Z., Wang, X., Chen, H., Zhang, R., Wei, T., Yao, L., & Wang, P. (2023). Review on performance of chromium-coated zirconium and its failure mechanisms. *Frontiers in Nuclear Engineering*, *2*, 1212351. Advance online publication. DOI: 10.3389/fnuen.2023.1212351

Nelson, A. T. (2023). Prospects for additive manufacturing of nuclear fuel forms. *Progress in Nuclear Energy*, *155*, 104493. DOI: 10.1016/j.pnucene.2022.104493

Nickel, H., Nabielek, H., Pott, G., & Mehner, A. (2002). Long time experience with the development of HTR fuel elements in Germany. *Nuclear Engineering and Design*, *217*(1–2), 141–151. DOI: 10.1016/S0029-5493(02)00128-0

Parisi, C., Ma, Z., Mandelli, D., Anderson, N., & Zhang, H. (2020). Risk-Informed Safety Analysis for Accident Tolerant Fuels. *Nuclear Science and Engineering*, *194*(8–9), 748–770. DOI: 10.1080/00295639.2020.1732699

Petrie, C. M., Linton, K. D., Vasudevamurthy, G., Schappel, D., Seibert, R. L., Carpenter, D., Nelson, A. T., & Terrani, K. A. (2023). Fission gas retention of densely packed uranium carbonitride tristructural-isotropic fuel particles in a 3D printed SiC matrix. *Journal of Nuclear Materials*, *580*, 154419. DOI: 10.1016/j.jnucmat.2023.154419

Rebak, R. B. (2023). Improved and Innovative Accident-Tolerant Nuclear Fuel Materials Considered for Retrofitting Light Water Reactors—A Review. *Corrosion and Materials Degradation*, *4*(3), 466–487. DOI: 10.3390/cmd4030024

Sanchez-Torrijos, J., Aragón, P., Queral, C., Feria, F., Schubert, A., & Van Uffelen, P. (2023). A review on the thermo-mechanical modelling needs in system codes. *Nuclear Engineering and Design*, *406*, 112243. DOI: 10.1016/j.nucengdes.2023.112243

Setty, D. S., Kapoor, K., & Saibaba, N. (2017). Nuclear fuel cycle - developments and challenges in fuel fabrication technology in India. *Progress in Nuclear Energy*, *101*, 100–117. DOI: 10.1016/j.pnucene.2017.03.005

Yin, L., Jurewicz, T. B., Larsen, M., Drobnjak, M., Graff, C. C., Lutz, D. R., & Rebak, R. B. (2021). Uniform corrosion of FeCrAl cladding tubing for accident tolerant fuels in light water reactors. *Journal of Nuclear Materials*, *554*, 153090. DOI: 10.1016/j.jnucmat.2021.153090

Yun, D., Lu, C., Zhou, Z., Wu, Y., Liu, W., Guo, S., Shi, T., & Stubbins, J. F. (2021). Current state and prospect on the development of advanced nuclear fuel system materials: A review. *Materials Reports: Energy*, *1*(1), 100007. DOI: 10.1016/j.matre.2021.01.002

Chapter 6
Small Modular Reactor: Road to Carbon Free and Sustainable Energy

Kashvi Chaturvedi
https://orcid.org/0009-0005-3192-3400
Ajeenkya D.Y. Patil University, India

Atharva Saraf
https://orcid.org/0009-0006-3842-5508
Ajeenkya D.Y. Patil University, India

Sunil Sankathala
https://orcid.org/0009-0004-7809-7647
Ajeenkya D.Y. Patil University, India

Shaurya Talewar
https://orcid.org/0009-0006-7282-1547
Ajeenkya D.Y. Patil University, India

Aditya Shrivastav
https://orcid.org/0009-0000-8461-8171
Ajeenkya D.Y. Patil University, India

Susanta Das
https://orcid.org/0000-0002-9314-3988
Ajeenkya D.Y. Patil University, India

Yadnyesh Khapekar
Ajeenkya D.Y. Patil University, India

Shrikaant Kulkarni
Sanjivani University, India

ABSTRACT

Small Modular Reactors (SMRs) are safer, more flexible, & more affordable than big reactors. They are becoming popular as an affordable, scalable, & adaptable substitute for conventional large-scale nuclear power. SMRs are positioned to be a key component of low-carbon, sustainable energy systems due to growing international interest & regulatory support. Rapid deployment in remote locations & integration with renewable energy sources are made possible by their factory-based modular construction. SMRs contribute to clean energy objectives by offering dependable electricity & purification. Their scalability & passive safety features lower capital risks & improve resilience. However, there are still issues such as public awareness, economic viability, & licensing barriers. This study investigates how SMRs can work together to create a future energy landscape that is secure, varied, & sustainable.

DOI: 10.4018/979-8-3373-0685-8.ch006

1. INTRODUCTION

The world must immediately lessen its reliance on fossil fuels while guaranteeing universal access to power in order to accomplish the Sustainable Development Goals of the UN, including SDGs 7 (cheap, clean energy) and 13 (climate action). Clean and scalable alternatives are crucial since 733 million people do not have access to electricity, and many emerging nations need to expand their energy supplies quickly (Vinoya et al., 2023).

For more than 70 years, nuclear energy has been utilized to provide power in a way that is dependable, economical, and emits few greenhouse gases. Large nuclear power facilities have historically provided the majority of baseload electricity for nations since they were frequently the most affordable choice (Lee, 2024).

Although just a few nations employ nuclear power, it nonetheless provides a sizable portion of the world's electricity since large energy-consuming nations depend on it (Daigle et al., 2024).

Despite their growing popularity, renewable energy sources like solar and wind power have drawbacks, chief among them being their low efficiency rates; solar normally operates between 10 and 25 percent and wind between 21 and 43 percent. They therefore function better when paired with consistent, dependable energy sources (referred to as baseload providers), which are frequently derived from fossil fuels. However, fossil fuels like gas, coal, and diesel emit a lot of carbon dioxide and play a big part in climate change (Vinoya et al., 2024).

There is a global movement to use more low-carbon energy in order to fight climate change. Consistent energy sources are still required to ensure dependable electricity output, even while renewable energy sources like solar and wind are growing. Nuclear energy, which currently accounts for 10% of electricity worldwide, is crucial to this shift. Microreactors and SMRs are examples of a new nuclear technology that could increase this function (Black et al., 2022).

Nuclear energy must be a major component of the world's energy mix in order to drastically reduce greenhouse gas emissions. It has, however, grown slowly and far less quickly than is required. This is mostly because building new large-scale nuclear facilities is expensive and takes a long time, especially in nations that haven't constructed any in recent decades. These problems have undermined investor and public trust, making it more challenging to finance new initiatives (OECD Nuclear Energy Agency, 2021). Lee (2024) mentions in his paper about issue brought forward by large plants like serious accidents as they retain more radioactive material.

SMRs, a more recent kind, are gaining attraction. Compared to conventional nuclear facilities, these are simpler, less expensive, and smaller to construct. They are also thought to be safer and more adaptable, which makes them ideal for isolated locations, islands, or locations that want clean energy fast (Daigle et al., 2024).

They are becoming more and more popular because of their compact size, versatility, and possible applications outside power generation, such as purifying saltwater or making hydrogen. They come in a variety of designs, from water-cooled to molten salt (Mignacca & Locatelli, 2019).

SMRs are useful in the fight against climate change since they can be a dependable source of low-carbon electricity. They provide not just electricity but also heat and desalinated water, which makes them particularly helpful in isolated or underdeveloped locations (Hussein, 2020).

They have the capacity to generate up to 300 megawatts of electricity. They are simpler and faster to construct than conventional big reactors since they are manufactured in factories and delivered in modules to their ultimate location (Mignacca & Locatelli, 2019).

SMRs are constructed in factories and moved to locations, as opposed to conventional huge nuclear facilities. This shortens building times, lowers initial costs, and facilitates expansion. They employ cutting-edge safety technologies that increase dependability and lower the danger of accidents without the need for human intervention (Josephs et al., 2025).

From water-cooled land or marine reactors to advanced gas, metal, and molten salt systems, SMRs come in a variety of technological configurations, each with special advantages for application type, efficiency, and safety. By facilitating mass production, quicker deployment, and learning curve effects—all of which progressively lower construction and operating costs—modularization and co-siting further improve cost-efficiency (Vinoya et al., 2023).

To put it briefly, SMRs have the ability to promote the social, environmental, and economic pillars of sustainable development (Hussein, 2020). Daigle et al. (2024) states that China and Russia are currently the only 2 countries operating SMRs, despite many nations' interest on constructing and optimising SMRs.

The International Atomic Energy Agency (IAEA) is currently monitoring 72 distinct SMR concepts. SMR supporters now assert that these reactors can address a number of problems that have hindered the expansion of nuclear power, including their capacity to generate high-temperature heat for a range of uses and adapt to fluctuating energy demands (load following) (Ramana, 2021).

Microreactors are extremely compact nuclear power plants with flexible and effective energy production capabilities; they usually produce less than 20 MW of electricity. They can be useful in situations when storage, weather, or space constraints make renewable energy unfeasible (Black et al., 2022).

Microreactors might be financially feasible in specialized markets where big nuclear plants aren't appropriate, but cost is still an obstacle, particularly when contrasted with other clean energy sources. Microreactors are more suited for small-scale or microgrid applications because they don't require massive emergency zones like large nuclear plants do. These comprise semi-remote or transportable configurations and applications that call for more compact, adaptable power sources (Black et al., 2022).

Their modular architecture reduces construction time and cost by enabling off-site production and quicker on-site assembly. They are perfect for deployment in rural locations, industrial zones, or in conjunction with renewable energy systems due to their lower size and improved built-in safety features. These reactors are particularly useful in areas with increasing energy demands and where massive nuclear plants are impractical (Tan et al., 2023).

The high expenses of fuel transportation and grid extension make it difficult to supply electricity to isolated or off-grid areas. Tiny-scale solutions such as smart microgrids (SMGs) and very tiny modular reactors (vSMRs) have been proposed to address this. vSMRs are small, portable nuclear reactors that are perfect for remote locations because they can generate less than 20 MW of power. They create dependable and sustainable microgrids when combined with energy storage devices and renewable energy sources (Raza et al., 2024).

By offering clean, dependable, and scalable energy solutions, SMRs and microreactors also closely align with UN Sustainable Development Goals 7 and 13, which are part of the global climate goals. They provide the chance to power off-grid populations, replace coal facilities, and assist industrial processes like desalination and hydrogen manufacturing (Tan et al., 2023).

As renewable energy sources thrive, these reactors are increasingly being viewed as essential to maintaining grid resilience and energy security as part of Advanced Nuclear Power Plants (aNPPs). Hybrid Energy Systems (HESs) and Integrated Energy Systems (IESs) are ideal for SMRs because they mix various energy sources (nuclear, fossil, and renewables) and outputs (clean water, fuel, and electricity) to increase total flexibility and efficiency. This is especially crucial as extreme events are occurring more frequently due to climate change, which puts pressure on energy systems to become more resilient and adaptive (Di Maio et al., 2022).

There are many advantages to SMRs. They are perfect for nations or regions that cannot support large-scale reactors because of their smaller size and lower cost. Compared to conventional nuclear projects, they can be built in three to five years. Multiple SMRs can be installed in tandem to meet power requirements due to their scalability and modularity. Passive safety features that reduce the possibility of radiation leakage are one way that their small, factory-tested design enhances safety. Additionally, SMRs are very adaptable and can power industrial sites, isolated or off-grid settlements, and other places where conventional nuclear power is impractical. They also complement renewable energy sources well, providing dependable backup power and aiding in grid stabilization in the event of variations in solar or wind output (Shobeiri et al., 2023).

Deployment is hampered by issues like public perception, cost scalability, license complexity, and a lack of operational experience, notwithstanding its potential. However, governments and business are becoming more interested in and supportive of them, which is speeding up their development (Tan et al., 2023).

Public education, open governance, and consistent policy backing are necessary for effective deployment. Furthermore, SMRs must address concerns like proliferation hazards, particularly for specific reactor types that require higher enriched uranium, even though they can encourage the development of jobs and local advantages (Vinoya et al., 2023).

While some countries, like Germany, Switzerland, and Italy, have paused or withdrawn from nuclear energy due to incidents like the Fukushima disaster, many others, like China, Russia, India, the United Arab Emirates, and to some extent the United States and the United Kingdom, continue to invest heavily in nuclear programs. Global energy consumption is expected to more than double by 2035 compared to 1995, largely due to rapid growth in non-OECD countries, especially China. During that time, electricity generation alone is expected to triple, with non-OECD countries increasing their energy use five to six times (Locatelli et al., 2014).

The U.S.-led FIRST Program, which assists countries interested in implementing American SMR technology, has extended an invitation to Southeast Asian nations like Thailand, the Philippines, and Indonesia to join. This occurs at a time when the region's energy demand is expected to rise quickly, reaching a combined GDP of $20 trillion by 2040. Energy security and carbon emissions are key problems because the region is heavily dependent on fossil fuels, particularly coal, which is predicted to nearly double in capacity by 2040 (Trajano et al., 2023).

Despite their flexibility and scalability, SMRs have drawbacks over big reactors, including greater initial costs per megawatt, less operational experience, and complicated multi-unit regulation. To address risks unique to SMRs and guarantee safe, effective deployment in the dynamic energy market, regulators must modify current frameworks (Josephs et al., 2025).

SMRs also encounter developmental and regulatory obstacles. Shobeiri et al.(2023) studies about few designs that have been approved by regulators or started construction. As of yet, CAREM in Argentina, SMART in South Korea, and NuScale in the United States have been approved. Countries such as Canada have implemented pre-licensing schemes to assist developers in adhering to safety regulations in order to foster innovation. In light of the global energy shift, SMRs show great promise.

This paper mentions the use of a hybrid energy system that consists of the main grid, solar panels, wind turbines, battery storage, and SMRs. It compares the current state of microreactor development to that of SMRs and major nuclear plants. It looks at their market potential, financial difficulties, and legal restrictions that need to be removed in order for them to be widely used. The paper also evaluates the impact of SMR deployment on grid stability.

2. LITERATURE REVIEW

2.1 Overview and History of SMR

A. Overview

SMRs are defined by the International Atomic Energy Agency (IAEA) as nuclear reactors having a maximum capacity of 300 MWe. Early nuclear development started with low-power designs, but contemporary SMRs purposefully take use of their tiny size to increase economic viability, safety, construction feasibility, and operating simplicity (Hussein, 2020).

SMRs are a new class of nuclear energy products that are different from big reactors, however they are not categorized by generation. SMRs can be cooled with light water or with other coolants, such as gasses or liquid metals. They are made to be affordable, safe, and small (Lee, 2024).

They are made to be modular, which allows them to be installed on-site after being manufactured in factories. Many reactor designs, many of which are still in the conceptual stage, can be categorized as "advanced" due to the term's uncertainty. Due to economic considerations, nuclear power's share of the world's electricity output has decreased from 17.5% in 1996 to roughly 10% in 2019. According to a 2003 MIT research, nuclear power lacked economic viability. After the Energy Policy Act of 2005, there was initial hope for a nuclear comeback, but many new projects experienced severe delays and cost overruns. For instance, the Vogtle project in Georgia has seen delays in its operational timetables and cost increases from its original projection of $14 billion to $29 billion (Ramana, 2021).

The nations that now own nuclear reactors or intend to construct them utilize the majority of the world's electricity. This implies that there is a sizable market for SMRs. Growth, however, is depending on the rate of increase in energy demand. It is anticipated that nations with rapidly increasing demand, such as China and India, will lead the world in new nuclear installations (Daigle et al., 2024).

It takes a lot of preparation, public involvement, and financial resources to launch a new nuclear power program, especially for SMRs. The economic feasibility of SMRs depends on a sizable market, and quick action is required to guarantee there is enough nuclear power to meet future energy demands (Shobeiri et al., 2023).

Six key areas, technology selection, licensing and regulations, supply chain management, public image, financial support, and proliferation risks, are recognized as barriers to the widespread adoption of SMRs (Shobeiri et al., 2023).

Choosing the right technology is essential when investing in SMRs. No SMRs are presently being built, despite more than 70 designs being submitted worldwide, and it is challenging for any one manufacturer to control the market due to design competition (Shobeiri et al., 2023).

B. Nuclear Reactor Generations:

- Generation I: Reactor prototypes used to validate ideas and technical procedures (Lee, 2024).
- Generation II: The market-dominant commercial power plants, renowned for their affordability and high capacity (Lee, 2024).
- Generation III: These reactors, which prioritize safety and incorporate cutting-edge technologies, were created in the wake of significant catastrophes (such as Chernobyl). Light-water reactors make up the majority of them (Lee, 2024).
- Generation III+: An upgrade with better safety features than Generation III (Lee, 2024).
- Generation IV: Currently undergoing research, these reactors depart from light-water technologies and strive for safety and sustainability (Lee, 2024).

C. In Present

SMRs are being developed with an emphasis on modularity, small-scale energy production, and shorter building timeframes in order to solve the drawbacks of traditional nuclear power stations. Two businesses, mPower America Partnership and NuScale Power Partnership, have been awarded funds to design and create functional models of SMRs as part of the $472 million allotted by the U.S. Department of Energy (DOE) to support the development of SMRs. They have the potential to be utilized in remote locations and smaller towns because of their increased efficiency and flexibility (Richter, 2014).

1. The Reasons SMRs Could Take Off:

SMRs are adaptable and have a wide range of applications, such as:

- It is difficult to develop large power plants or expand the system in remote areas especially on islands (Daigle et al., 2024).

- Installing smaller, cleaner SMRs in place of outdated coal and gas facilities (Daigle et al., 2024).
- Desalination facilities to produce drinking water from saltwater, particularly in arid areas like the Middle East (Daigle et al., 2024).

2. SMR examples include:

- SMART: A design-approved integral type PWR from Korea that prioritizes safety through small size (Lee, 2024).
- NuScale Power Module: A modular reactor that lowers costs and improves safety by enabling the operation of numerous units with fewer operators (Lee, 2024).
- ACP100: A Chinese SMR that is currently being built and has obtained international safety approval (Lee, 2024).
- NUWARD: A French SMR that is presently under conceptual design and should be finished by 2025 (Lee, 2024).
- A Korean initiative called i-SMR aims to significantly lower the frequency of core damage and is intended for a number of uses, such as desalination and hydrogen production (Lee, 2024).

2.2 Advantages

A. Economic and Technical Advantages
    - Improved Safety: The risk and complexity of failure modes are decreased by smaller fuel stockpiles and passive safety measures (such as natural circulation) (Hussein, 2020).
    - Robustness and Simplicity: Design simplicity reduces fabrication and construction difficulties (Hussein, 2020).
    - Grid Compatibility: Perfect for small-scale or remote applications, such as micro-grids, industrial facilities, and off-grid settlements (Hussein, 2020).
    - Standardization makes it possible to share operating experience, lower capital expenditures, and simplify licensing (Hussein, 2020).
    - Financial viability: Investment risk is decreased by lower upfront expenses. The labor inefficiencies that tend to favor bigger reactors can be mitigated by automation (Hussein, 2020).
    - Scalability: Modular deployment makes it possible to grow capacity incrementally, which enhances cost control and demand alignment (Hussein, 2020).

- SMRs are less likely to suffer catastrophic incidents because of their stringent safety requirements and design elements (Tan et al., 2023).
- Their passive safety measures can lower the chance of radioactive contamination and prevent mishaps (Tan et al., 2023).

B. On Environment

- Life cycle assessments, which contrast the greenhouse gas emissions of SMRs with those of other energy sources, are used to evaluate their environmental sustainability (Vinoya et al., 2023).
- Research suggests that SMRs are less harmful to the environment than fossil fuels, but more study is required to fully comprehend their effects (Vinoya et al., 2023).
- Since there is now just one SMR operating commercially, the majority of LCAs (Life Cycle Analaysis) for SMRs are based on theoretical models. Data from reference LRs is frequently scaled in these analyses (Vinoya et al., 2024).
- Research suggests that SMRs and LRs have similar global warming potentials (GWPs). Nuclear, for instance, generates about 12.4 grams of CO_2 per kilowatt-hour (g CO_2-eq/kWh), but wind and hydropower have the highest and lowest GWPs, at 28.6 and 3.5 g CO_2-eq/kWh respectively (Vinoya et al., 2024).
- A useful method for evaluating how technologies affect the environment is LCA. It takes into account variables including greenhouse gas emissions, influence on water resources, and toxicity from the extraction of raw materials until decommissioning (Vinoya et al., 2024).
- Using kilowatt-hours (kWh) as a functional unit of comparison, this study examines the environmental effects of SMRs on the production and transmission of electricity (Vinoya et al., 2024).
- Its LCA assesses SMRs' effects on the environment at every stage and demonstrates that they can have less of an impact than conventional big reactors (Josephs et al., 2025).
- SMRs with advanced fuel cycles can cut waste, but they may also present new difficulties. SMR decommissioning has benefits and drawbacks, and creative solutions are required for efficient waste management (Josephs et al., 2025).
- Studies show that nuclear power produces fewer employment per unit of energy than renewable sources like solar and wind, despite the fact that SMRs are predicted to create jobs. Further restricting the creation of jobs is the fact that many SMR designs seek to lower the number of operators (Ramana, 2021).

C. The Reasons SMRs are Helpful

- Flexible Market Demand: SMRs are able to combine with renewable energy sources and adjust to changing energy demands. They can coexist with different energy sources and are thought to be a good alternative to outdated fossil fuel facilities (Locatelli et al., 2014).
- Modular Design: SMRs can be transported and installed more easily because of their modular design. They can be deployed in a variety of configurations due to their design, which makes them appropriate for a range of energy requirements (Locatelli et al., 2014).

2.3 Disadvantages/ Difficulties

A. Principal Difficulties
  - Regulations: SMRs are nevertheless subject to the same regulations as large nuclear reactors, which results in a drawn-out and costly approval procedure (Daigle et al., 2024).
  - Policy uncertainty: Investors are uneasy as several nations, such as Japan and Germany, are abandoning nuclear power. Others are just getting started, such as the UAE and Turkey (Daigle et al., 2024).
  - Due of the possibility of accidents and the requirement for strong safety measures, safety issues are a significant problem with SMRs (Richter, 2014).
  - The possibility that terrorists might attack SMRs and the requirement for safe nuclear material storage and transportation raise further security considerations (Richter, 2014).
  - Another problem is waste; SMRs generate radioactive waste, which needs to be properly stored and disposed of (Richter, 2014).

B. Challenges with Licensing and Regulation
  - Adoption of SMR depends on efficient licensing and regulatory frameworks. Standardization is made more difficult by the variety of designs and the novelty of SMR technology (Josephs et al., 2025).
  - Long-term obstacles like regulatory fragmentation and short-term difficulties like exorbitant fees and protracted licensing periods are examples of regulatory concerns. There are chances to expedite the licensing procedure by means of global cooperation and flexible regulatory structures (Josephs et al., 2025).
  - SMR design and licensing procedures are still in their infancy, and there is a great deal of uncertainty around their budgets and schedules (Rosner et al., 2011)

- Since each SMR design needs to be licensed in its home nation, regulatory approval is a significant obstacle that can be time-consuming and complicated (Rosner et al., 2011).

C. Obstacles
  - There is currently no established manufacturing infrastructure for SMRs, necessitating a significant financial outlay to construct specialized production facilities (Rosner et al., 2011).
  - Since each SMR design needs to be licensed in its home nation, regulatory approval is a significant obstacle that can be time-consuming and complicated (OECD Nuclear Energy Agency, 2021).
  - Public acceptability: Given previous nuclear mishaps, social acceptability of nuclear energy is crucial. The effective implementation of SMRs will depend on interacting with communities and resolving safety issues (Locatelli et al., 2014).
  - Economic and Regulatory Aspects: Issues with licensing, public opinion, and the supply chain could prevent SMRs from being widely used. The development and implementation of SMRs require a stable regulatory environment (Locatelli et al., 2014).

2.4 Fuel

Fuel Innovation and Special Configurations

- High neutron economy, flexibility in fuel types, including thorium, and on-power refueling are all made possible by Heavy Water Reactors (HWRs) with Natural Uranium CANDU-derived SMRs (Hussein, 2020).
- Fuel recycling capability and reduced enrichment requirements provide advantages for resource and economic use (Hussein, 2020).
- For instance, CANDU SMR and AHWR-LEU permit the use of thorium, which lowers the waste of long-lived actinides (Hussein, 2020).
- Reactors that use particle fuel (TRISO)
  - o Fuel in small coated particles (TRISO) permits high-temperature operation and improves safety by containing fission products (Hussein, 2020).
  - o High efficiency and low pressure operation make it appropriate for both gas and molten salt coolants (Hussein, 2020).
  - o Fuel can withstand high temperatures without active cooling thanks to passive safety (Hussein, 2020).

Fuel Composition in SMRs

a. TRISO Particle Fuel: High-temperature reactors (HTRs) use TRISO (Tristructural Isotropic) fuel, which has the following characteristics:
   - o UO_2, UC, UN, and UCO have high ceramic-based temperature tolerance (Hussein, 2020).
   - o Fission products are retained by multilayer carbon and SiC coatings (Hussein, 2020).
   - o Because of its exceptional high-temperature performance (20% burnup, 1400°C irradiation, and up to 1800°C accident tolerance), UCO (uranium oxycarbide) is utilized in SC-HTGR and Xe-100 (Hussein, 2020).
   - o HTMR-100 and GTHTR300 employ TRISO based on UO_2 (Hussein, 2020).
   - o In order to provide high fuel loading, EM2 uses UC with SiGATM cladding (Hussein, 2020).

b. Advanced Fuel Types:
   - Uranium Silicide (U_3Si_2)
     - o denser uranium (11,300 kg-U/m^3 for UO_2 compared to 9,700 kg-U/m^3 for UO_2) (Hussein, 2020).
     - o Improved heat capacity and thermal conductivity increase accident tolerance (Hussein, 2020).
     - o allows for the use of alternate cladding (such as FeCrAl, SiC/SiC, and Ti_3SiC_2) (Hussein, 2020).
     - o taken into account for advanced PWRs and maritime SMRs (Hussein, 2020).
   - Uranium Nitride (UN)
     - o high thermal conductivity and density (Hussein, 2020).
     - o Nevertheless, it is chemically reactive and dissociates at about 1600°C (Hussein, 2020).
     - o To lessen the formation of 14C from neutron interactions, 15N enrichment is required (Hussein, 2020).
     - o utilized in the SVBR-100, eVinci, LFR-TL-X, and, Westinghouse LFR; the BREST-OD-300 uses mixed UN-PuN (Hussein, 2020).
   - Metallic Fuel (U-Zr, U-Pu-Zr)
     - o Despite having lower melting points, fast reactors are preferred for neutron economy (Hussein, 2020).
     - o High heat conductivity is advantageous (Hussein, 2020).
     - o utilized in SMRs such as BREST and 4S (Hussein, 2020).
   - Molten Metal Fuel
     - o LAMPRE-1 was tried in the past, but because of its high enrichment requirements, contemporary SMRs avoid it (Hussein, 2020).

- offers high burnup, online reprocessing potential, and passive safety (Hussein, 2020).

c. Limitations
- Material problems arise from the high reactivity and water sensitivity of carbide, nitride, and silicide fuels (Hussein, 2020).
- Fast SMRs with small cores suffer restrictions in enrichment and neutron leakage, while thermal SMRs use LEU (<20%) for proliferation resistance (Hussein, 2020).

d. Refueling Strategies
- Batch Refueling
    - Standard for most commercial reactors (Hussein, 2020).
    - Light water SMRs show extended cycles: e.g., RITM-200M (120 months), ABV-6E (120–144) (Hussein, 2020).
    - High-density fuels (UN, UC, U-Zr) enable longer cycles: e.g., SEALER (336 months), SUPERSTAR (180) (Hussein, 2020).
    - Downside: shutdown periods (~1 month), added O&M costs (Hussein, 2020).
- Online/On-Power Refueling
- Used in:
    - Channel-type reactors: AHWR-300, CANDU SMR (Hussein, 2020).
    - Stable Salt Reactors: fresh pellets inserted, spent fuel removed hot (Hussein, 2020).
    - CMSR (Molten Salt): continuous feed of SNF/thorium, 12-year lifetime (Hussein, 2020).
    - Pebble-bed reactors (e.g., Xe-100): continuous fuel cycling (Hussein, 2020).
- Advantages:
    - Higher fuel utilization (Hussein, 2020).
    - No refueling downtime (Hussein, 2020).
    - Can also support online fission product removal (Hussein, 2020).
- Battery-Powered Function
    - Made for transportable, sealed use (navy, remote areas) (Hussein, 2020).
    - Reactors are returned for waste disposal and refueling after a lifespan of 10 to 30 years (Hussein, 2020).

For instance:

- U-battery, eVinci, and MMR: 10–20 years with LEU (Hussein, 2020).
- 4S (30 MWt): fast-spectrum with moveable reflector-based burnup compensation (Hussein, 2020).

- Limitations:
  - limited power because of neutron leakage and fuel enrichment limitations (Hussein, 2020).
  - need either neutron economy improvements (reflectors, blankets) or high-density fuel (Hussein, 2020).

e. Breeder, Converter, and Burner Reactors Breeders
   - converters produce fissile fuel from Th-232 or U-238, which are fertile. For instance, TerraPower TWR (breed-and-burn), BREST, and EM2 (convert-burn core) (Hussein, 2020).
   - Use weapons-grade plutonium or SNF for burners. For instance, rapid reactors, molten salt, and stable salt wasteburners (Hussein, 2020).
   - The two main benefits are waste reduction and fuel cycle closure (Hussein, 2020).

f. A wide range of technologies are used in reactors, including:
   - Light Water Reactors, or LWRs (Black et al., 2022).
   - TRISO-fueled high-temperature gas reactors, or HTGRs (Black et al., 2022).
   - Melted Salt Reactors, or MSRs (Black et al., 2022).
   - Sodium-cooled Fast Reactors, or SFRs (Black et al., 2022).
   - Reactors with Heat Pipes (Black et al., 2022). (Black et al., 2022).

g. Types of fuel:
   - LEU (3–5% U-235) is typically used in SMRs. (Black et al., 2022).
   - Microreactors: For improved performance and longer core lifetimes, choose HALEU (5–19.75% U-235). (Black et al., 2022).
   - Particularly with microreactors, design simplicity and passive cooling (such as heat pipes) are important differentiators (Black et al., 2022).

2.5 Technical Specifications

A. Modularity of SMR

Modularity in the context of SMR comprises modules manufactured in a factory and delivered to the installation location. Modules, as mentioned by Hussein (2020), may include:

- Vessels for reactors
- Coolant systems
- Turbines
- Controls and Instrumentation
- Components that are structural and electrical

Hussein (2020) in his paper also listed the following types of modularity:

- Integrated modularity reduces pipe hazards by combining the reactor and cooling system into a single unit.
- Close-Coupled Modularity: submarines with compact designs that require less containment.
- Combined Modularity: When many SMR units are combined, larger plants are produced.

Hussein (2020) has mentioned many details regarding modularity. IAEA defines:

- Depending on need, a single or multiple module deployment
- Modularity in scale:
    - Combined scale: a single facility with several reactors(Hussein, 2020)
    - Standardized scale: Using the same units repeatedly(Hussein, 2020)

A modular design highlights:

- Partitioning systems into independently functional units is known as functional partitioning (Hussein, 2020)
- Standard interfaces guarantee interchangeability and compatibility (Hussein, 2020)
- Scalability: The capacity to reuse modules across applications and designs (Hussein, 2020)

There are three kinds of modules:

- Common modules are utilized by many kinds of plants (Hussein, 2020)
- Self-contained modules that require little setup and are "plug-and-play" (Hussein, 2020)
- Modular installations are platforms that can hold different modules (Hussein, 2020)

Modular Design Benefits

- cuts down on the complexity and duration of building (Hussein, 2020)
- improves recyclability, upgradability, and maintainability (Hussein, 2020)
- encourages the advancement of technology (constructal law) (Hussein, 2020)
- increases assembly-line efficiency and mass production (Hussein, 2020)

Phases of Modularization

- Design: Construct subsystems that operate independently (Hussein, 2020)
- Manufacturing: Produce in regulated settings (Hussein, 2020)
- Building: Quickly assemble modules on location (Hussein, 2020)
- Operations & Maintenance: Make it possible to upgrade or service individual units (Hussein, 2020)
- Decommissioning: Streamlined waste management and dismantling (Hussein, 2020)

B. Reactor Design

In order to improve safety, dependability, and economies, modern reactor designs—particularly those in the Generation IV category—place a strong emphasis on integral and passive safety, modular integration, and creative fuel systems. These concepts are used by microreactors and SMRs to increase deployability, decrease operating complexity, and save capital expenses (Hussein, 2020).

1. Passive and Inherent Safety Mechanisms
    - Integral safety: During transients, the reactor is stabilized or shut down using physical rules (such as negative reactivity feedback) rather than engineering controls (Hussein, 2020).
    - Using natural convection, gravity, and radiative heat transfer in the absence of active parts or external power is known as passive safety. While not infallible, this is more dependable in an emergency (Hussein, 2020).
2. Configurations of Integral Reactors
    - Important innovation: combining the reactor core, steam generator, pumps, and pressurizer—the main coolant components—into a single pressure vessel (Hussein, 2020).
    - Organic circulation Boiling water reactors, or BWRs, simplify plant design and lessen the need for pumps (Hussein, 2020).
    - Efficiency is increased by direct steam generation, which eliminates the need for intermediary heat exchangers (Hussein, 2020).
    - Safety and Financial Benefits:

    - reduces the possibility of big break loss-of-coolant (LOCA) accidents (Hussein, 2020).
    - improves seismic resistance and streamlines system design (Hussein, 2020).
    - simplifies modular deployment and factory fabrication (Hussein, 2020).
- Limitation: Compactness increases waste volume and decreases thermal efficiency by increasing neutron leakage and decreasing fuel burnup (Hussein, 2020).

C. Integration with Renewable Sources

Hybrid renewable energy systems are growing in popularity as the globe shifts to cleaner, low-carbon energy to combat climate change. Instead of depending on just one energy source, these systems integrate sources like sun and wind to produce a more dependable and economical energy supply. Due to their complementary nature and ability to produce more consistent power, solar-wind combinations are particularly popular (Paisiripas et al., 2024).

Reliability can be further increased by including nuclear energy in the mix. Hybrid nuclear-renewable systems combine the variable character of renewable energy sources with a stable nuclear power output. This lowers pollution and fuel expenses. Additionally, battery storage can be used to manage the fluctuations of renewable energy sources and guarantee a consistent power source. According to some research, these hybrid systems can lower overall energy expenses and carbon emissions by up to 40% (Paisiripas et al., 2024).

Cities and other places with significant energy demands benefit greatly from these systems. They promote a cleaner, more sustainable energy mix and lessen reliance on the grid by integrating nuclear and renewable energy. All things considered, nuclear-powered grid-connected hybrid systems can greatly cut greenhouse gas emissions, increase energy reliability, and minimize electricity costs (Paisiripas et al., 2024).

a. Technical Synopsis

Micro-reactors

- SMR subsets with outputs that normally range from less than 1 MW(e) to 20 MW(e), with a maximum of about 50 MW(e). (Black et al., 2022).
- Designed for remote industrial applications, off-grid use, microgrids, renewable energy integration, and portability. (Black et al., 2022).
- Make it possible for distributed energy applications to replace fossil fuel systems with cogeneration. (Black et al., 2022).

- Stress factory-based manufacture, modularity, and compactness to enable transportability and possibly mobile deployment. (Black et al., 2022).
- Reduce staffing needs and aim for semiautonomous operation (Black et al., 2022).

SMR (Black et al., 2022).

- SMRs may produce between 50 and 300 MW(e) of power per module, with multi-unit systems capable of producing up to 900 MW(e). (Black et al., 2022).
- Represent a change from conventional large nuclear plants (600–1750 MW(e)) to alternatives that are modular, standardized, and easier to build. (Black et al., 2022).
- Permit decreased emergency planning zones, shorter building schedules, lower footprints, and co-siting with industrial loads or renewable energy sources (Black et al., 2022).

b. Advanced Technology

- SMR efficiency and safety can be increased using advanced materials like accident-tolerant fuel and TRISO fuel (Tan et al., 2023).
- Systems for digital instrumentation and control (DI&C) can enhance SMR performance and safety (Tan et al., 2023).
- SMRs must be protected from cyber threats by cybersecurity measures (Tan et al., 2023).
- Operators can get help from artificial intelligence (AI) in identifying and detecting faults (Tan et al., 2023).
- Complex issues in SMR design and operation can be simulated using the high-performance computing framework MOOSE (Tan et al., 2023).

c. Advancements in Technology

- Currently, 72 SMR development projects are underway, with Russia, China, and Argentina leading the way (Vinoya et al., 2023).
- Modular construction is made possible by SMR designs, making installation and transit simpler. By reducing the number of components, integrated designs lower the chance of accidents (Vinoya et al., 2023).
- Microreactors, which may meet smaller energy needs, and offshore and underwater reactors, which improve safety against natural calamities, are examples of innovations (Vinoya et al., 2023).

2.6 Reactor Vessel

a. Integral Design of Reactor Vessels
   - By reducing the quantity and size of penetrations, SMRs with integral reactor vessel layouts eliminate the possibility of control rod ejection accidents (CREAs) and big break loss of coolant accidents (LBLOCAs) (Di Maio et al., 2022).
   - This design element lowers the possibility of serious reactor core accidents and increases safety margins (Di Maio et al., 2022).

b. Higher Volume of Relative Coolant and Relative Pressurizer
   - Larger relative coolant stocks and pressurizer volumes in SMRs with integrated designs increase thermal inertia and coping time during transients of loss of ultimate heat sink (LUHS) (Di Maio et al., 2022).
   - SMRs can satisfy the functional requirement (FR) of minimal sensitivity to external risks because to this characteristic (Di Maio et al., 2022).

c. High Reactor Vessel with Small Diameter
   - Safe cooldown without the need for pumps and motors is made possible by SMRs with taller reactor vessels and smaller diameters, which improve heat removal and allow for natural circulation (Di Maio et al., 2022).
   - SMRs can meet FRs for island mode operation, natural circulation, and rapid and infinite core cooling because to this design feature (Di Maio et al., 2022).

d. The reactor building's subpar construction and small fuel inventory
   - Because SMRs have smaller fuel stockpiles, there is less chance of radioactive contamination and core decay heat (Di Maio et al., 2022).
   - Reactor buildings built below grade minimize discharge routes and strengthen the system against natural disasters (Di Maio et al., 2022).

e. Tiny Size
   - Because SMRs are compact, they require less power for shutdown cooling, which makes passive cooling systems possible and lessens the need for emergency power sources (Di Maio et al., 2022).
   - Because of their small size, SMRs may also function in island mode, providing other reactor units with core self-cooling power and housekeeping power (Di Maio et al., 2022).

f. Adaptability
   - Because SMRs are modular, they may be deployed incrementally, which increases operational flexibility and permits load-following (Di Maio et al., 2022).

- Because it is extremely improbable that an accident scenario will arise when fuel failure occurs in all SMR units simultaneously, modularity further lowers the likelihood of a significant radiation emission (Di Maio et al., 2022).

g. Produced within the factory
    - Because SMRs may be produced in factories, building time and expenses can be decreased (Tan et al., 2023).
    - Factory production can cut waste and increase quality (Tan et al., 2023).
    - Reducing the number of components and simplifying construction are two benefits of modular design (Tan et al., 2023).

h. The Chain of Supply
    - The supply chain for nuclear fuel is intricate and needs to be managed carefully (Tan et al., 2023).
    - While SMRs can construct new reactors using conventional techniques, several features are given additional attention (Tan et al., 2023).
    - The mining, conversion, enrichment, and fuel production of uranium are all part of the nuclear fuel cycle (Tan et al., 2023).

2.7 vSMRs

Micro-reactors are defined as those that produce 1–10 MWe.

- Portable, low-maintenance, and self-contained (Hussein, 2020).
- Categorized as a Category 2 Hazard (effects only on-site) (Hussein, 2020).
- Plug-and-play options for desalination, emergency power, hydrogen generation, and distant locations (Hussein, 2020).
- Demonstrated in space applications (e.g., Romashka, BUK, and SNAP-10A systems) (Hussein, 2020).
- Unlike radioisotope thermoelectric generators (RTGs), which use decay heat rather than fission heat (Hussein, 2020).

Additionally, SMRs and vSMRs are physically smaller that permit deployment on offshore platforms, industrial ships, and restricted or urban areas (Hussein, 2020).

a. Examples of designs:
    - SEALER (3 MWe): 2.748 m in diameter, 6 m in height, 30 tons (Hussein, 2020).
    - mPower (190 MWe): 27.4 m height, 4.15 m diameter (Hussein, 2020).
    - Shippable in sections, ELENA (68 kWe) has a diameter of 1.25 meters (Hussein, 2020).

- CMSR (100–115 MW): vessel height of 2.5 m (Hussein, 2020).
- Westinghouse SMR, eVinci, SUPERSTAR, and UK SMR are all made for modular deployment and transportation (Hussein, 2020).

b. System Microgrid
- The microgrid system, which consists of a vSMR, PV system, WTS, and BESS, is intended to run at 11 kV (Raza et al., 2024).
- The baseload power source is the vSMR, and the power supply is supplemented by the PV system and WTS (Raza et al., 2024).
- When energy generation is minimal, the BESS is utilized to supply power and store extra energy produced by the PV system and WTS (Raza et al., 2024).

2.8 Safety Considerations

Compared to conventional big reactors, SMRs have a stronger safety argument thanks to their enhanced safety measures. These consist of:

- Passive Safety Mechanisms: By using natural cooling processes, SMRs eliminate the requirement for active systems and streamline safety assessments (OECD Nuclear Energy Agency, 2021).
- Fewer Failure Modes: They are more resilient to mishaps like loss-of-coolant occurrences because the design is simpler, which reduces the number of possible failure points (OECD Nuclear Energy Agency, 2021).
- Decreased Areas for Emergency Planning (EPZ): Public safety perceptions may be improved by smaller EPZs made possible by lower inventory and better passive safety features (OECD Nuclear Energy Agency, 2021).

The following intrinsic safety attributes of SMRs strengthen their safety case:

- By lowering dependency on active systems, passive safety mechanisms enhance decay heat removal and streamline safety assessments (Vinoya et al., 2023).
- Fewer Failure Modes: SMRs are more accident-resistant due to the design simplification, which reduces the number of possible failure locations (Vinoya et al., 2023).
- Decreased Areas for Emergency Planning (EPZ): Public safety perceptions may be improved by smaller EPZs made possible by lower inventory and better passive safety features (Vinoya et al., 2023).
- Advanced Safety systems: SMRs reduce the possibility of catastrophic accidents by incorporating passive safety systems. The KLT-40S and the mPower

reactor are two examples of integrated designs that improve safety and lower the risk of accidents (Locatelli et al., 2014).

- Waste and Safety Issues

SMRs nevertheless carry the same risks as nuclear power, despite being marketed as safer and generating less waste. The features of waste management and safety in various SMR designs range greatly (Ramana, 2021).

- SMRs have major design improvements that increase safety while using the same fuel and cooling systems as conventional big light water reactors (LWRs) (Rosner et al., 2011).
- Important safety characteristics include integral designs that reduce the chance of coolant leakage, passive safety systems that use natural cooling processes, and lower decay heat levels that require less cooling after shut-down (Rosner et al., 2011).
- By locating reactor components in subterranean pools, these designs seek to improve earthquake resilience and do away with the requirement for backup power systems (Rosner et al., 2011).
- Nuclear governance frameworks, including safety, security, and non-proliferation procedures, need to be reviewed and strengthened as SMR projects progress in the Asia-Pacific area (Trajano et al., 2023).
- It might be necessary to modify current international nuclear conventions in order to provide complete safety criteria and sufficiently include SMRs, especially floating designs (Trajano et al., 2023).
- Strong security measures are required since the digitalization of SMRs raises cybersecurity concerns. The scattered nature of SMR deployment necessitates the adaptation of physical protection techniques (Josephs et al., 2025).
- "Safeguards by Design" ensures efficient nuclear material monitoring and verification by including non-proliferation measures into reactor design (Josephs et al., 2025).

2.9 Case Studies

The study shows that aNPPs (alternative advanced nuclear power plants) with design features appropriate to withstand natural hazards and meet functional requirements can enhance IES resilience against NaTech events. The case study compares the resilience of an IES with a conventional nuclear power plant (cNPP) and three aNPPs comprising multiple SMR units.

- Unites States: In order for the United States to share its nuclear technology, the Philippines and the United States announced bilateral negotiations for a civilian nuclear energy cooperation agreement in May 2023. President Ferdinand Marcos Jr. indicated interest in adding nuclear energy to the Philippines' power mix, and U.S. companies like NuScale Power, which plans to build SMRs in the country by 2031, are potential suppliers. The United States will also support the development of a national nuclear engineering workforce focused on SMR technology (Trajano et al., 2023).
- Thailand: As part of Thailand's efforts to reach net-zero emissions by 2065, a clean energy collaboration was established to encourage the deployment of advanced nuclear reactor technologies, including SMRs, during a visit by U.S. Vice President Kamala Harris in November 2022 (Trajano et al., 2023).
- Indonesia: The United States and Indonesia signed an agreement in March 2023 to support the development of Indonesia's nuclear energy program, with a special emphasis on SMRs. To assess the viability of Indonesia's first SMR plant, the United States will offer capacity-building assistance (Trajano et al., 2023).
- Myanmar: In an effort to alleviate energy constraints brought on by depleting natural gas reserves, Myanmar's military government is exploring SMRs constructed in Russia. A memorandum of understanding with Rosatom permitted the start of a feasibility study on SMRs (Trajano et al., 2023).
- South Korea: South Korean enterprises and U.S. SMR companies are collaborating on the development of micro-modular reactors. Among the noteworthy partnerships is Hyundai's development of SMR prototypes with Holtec International and Ultra Safe Nuclear Corporation (Trajano et al., 2023).
- Japan: The SRZ-1200 is an integrated compact reactor created by Mitsubishi Heavy Industries for use in emergency power supplies and small grids. In order to obtain access to cutting-edge SMR technology, Japan is also working with American businesses (Trajano et al., 2023).
- China: Through state-owned companies, China is aggressively developing SMRs. Examples of these include the demonstration of a floating nuclear reactor intended for offshore power generation and a high-temperature gas-cooled reactor (Trajano et al., 2023).
- There is also potential in areas of Europe, Africa, Latin America, and North America (Black et al., 2022).
- Distributed energy and isolated operations are the most potential markets (Black et al., 2022).
- Human Resources and the Supply Chain

For SMR deployment, a dependable supply chain with the required technical know-how must be established. Cost savings are possible thanks to the modular architecture, but order disruptions can be expensive. To develop a skilled workforce, cooperation with suppliers and academic institutions is required (Shobeiri et al., 2023).

- Perception and Acceptance by the Public

Historical catastrophes and cultural perceptions of risk are frequently the root causes of public opposition to nuclear power. But there has been a rise in support for nuclear energy, especially as worries about climate change increase. More people might accept SMRs with improved safety measures (Shobeiri et al., 2023).

- Political and Financial Assistance

SMRs' modular deployment, which lowers risk and permits incremental investment, is what makes them financially appealing. In order to finance research and development and maintain public safety, government support is essential (Shobeiri et al., 2023).

- Risks of Explosion

Concerns regarding nuclear explosion and safety measures will need to be addressed once SMRs become economically feasible. The possible hazards related to the nuclear fuel cycle must be taken into account when designing and implementing SMRs (Shobeiri et al., 2023).

- Additional Concerns

Concerns about spent fuel management and the nuclear fuel cycle persist with SMRs. The isotopic makeup of spent fuel must be confirmed, and alternatives for ending the nuclear fuel cycle must be investigated (Shobeiri et al., 2023).

2.10 Grid Specifications

- For SMRs to be successful, they must be integrated into current electricity grids. Because of their modularity, operational flexibility, and reduced size, SMRs are attractive because they can be deployed gradually to meet the rising need for energy (Josephs et al., 2025).

- Because of their variable output capabilities, they can be used to supplement irregular renewable energy sources like solar and wind, improving grid stability (Josephs et al., 2025).
- Modular construction, scalability, simplified design, cutting-edge safety technology, and accelerated regulatory approvals are important elements that support SMR integration. Financial difficulties still exist, nevertheless, especially in light of the significant initial capital expenditures (Josephs et al., 2025).

2.11 Economy and Finance

One important factor influencing nuclear energy projects' viability is their financial and economic competitiveness, particularly in a power market that has been liberalized. Due to a number of comparative benefits relating to its modular design, variable deployment techniques, and life-cycle costs, SMRs have become a financially feasible substitute for LRs (Mignacca & Locatelli, 2019).

SMRs are made for factory-based modular construction, which enables reproducible construction methods, standardized design, and large production. With the use of automation, supply chain optimization, and learning-by-doing, economies of multiples are made possible by this industrial method. Over time, the cost of each unit is decreased by fewer work hours, improved designs, and quality assurance. Similar modular industries have shown a learning rate of 5–10% (cost reduction per doubling of cumulative production), which is anticipated to hold true for SMR production. Unlike LRs, which are usually custom-built with little room for replication benefits, these learning effects help to increase economic competitiveness over a fleet deployment strategy (Mignacca & Locatelli, 2019).

The flexibility of SMRs to scale and expand capacity incrementally is one of their main economic benefits. To meet demand right away, utilities can implement a single SMR module; if need increases or funding becomes available, they can add more modules. This method more closely matches generation capacity with changing grid demands while lessening the burden of significant upfront capital costs. LRs, on the other hand, necessitate significant upfront investments, with entire capacity committed. This frequently results in overcapacity, stranded investments, or underutilization during the initial years. A more adaptable and financially viable expansion route is provided by SMRs for developing economies or areas with moderate load increase (Mignacca & Locatelli, 2019).

Since shared infrastructure (cooling systems, grid connections, emergency response systems) can drastically save costs, SMRs are especially well-suited for deployment at brownfield sites or existing nuclear reactors. To further cut costs per MW, multi-module SMR plants might share security systems, staffing, and control rooms. Additionally, revenue can be generated from early modules while later ones are still being built thanks to phased licensing and staged construction of numerous modules, which enhances project cash flows and financing viability. Since LRs function as a single, enormous unit with little room for infrastructure reuse, they are mainly unable to take advantage of these co-siting and multi-module synergies (Mignacca & Locatelli, 2019).

1. Financial Aspects
    - Fuel Costs: May go down with more operating experience, but slightly costlier because of longer refilling cycles (Hussein, 2020).
    - Operating Costs: Automation and modular O&M techniques can help reduce higher labor costs per MW (Hussein, 2020).
    - Savings on Co-siting and building: Co-location, or the deployment of several units on one site, and shorter building schedules improve economics (Hussein, 2020).
2. Traditional Cost Estimation Issues: Top-Down Method (TDA)
    - Large nuclear power plants' (NPPs') historical costs are reduced via TDA. (Black et al., 2022).
    - tends to overstate SMR and microreactor expenses because:
        - incorrect assumptions about the scale. (Black et al., 2022).
        - disregarding modular efficiency and design simplifications. (Black et al., 2022).
        - ignoring the benefits of industrial fabrication. (Black et al., 2022).
    - results in negative competitiveness implications and exaggerated per-MW capital costs (Black et al., 2022).
3. Improved Estimation: Bottom-Up Method (BUA)
    - BUA assesses system and component costs unique to new designs. (Black et al., 2022).
    - Reflects:
        - fewer pieces of equipment as a result of integration. (Black et al., 2022).
        - reduced indirect costs due to fabrication in the facility. (Black et al., 2022).
        - smaller buildings and easier logistics for construction. (Black et al., 2022).
    - Studies that compare NuScale to normal PWR-12, for example, demonstrate significant reductions in direct and indirect costs per kW (Black et al., 2022).
4. Data Availability and Design Maturity

- For more developed SMR designs (such as NuScale and Westinghouse SMR), BUA is more practical. (Black et al., 2022).
- Microreactors are still restricted because of:
    - o fewer intricate public designs. (Black et al., 2022).
    - o continuous regulatory development and design iteration (Black et al., 2022).
- A broader measure of competitiveness, incorporates:
    - o Direct and indirect capital costs. (Black et al., 2022).
    - o Discount rates and financing charges. (Black et al., 2022).
    - o Plant life, decommissioning, fuel and O&M expenses, and capacity factor. (Black et al., 2022).
- Capital costs are the primary factor for nuclear technology, particularly in the United States, where indirect costs have averaged more than 50%.
- Indirect costs per kW could be greatly reduced by small reactor designs, indicating greater competitiveness if LCOE is evaluated comprehensively.
- It's important to understand that value is not solely determined by costs; rather, market viability depends on features like dispatchability, reliability, and non-carbon emissions (Black et al., 2022).

5. Economic and Safety Considerations:
    - Because of their smaller size, which enables improved passive safety mechanisms and less radioactive fallout in the event of an accident, SMRs are thought to be safer than big reactors (Lee, 2024).
    - SMRs are more economically viable than conventional big nuclear facilities because of their modular architecture, which attempts to cut down on construction time and expenses (Lee, 2024).
    - Large-scale reactors' overnight costs have gone up dramatically, and although SMRs might be less expensive up front, they might nevertheless have greater costs per kilowatt-hour than larger reactors (Rosner et al., 2011).
    - The current natural gas market, which is less expensive than nuclear energy without carbon pricing, has an impact on the financial feasibility of SMRs. On the other hand, changes in gas costs can make SMRs more appealing (Rosner et al., 2011).
    - Because SMRs may increase capacity incrementally and have smaller pre-completion risks, their weighted average cost of capital (WACC) may be lower (Rosner et al., 2011).
    - Large-scale reactors' overnight costs have gone up dramatically, and although SMRs might be less expensive up front, they might neverthe-

less have greater costs per kilowatt-hour than larger reactors (OECD Nuclear Energy Agency, 2021).

- The current natural gas market, which is less expensive than nuclear energy without carbon pricing, has an impact on the financial feasibility of SMRs (OECD Nuclear Energy Agency, 2021).
- Metrics like as Total Capital Investment Cost (TCIC) and Levelized Cost of Electricity (LCOE) are used to evaluate the economic viability of SMRs (Vinoya et al., 2023).
- Even while SMRs could cost more up front, their modular design makes it possible to invest incrementally and lowers financial risk (Vinoya et al., 2023).
- According to studies, SMRs may be able to compete with traditional energy sources, particularly in markets where larger reactors are not feasible (Vinoya et al., 2023).

6. Aspects of Socio-politics
   - Because of past mishaps and risk concerns, public acceptability of nuclear power is vital. Improving public awareness of SMRs' safety characteristics can aid in increasing their acceptance (Vinoya et al., 2023).
   - SMRs may not be as socially accepted as conventional reactors, necessitating the use of specialized public engagement techniques (Vinoya et al., 2023).
   - SMRs are dependable energy sources that can adapt to changing market demands (Tan et al., 2023).
   - They can be utilized in hybrid energy systems and combined with renewable energy sources (Tan et al., 2023).
   - Compared to conventional fossil fuel-based power plants, SMRs have a smaller environmental impact (Tan et al., 2023).
   - They offer social and economic advantages in isolated and off-grid areas (Tan et al., 2023).
7. Analysis of the Economy
   - A conventional technique for assessing and cataloging dispatchable power producing assets based on yearly production costs is the levelized cost of electricity (LCOE) (Tan et al., 2023).
   - SMRs are a competitive alternative because of their lower LCOE when compared to other renewable energy sources (Tan et al., 2023).
   - Cost Competitiveness: When compared to other renewable energy sources, SMRs' LCOE is competitive. Modular construction and factory manufacturing, which cut down on building time and costs, can help SMRs reach reduced costs (Locatelli et al., 2014).

- Life Cycle Costs: Capital, operating, maintenance, and decommissioning expenses are all included in the economic assessment of SMRs. SMRs may be economically feasible, according to studies, particularly when taking into account their potential for mass manufacturing and lower financial risks (Locatelli et al., 2014).
- Software called the System Advisor Model (SAM) is used to conduct the economic analysis (Raza et al., 2024).
- As part of the analysis, the budget and financial aspects of the vSMR, renewable energy, and BESS installation project are evaluated (Raza et al., 2024).
- The project's investment viability is assessed using the cash flow for each year, the simple payback period (SPP), net present value (NPV), and internal rate of return (IRR) (Raza et al., 2024).

8. Economic Difficulties
    - Since many nations are diversifying their energy sources rather than concentrating only on nuclear power, the building of new nuclear reactors has stalled globally (Ramana, 2021).
    - Although they have financial difficulties, SMRs are thought to be a viable way to revive nuclear energy. Because larger reactors benefit from economies of scale, smaller reactors are often more expensive per unit of output (Ramana, 2021).
9. Economic Consequences of Lower Production
    - Because of the economies of scale principle, which states that larger reactors have lower costs per megawatt, smaller reactors should be more expensive (Ramana, 2021).
    - Inexperience with SMR designs causes uncertainty in construction time and expense, which frequently contributes to the overruns that are common in "First of a Kind" projects (Ramana, 2021).
    - For SMRs to be accepted by the market, their economic competitiveness is essential. They must be reasonably priced compared to alternative energy sources, particularly when considering carbon-free electricity (Josephs et al., 2025).
    - Because of their modular architecture and factory-based production, which enable phased expenditures, SMRs have reduced upfront capital costs. They can offer reasonable energy pricing and are especially well-suited for isolated locations (Josephs et al., 2025).
    - Because smaller reactors are typically more expensive per unit of output, SMRs confront difficulties with economies of scale notwithstanding their benefits. To promote their viability and reduce expenses, government investment is necessary (Josephs et al., 2025).

2.12 Public Perception

- Risk perceptions and past tragedies have an impact on public opposition to nuclear power. Nonetheless, there is growing support for nuclear energy, especially as worries about climate change increase (Rosner et al., 2011).
- More people may embrace SMRs with improved safety features, particularly if they help boost local employment and energy security (Rosner et al., 2011).
- Despite the minimal likelihood of accidents, public opposition to nuclear power frequently arises from these worries. Acceptance will depend on addressing safety issues and educating the public about the advantages of SMRs (OECD Nuclear Energy Agency, 2021).
- Consideration of local community preferences and employment possibilities will be necessary for the successful siting of SMR sites (OECD Nuclear Energy Agency, 2021).

2.13 Laws and Policies

a. Aspects of Policy
   - Governments must assist the deployment of SMR financially, develop domestic programs, evaluate licensing arrangements, and offer long-term policy support (OECD Nuclear Energy Agency, 2021).
   - Harmonizing licensing systems and resolving regulatory issues require international cooperation (OECD Nuclear Energy Agency, 2021).

b. Framework for Law and Regulation
   - Navigating intricate legal and regulatory environments, including as international conventions on nuclear safety and liability, is necessary for the implementation of SMRs (Vinoya et al., 2023).
   - SMR deployment could be facilitated and regulatory issues could be resolved by harmonizing licensing procedures across nations (Vinoya et al., 2023).

2.14 Future Ahead

SMRs could contribute to the achievement of global clean energy targets, particularly in regions that require dependable, emission-free electricity. But in order to get there, we must complete following points:

- Improved and expedited procedures for regulatory approval (Daigle et al., 2024).

- Both public and private funding (Daigle et al., 2024).
- Public backing and steady, unambiguous policy (Daigle et al., 2024).
- Addressing a number of issues, such as public opinion, regulatory barriers, and financial concerns, will be necessary for the effective implementation of SMRs (Vinoya et al., 2023).
- To manage these challenges and guarantee a sustainable future for nuclear energy, cooperation between governments, businesses, and communities will be crucial (Vinoya et al., 2023).

Following are some problems and prospects for the future of SMR that it can overcome in future:

- SMRs need to overcome obstacles pertaining to building costs, operational efficiency, and licensing in order to succeed (Lee, 2024).
- Modular construction methods are being promoted as a way to cut expenses and expedite manufacturing (Lee, 2024).
- It also emphasizes the potential applications of SMRs in fields other than power generation, like industrial heating and maritime transportation (Lee, 2024).
- Regulations and Licensing

Since new designs must be assessed and approved by national regulatory agencies, licensing is a major obstacle. Because authorities must guarantee safety and traceability across the supply chain, the modular design of SMRs makes this procedure more difficult (Shobeiri et al., 2023).

2.15 Examples on few fields

a. Resilient Urban
    - Megacities and other densely populated emerging cities with inadequate or non-existent power infrastructure are examples of resilient urban use cases. (Black et al., 2022).
    - Key characteristics include co-location with energy users, plug-and-play capability, and a tiny footprint. (Black et al., 2022).
    - Possibility: Without requiring extensive national grid infrastructure, microreactors might provide both industrial heat and electricity. (Black et al., 2022).

- Support: To fulfill the energy demands of cities, technologies such as nuclear batteries and innovative materials are being developed (Black et al., 2022).

b. Disaster Relief
   - Use Case for Disaster Relief: Emergencies like natural catastrophes (wildfires, earthquakes, and hurricanes). (Black et al., 2022).
   - Key features include off-grid operation, quick setup (within 3 days), and mobile deployment. (Black et al., 2022).
   - Current Options: Diesel generators are the most common, however they have carbon emissions and fuel supply issues. (Black et al., 2022).
   - Microreactor Advantage: Long-lasting, cleaner electricity for communications, hospitals, emergency services, and desalination; can be given rapidly after disasters or pre-deployed. (Black et al., 2022).
   - Packaging: Made to fit into common shipping containers for transportation by land, sea, or air (Black et al., 2022).

c. Marine Population
   - Use Case for Marine Propulsion: Providing energy to ships to lessen dependency on fossil fuels. (Black et al., 2022).
   - Long refueling intervals, lower emissions, more cargo room, and the possibility of standardized sizing are some of the salient features. (Black et al., 2022).
   - Current alternatives include hybrid systems and wind-assisted sails, although nuclear provides more power and range. (Black et al., 2022).
   - Challenges include the requirement for secure waste management, port-based refueling stations, and qualified nuclear operators. (Black et al., 2022).
   - For instance, nuclear-powered vessels such as the NS Savannah show promise (Black et al., 2022).

2.16 Analysis

a. On SMR

For the bibliometric analysis on SMR and Nuclear research, data was taken from PubMed. The following displayed graphs, Fig. 1 and Fig. 2, give a chronological summary of research publishing patterns. They were created using Python's plotting tools.

Figure 1. Graph showcasing number of research papers published on SMRs between 1984 to 2025 from PubMed

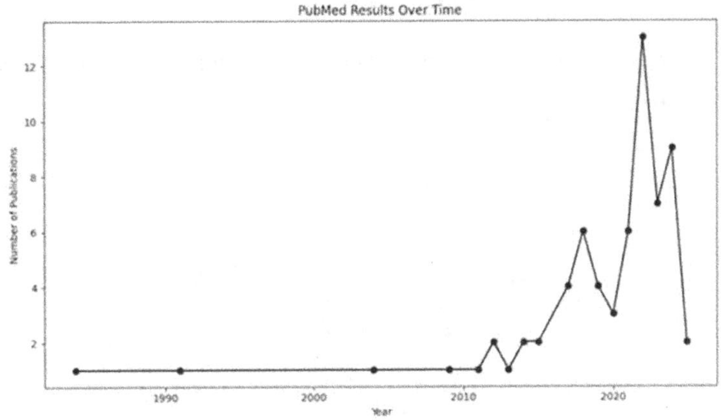

Figure 2. Graph showcasing number of research papers published on Nuclear Research between 2010 to 2025 from PubMed

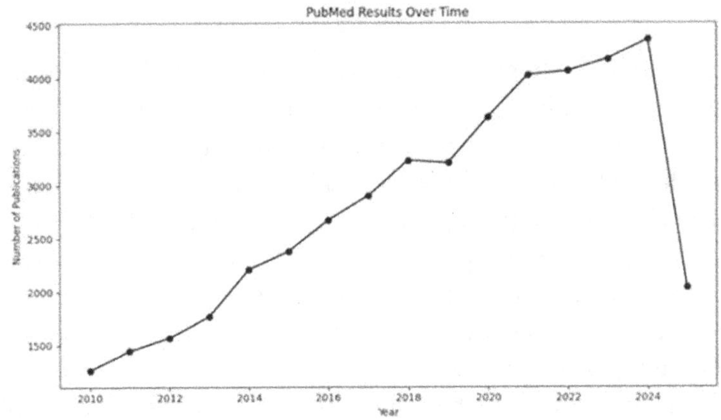

Fig. 3 shows bibliometric analysis, which was produced with VOSviewer using data from PubMed. It provides a thorough visual depiction of the state of research on nuclear energy and its multidisciplinary uses. Every node in the network represents a term or keyword from scientific papers; the lines (or, edges) show co-occurrence associations, and the node size corresponds to the term's frequency of occurrence. In order to provide a thorough picture of how nuclear energy-related research crosses areas such as biology, physics, medicine, and the environment, clusters are organized and color-coded to represent discrete but linked subject topics.

Figure 3. VOSviewer-generated keyword co-occurrence network based on PubMed Database on SMR

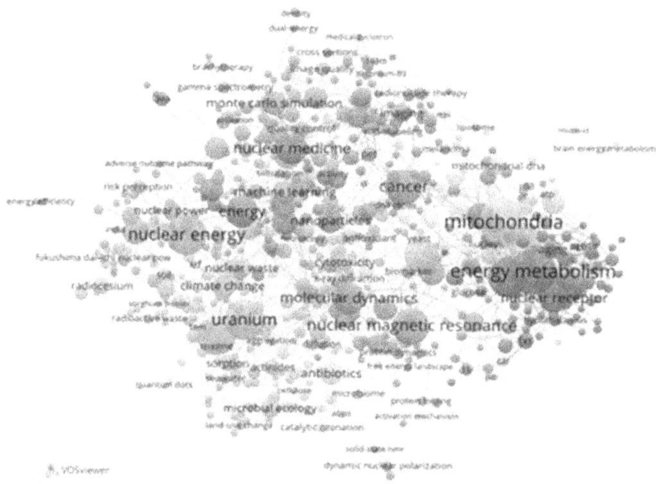

The phrase "nuclear energy" appears as a dominating node near the map's centre, indicating its importance in this mass of work. Related terms such as "nuclear power," "nuclear waste," "nuclear medicine," and "nuclear magnetic resonance" surround it. This core cluster (orange and green) highlights concerns including radioactive waste, climate change, risk perception, and radiocesium, indicating a heavy focus on nuclear energy generation and its effects. Words like "soil," "energy efficiency," and "Fukushima Daiichi nuclear power" imply the inclusion of environmental impact studies and public perception, indicating the sustainability issues and societal concerns related to nuclear power.

The blue and red clusters to the right of the central node represent biochemical and biomedical research that converges with nuclear technology. Names like "mitochondria," "energy metabolism," "nuclear receptor," "cancer," and "ATP" point to a substantial amount of research that links nuclear-related technologies and processes to cellular biology and health sciences. The use of terms like "aging," "p53," and "hypothalamus" indicates that there is ongoing study into the use of nuclear energy techniques, especially in imaging and radiation therapy, to comprehend and treat metabolic problems and age-related diseases. A prominent research domain in nuclear medicine and diagnostics, including PET scans, MRIs, and radiolabelling techniques, is also indicated by the strong occurrence of the terms like "cancer" and "imaging."

Environmental and ecological issues are the subject of the green and light blue clusters in the left-bottom section, which include recurrent topics like "uranium," "antibiotics," "microbial ecology," "sorption," "actinides," and "seawater." This suggests that uranium's effects on the environment, control of contamination, and interactions with microbial populations are gaining attention. In order to evaluate ecological dangers and create bioremediation plans around nuclear facilities, such studies are crucial. Furthermore, phrases like "radioactive waste," "soil," and "land-use change" imply interdisciplinary interaction with environmental science and policy.

The purple and blue clusters in the centre-right investigate the science related to nuclear research at the molecular and atomic level. In structural biology, materials science, and quantum chemistry, terms such as "molecular dynamics," "nanoparticles," "protein folding," and "dynamic nuclear polarization" denote the use of nuclear magnetic resonance (NMR) and associated simulations.

In conclusion, this illustration shows the diverse character of nuclear energy research, ranging from technological and environmental issues to molecular science and biomedical innovation. It emphasizes how nuclear technologies have an impact on many fields.

b. On Clean Energy

The following graph, Fig. 4, is a Python-generated graph which provides a summary at the growing trend of biomedical research and publication efforts throughout the previous ten years, emphasizing the notable expansion in recent years.

Figure 4. Graph showcasing number of research papers published on Clean Energy between 2010 to 2025 from PubMed

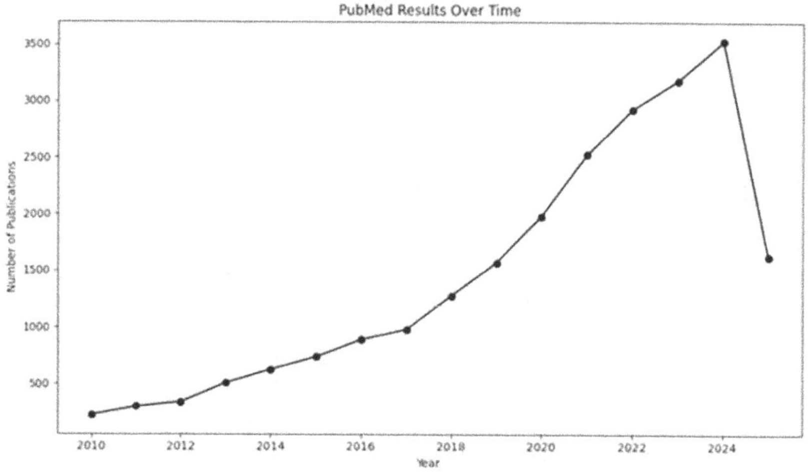

Fig. 5 shows bibliometric analysis, which was produced with VOSviewer using data from PubMed. It provides a thorough visual depiction of the major themes and research trends in the field of renewable energy and related fields. Each node (or bubble) represents a keyword, and the size of the node shows the frequency of occurrence of that keyword in the dataset. This map is created using co-occurrence data of keywords from scientific literature. The strength of co-occurrence links is reflected in the proximity between nodes; nodes that are closer together are more likely to occur together in the same document. These nodes are grouped into clusters based on color, each of which represents a thematic collection of linked subjects

Figure 5. VOSviewer-generated keyword co-occurrence network based on PubMed Database on Clean Energy

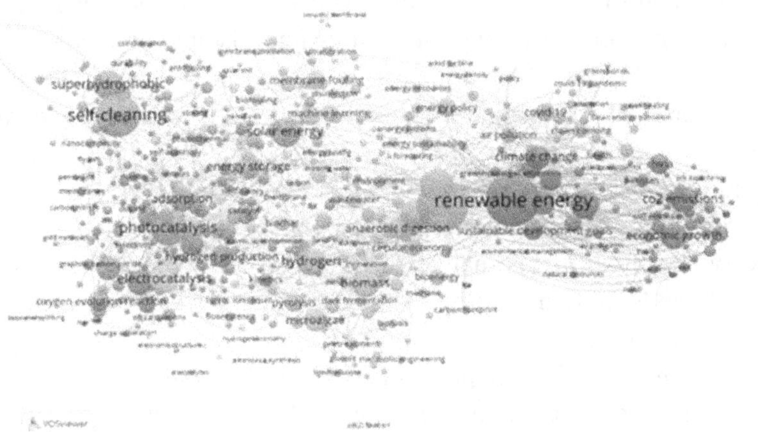

"Renewable energy" is the most noticeable and dominant term on the map, and it is symbolized by a big red bubble in the middle. Its size and central location demonstrate its broad applicability and frequent correlation with several subtopics in a wide range of research fields. "Renewable energy" is closely related to major global issues like "climate change," "sustainable development goals (SDGs)," "co2 emissions," "economic growth," and "energy policy." The red and orange clusters on the right side of the map represent these concepts, emphasizing their close relationship and shared emphasis on sustainability, environmental effect, and policy implications. This shows that research on renewable energy is strongly aligned with socioeconomic factors and climate mitigation initiatives.

Numerous more specialized clusters that represent various research fronts surround the centre cluster. Terms like "biomass," "bioenergy," "anaerobic digestion," "circular economy," and "microalgae" are included in a noticeable cyan cluster to the left and bottom centre. Growing interest in renewable energy produced from organic and waste materials is indicated by this cluster's concentration on biological and waste-to-energy technologies. Additionally, it touches on subjects like "dark fermentation" and "wastewater treatment," demonstrating interdisciplinary study between environmental engineering and sustainable energy generation.

A green cluster with phrases like "self-cleaning," "superhydrophobic," and "membrane fouling" appears in the upper left corner of the map. These words refer to research in surface engineering and material science that is relevant to renewable energy systems, such as enhanced solar panel materials or water purification membranes. The purple cluster next to this is focused on "photocatalysis," "electrocatalysis," "oxygen evolution reaction," and "hydrogen production," indicating a high level of activity in the fields of water splitting, hydrogen energy, and catalysis—all essential technologies for sustainable energy conversion and storage.

Near the centre-top, the yellow cluster centres on "solar energy," "machine learning," and "membrane fouling," indicating interdisciplinary research combining renewable technology and AI to maximize system performance. Here, the terms "energy storage" and "adsorption" connect these initiatives to more general concepts of effectiveness and system integration.

The orange cluster, which includes terms like "covid-19," "green bonds," and "energy transition," represents recent global events like the COVID-19 pandemic that have found a place in this analysis. This suggests that researchers are starting to investigate how disruptions caused by the pandemic have affected energy policies and sustainability objectives.

Along with highlighting the prevalence of important themes like sustainability, climate change, and hydrogen energy, this VOSviewer illustration, also identifies new interdisciplinary approaches that combine data science, materials science, and policy studies—all of which are crucial in determining how clean energy transitions will develop in the future. Based on data taken from PubMed, the following displayed graph provides a clear visual representation of the quantity of research papers published over time. The vertical axis (y-axis) shows the quantity of publications, while the horizontal axis (x-axis) shows the years, from 2010 to 2025. It is simple to follow the trend in publishing volume over time thanks to the plotting of the data points using Python.

The graph shows that, from 2010 and 2024, publishing production increased steadily. Beginning with about 250 publications in 2010, the number of papers progressively rises annually, with a discernible uptick in growth starting in 2018. Around 2017–2018, the number of publications exceeds 1,000, and by 2021, it has rapidly increased to over 2,500. This rapid rise, which peaked in 2024 at over 3,500 papers, suggests a time of high scholarly output and research activity.

Nonetheless, there is a notable decline in publications in 2025, with the total dropping to about 1,600. Since 2025 is still in progress, this decline is probably the consequence of insufficient data gathering for the present year rather than a real decline in research activities. Real-time publishing analysis frequently shows this kind of decline, which needs to be taken with caution.

3. CONCLUSION

SMRs are at the front of a nuclear energy revolution that could address the twin problems of energy accessibility and climate change. SMRs have several advantages over traditional large-scale nuclear power plants, as discussed in the introduction and literature review. These advantages include scalability, improved safety features, and modular design. These features make them particularly well-suited for deployment in urban microgrids, rural locations, and developing countries looking to reduce greenhouse gas emissions while meeting growing energy demands.

The significance of SMRs in future energy systems is highlighted by the growing demand for dependable, carbon-free energy sources, which is being driven by international commitments like the UN Sustainable Development Goals (especially SDG 7 and SDG 13). They are valuable assets in a varied energy mix because of their capacity to function independently or in conjunction with renewable energy sources, which improves grid stability. The multifunctionality of SMRs is highlighted by their passive safety systems, lower fuel consumption, and versatility in a range of applications, such as desalination, hydrogen production, and industrial heat.

Despite their potential, social, economic, and regulatory barriers stand in the way of SMR's broad adoption. Significant obstacles still include difficult licensing procedures, expensive initial capital expenditures per megawatt, public mistrust stemming from past nuclear accidents, and a lack of operational experience. Nonetheless, these problems are gradually being addressed by global cooperation, encouraging legislative frameworks, and continuous technology advancement. While emerging nations in Southeast Asia are looking at partnerships for future deployment, countries such as the United States, China, Russia, and Canada are spearheading efforts to demonstrate and market SMR technologies.

Phased construction, modular scalability, and co-siting with existing infrastructure make SMRs an economically attractive option. These elements support gradual capacity increase and lower investment risk, which is especially advantageous in emerging nations. With fewer greenhouse gas emissions than fossil fuels and the possibility for cutting-edge waste reduction technologies, SMRs also exhibit good environmental performance, as shown by life-cycle assessments.

According to the literature, SMRs have the potential to significantly contribute to the achievement of global clean energy goals, particularly when combined with renewable energy sources in hybrid energy systems. They are a technological innovation as well as a sociotechnical solution that satisfies changing global energy demands and environmental imperatives because of their distinctive blend of safety, adaptability, and sustainability.

In conclusion, SMRs are a redesigned approach to nuclear energy, not just a smaller version of conventional reactors. SMRs have the potential to provide equitable, clean, and resilient energy solutions for the twenty-first century and beyond with sustained investment, regulatory adaption, and public involvement.

REFERENCES

Black, G., Shropshire, D., Araújo, K., & Van Heek, A. (2022). Prospects for Nuclear Microreactors: A review of the technology, economics, and regulatory considerations. *Nuclear Technology, 209*(sup1), S1–S20. DOI: 10.1080/00295450.2022.2118626

Daigle, B., DeCarlo, S., & Lotze, N. & Office of Industry and Competitiveness Analysis. (2024). Big change goes small: Are small modular reactors (SMRs) the future of nuclear energy? In *Working Paper ICA-105.*

Di Maio, F., Bani, L., & Zio, E. (2022). The contribution of small modular reactors to the resilience of power supply. *Journal of Nuclear Engineering*, *3*(2), 152–162. DOI: 10.3390/jne3020009

Hussein, E. M. (2020). Emerging small modular nuclear power reactors: A critical review. *Physics Open*, *5*, 100038. DOI: 10.1016/j.physo.2020.100038

Josephs, R. E., Yap, T., Alamooti, M., Omojiba, T., Benarbia, A., Tomomewo, O., & Ouadi, H. (2025). Regulation of Small Modular Reactors (SMRs): innovative strategies and economic insights. *Eng—Advances in Engineering, 6*(4), 61. DOI: 10.3390/eng6040061

Lee, J. I. (2024). Review of Small Modular Reactors: Challenges in Safety and Economy to Success. *Korean Journal of Chemical Engineering*, *41*(10), 2761–2780. DOI: 10.1007/s11814-024-00207-0

Locatelli, G., Bingham, C., & Mancini, M. (2014). Small modular reactors: A comprehensive overview of their economics and strategic aspects. *Progress in Nuclear Energy*, *73*, 75–85. DOI: 10.1016/j.pnucene.2014.01.010

Mignacca, B., & Locatelli, G. (2019). Economics and finance of Small Modular Reactors: A systematic review and research agenda. *Renewable & Sustainable Energy Reviews*, *118*, 109519. DOI: 10.1016/j.rser.2019.109519

Montesclaros, J. M. L. (2023). The Future of Small Modular Reactors: Implications for Nuclear Governance Introduction.

ORGANISATION FOR ECONOMIC CO-OPERATION AND DEVELOPMENT & OECD Nuclear Energy Agency. (2021). Small modular reactors: challenges and opportunities. In ORGANISATION FOR ECONOMIC CO-OPERATION AND DEVELOPMENT & OECD Nuclear Energy Agency, *Nuclear Technology Development and Economics* [Report]. https://inis.iaea.org/collection/NCLCollectionStore/_Public/52/041/52041043.pdf?r=1

Paisiripas, D., Cho, K., & Park, S. (2024). Integration of Small Modular Reactors with Renewable Energy for Carbon Neutrality: A Case Study of Phuket, Thailand. *Energies, 17*(22), 5565. DOI: 10.3390/en17225565

Ramana, M. V. (2021). Small modular and advanced nuclear reactors: a reality check. In *IEEEAccess*. https://creativecommons.org/licenses/by/4.0/

Raza, M. K., Alghassab, M., Altamimi, A., Khan, Z. A., Kazmi, S. A., Ali, M., & Diala, U. (2024). Integration of very small modular reactors and renewable energy resources in the microgrid. *Frontiers in Energy Research, 12*, 1365735. Advance online publication. DOI: 10.3389/fenrg.2024.1365735

Richter, J. (2014). Small Modular Reactors: The future of nuclear energy? *ResearchGate, 1–4*, 1–4. Advance online publication. DOI: 10.1109/ETHICS.2014.6893449

Rosner, R., & Goldberg, S. Energy Policy Institute at Chicago, The Harris School of Public Policy Studies, Hezir, J. S., & EOP Foundation, Inc. (2011). *Small modular reactors – key to future nuclear power generation in the U.S.*https://www.energy.gov/sites/prod/files/2015/12/f27/ECON-SMRKeytoNuclearPowerDec2011.pdf

Shobeiri, E., Genco, F., Hoornweg, D., & Tokuhiro, A. (2023). Small modular reactor deployment and obstacles to be overcome. *Energies, 16*(8), 3468. DOI: 10.3390/en16083468

Tan, S., Cheng, S., Wang, K., Liu, X., Cheng, H., & Wang, J. (2023). The development of micro and small modular reactor in the future energy market. *Frontiers in Energy Research, 11*, 1149127. Advance online publication. DOI: 10.3389/fenrg.2023.1149127

Vinoya, C. L., Ubando, A. T., & Culaba, A. B. (2024). Life cycle analysis of a network of small modular reactors. *IOP Conference Series. Earth and Environmental Science, 1372*(1), 012059. DOI: 10.1088/1755-1315/1372/1/012059

Vinoya, C. L., Ubando, A. T., Culaba, A. B., & Chen, W. (2023). State-of-the-Art review of small modular reactors. *Energies, 16*(7), 3224. DOI: 10.3390/en16073224

ADDITIONAL READINGS

Liu, Z., & Fan, J. (2013). Technology readiness assessment of Small Modular Reactor (SMR) designs. *Progress in Nuclear Energy, 70*, 20–28. DOI: 10.1016/j.pnucene.2013.07.005

Locatelli, G., Bingham, C., & Mancini, M. (2014b). Small modular reactors: A comprehensive overview of their economics and strategic aspects. *Progress in Nuclear Energy*, *73*, 75–85. DOI: 10.1016/j.pnucene.2014.01.010

Rowinski, M. K., White, T. J., & Zhao, J. (2015). Small and Medium sized Reactors (SMR): A review of technology. *Renewable & Sustainable Energy Reviews*, *44*, 643–656. DOI: 10.1016/j.rser.2015.01.006

Vujić, J., Bergmann, R. M., Škoda, R., & Miletić, M. (2012). Small modular reactors: Simpler, safer, cheaper? *Energy*, *45*(1), 288–295. DOI: 10.1016/j.energy.2012.01.078

Wang, Y., Chen, W., Zhang, L., Zhao, X., Gao, Y., & Dinavahi, V. (2024). Small Modular Reactors: An overview of modeling, control, simulation, and applications. *IEEE Access : Practical Innovations, Open Solutions*, *12*, 39628–39650. DOI: 10.1109/ACCESS.2024.3351220

KEY TERMS AND DEFINITIONS

Cost-effective: Delivering good value or outcomes relative to the amount of money invested, particularly when it comes to obtaining cost-effective energy solutions.

Grid Integration: The process of effectively tying SMRs and other power sources to the electrical grid in order to preserve a steady and dependable supply.

Low-Enriched Uranium (LEU): Commonly used as fuel in nuclear reactors, it is uranium that has undergone processing to raise the concentration of the U-235 isotope to less than 20%.

Nuclear Licensing: The formal procedure by which governments assess and authorize the security, layout, and functioning of nuclear power plants is known as nuclear licensing.

Nuclear Reactor: It is a machine that creates and manages a long-term chain reaction of nuclear fission to produce heat and power.

Reducing GHG Emission: The process of reducing greenhouse gas emissions, such as CO2, in order to slow down global warming and climate change.

Regulatory Frameworks: A collection of laws, regulations, and policies put in place by authorities to guarantee the effective and safe use of technology like nuclear power.

SMR: A small, factory-built nuclear reactor with a power output of less than 300 MW that may be deployed in a variety of ways.

Chapter 7
Integrating AI and Nuclear Technology for Environmental Sustainability

Usharani Bhimavarapu
https://orcid.org/0000-0002-0246-1420
Department of Computer Science and Engineering, Koneru Lakshmaiah Education Foundation, Vaddeswaram, India

ABSTRACT

The integration of nuclear technology, ecosystem services, and sustainable development is faced with numerous challenges, including regional heterogeneities, environmental risks, and socio-economic constraints. There exist no models particularly to tackle localized sustainability change, transfer of technology, and long-term ecological impacts. The study discusses a more intricate framework that enhances the sustainability indicators by considering region-specific drivers, advanced risk assessment, and adaptive policy advice. The proposed algorithms improve upon the shortcomings of past studies by including predictive modeling of long-term material stability, improving life-cycle analysis of greenhouse gas emissions, and improving isotopic analysis for nutritional and environmental research. Through the integration of dynamic modeling techniques, multi-proxy data techniques, and case studies in real-world environments, this study improves the usability and applicability of nuclear-based environmental and energy solutions.

DOI: 10.4018/979-8-3373-0685-8.ch007

INTRODUCTION

Nuclear energy is world energy infrastructure with low-carbon, secure electricity. Climatically related disasters such as hurricanes, storms, flooding, and wildland fires are such gigantic risks to nuclear plants (Elmqvist et al., 2019). Weather storms and coastal erosion bring cooling systems, electric wire transmittals, and security hardware that are instrumental in plant functions into danger (IAEA, 2022). Nuclear facilities must be constructed capable of resisting such natural forces without sacrificing efficiency in the years to come (Freire and Andrade, 2021). The future reactor designs are with passive safety attributes that make them more robust at worst (Cizelj et al., 2024). Small modular reactors (SMRs) are an extensible and versatile approach to disaster-affected areas (Freire and Andrade, 2021). The new technology requires stringent engineering and strict regulatory master plans for ensuring security (IAEA, 2022). Climate risk assessment must be integrated into plant design and site selection by the nuclear industry (Bilgin and Kara, 2024). Nuclear power plants must be strengthened in an attempt to ensure energy reliability against climate change challenges (Allegretti et al., 2024).

Durable nuclear plants begin with planning and siting minimizing the effect of climate-related hazard (Elmqvist et al., 2019). Nuclear plants based on the seashore will have to plan for sea-level rise and storm surges, like reinforced fences and raised reactor component sections (Azam et al., 2021). Land vegetation will have to mitigate the risk of heatwaves and dryness to cold water supply (Balvanera et al., 2022). New reactors such as molten salt reactors and high-temperature gas reactors are being designed to utilize novel coolants, reducing water use (Cizelj et al., 2024). Seismic construction to withstand seismically active areas by means of robust structures and enhanced core stability is needed (Freire and Andrade, 2021). Duplicate power sources such as microgrids and batteries can prevent blackouts during extreme weather conditions (Ding and Shi, 2024). Cybersecurity must be enhanced to prevent climate-permitting cyberattacks on plant operation (Rajca, 2024). Emergency preparedness and response plans must be updated periodically to include new climate models (IAEA, 2022). Climate-resilient investment in nuclear infrastructure will ensure long-term energy security (Azam et al., 2021).

Nuclear power supports disaster recovery because it produces electricity at a constant rate when other sources of power fail (Fernández-Arias et al., 2024). Solar and wind, being weather dependent, are intermittently interrupted by means of extreme events, while nuclear power plants are not intermittently interrupted by means of extreme events (Allegretti et al., 2024). While wildfires or hurricanes hit, the nuclear energy powers the hospitals, emergency centers, and other infrastructures that are considered to be critical (Fernández-Arias et al., 2024). Distributed SMRs are sent out at the same time into disaster areas destroyed by such disasters in an attempt to restore power (Freire and Andrade, 2021). Invented portable microreactors can assist in delivering electricity to rural communities within days (Freire and Andrade, 2021). Nuclear power hydrogen also has the capability to facilitate recovery by providing electricity for backup energy and transportation (Fernández-Arias et al., 2024). Nuclear power's capability to provide grid stability minimizes cascading failure during peak demand (Ding and Shi, 2024). The future will be a combination of nuclear power and secure grids (Cizelj et al., 2024). Enhancing crisis recovery capability of nuclear enhances public safety and national security (IAEA, 2022).

Rising occurrences of natural disaster events illustrate the usefulness of nuclear energy versatility in suboptimal conditions (McDonald et al., 2020). Traditional large reactors are facing severe safety feature demands but must spend enormous resources in order to deliver operation in crisis situations. Nuclear technology that is new focuses on modularity and inherent autonomy for safety to respond in unexpected scenarios (Mondal et al., 2024). Passive cooling systems such as gravity-fed water supply reduce the requirement of off-site power in the case of crisis scenarios (Mondal et al., 2024). Reactors such as molten salt reactors achieve intrinsic safety, which does away with the meltdown possibility even in long blackouts (Mondal et al., 2024). Climate-resilient nuclear technology is among the utmost keys to duress-resilience in the grid (McCulloch et al., 2012).

Hybridization with other renewables of the nuclear offers a combined energy system that can adapt to the shifting nature of the climate (O'Reilly et al., 2003). Duress-resilience policies for resilient nuclear development need to be research, innovation, and partnership-based (Nicholson et al., 2021). A nuclear resilience strategy guarantees long-term nuclear energy access safely in a future climate change (McDonald et al., 2020). Climate change stresses necessitate regulation to evolve and strengthen nuclear energy resilience (Nisbet et al., 2020). Regulators and governments need to redesign safety standards in response to new climate threats (Nicholson et al., 2021). Precautionary risk management laws should include consideration of new and existing nuclear facilities' climate effects (Marx et al., 2021).

Licensing frameworks should facilitate innovation in high-temperature reactor technologies with inherent robustness characteristics (Mondal et al., 2024).

Climate-resilient nuclear infrastructure development programs should be funded through public-private partnerships (McDonald et al., 2020). There would be cross-border collaboration which would enable instant sharing of knowledge on best nuclear practices to curtail climate change (Nicholson et al., 2021). Climate-resilient innovation in nuclear will need to be directed towards marrying the research hubs in academia (Jez et al., 2021; Pereira et al., 2022).

Policy-makers should promote innovation but not to the detriment of rigorous safety and environmental standards (Nicholson et al., 2021; Newsome et al., 2007). Nuclear power plant disaster preparedness and risk reduction can be optimized using predictive analytics and AI in combination (Lu et al., 2024). Improved regulatory frameworks will make nuclear energy a manageable source of energy amidst an unpredictable climate (Mondal et al., 2024).

Disaster-resilient nuclear investment means investment in world future energy security (Mondal et al., 2024). Governments must invest in research and development of new reactor technology (Mondal et al., 2024). Climate resilience must be integrated into plans for modernization of nuclear power plants (McDonald et al., 2020). Installation of increased security and cooling systems makes the existing plants more resilient to undesirable climate (Mondal et al., 2024).

Decentralized nuclear power, such as microreactors and SMRs, enhances resilient energy supply access (Mondal et al., 2024). Nuclear and wind/solar capacity interconnectivity enhances overall energy security (Nisbet et al., 2020). Popular education can enhance support for the utilization of nuclear energy in making climate resilience (Nicholson et al., 2021; Newsome et al., 2007). Climatologists, nuclear engineers, and policymakers will work intersectorally to provide effective adaptation processes (McDonald et al., 2020; Jez et al., 2021). Green bonds can help provide resilient investment in nuclear infrastructure through sustainable funding processes (Nicholson et al., 2021). An anticipatory strategy of nuclear resilience safeguards energy infrastructure from expected future climate risks (McCulloch et al., 2012; Pereira et al., 2022).

The contribution of nuclear energy to climate resilience goes beyond electricity generation. Applications of nuclear technology to climate adaptation include desalination, clean hydrogen production, and industrial heat (Rodrigues et al., 2021). Nuclear power desalination offers clean water to arid and climate-modified regions. Nuclear-hydrogen production supports low-carbon economic transition through offering an alternative to fossil fuels. Low-carbon high-temperature nuclear reactors can replace industrial processes (Weisser, 2007). Nuclear utilization can be supplemented by carbon capture and storage for enhanced climate resilience. Multifunctional nuclear solutions are given a boost by innovation, which doubles its contribution to sustainable energy transition (Sachs et al., 2019). Climate adaptation policies enhance economic and environmental resilience with the addition of nuclear power (Salman et al., 2024). Nuclear development may be promoted by policy to maximize its contribution to mitigating the impacts of climate change (Yue et al., 2024). The versatility of nuclear technology places it at the forefront of international climate adaptation (Walsh et al., 2020).

Public trust and transparency in communication of safety procedures are required by a robust nuclear industry to facilitate public engagement in overcoming issues on nuclear power utilization in developing climate resilience (Rajca, 2024). Public education on the safety innovation of new reactors can help improve public acceptance of new technology (Rodrigues et al., 2021). Governments must provide clear guidelines on nuclear accident readiness in an attempt to provide an assurance of security to the public. Transparent reporting of plant safety, climate resilience, and regulatory compliance enhances nuclear power credibility. There is a need for cooperation with civil society actors in nuclear policymaking and the desire to be sustainable (Sachs et al., 2019). There should be encouraged by academics nuclear curricula to study in order to achieve an effective cadre of climate-resilient nuclear development (Walsh et al., 2020). Nuclear risk myths need to be dispelled to make it legitimate and continue to have political as well as public support. Public-private cooperation to success builds the credibility of nuclear energy towards climate resilience. Trust development in nuclear power fortifies nuclear energy's place in sustainable energy transformation.

The coming generation of nuclear power projects will have to resonate with global climate objectives and still offer long-term security and reliability (Salman et al., 2024). Fission for the future generation and technology for the future generation of fusion provide climate-resilient technologies for nuclear energy. Small modular and floating reactors provide space for decentralized, disaster-resistant energy facilities (Reddy et al., 2024). Nuclear fusion, if harnessed economically, will be an unlimited, carbon-free energy source for the next few centuries. Technological advances in nuclear fuel recycling and nuclear waste storage render nuclear power greener (Weisser, 2007). Nuclear integration in smart grids maximizes energy transmission efficiency under weather extremities. Artificial intelligence-based predictive maintenance reduces operation risk to zero and maximizes plant safety under climate change (Salman et al., 2024). Inclusion of nuclear in climate policy mechanisms makes nuclear a sustainable energy (Yue et al., 2024). Future innovation in resiliency for the use of nuclear globally is fostered by further research and collaboration globally (Sachs et al., 2019). Development in nuclear technology will have a fundamental role in determining a climate-resilient future for energy.

Nuclear power is international energy infrastructure that produces low-carbon electricity in a secure way (Weisser, 2007). Climatically caused disasters such as hurricane storms, floods, and wildfires create gargantuan threats to nuclear power plants. Weather and coastal erosion expose cooling systems, electric transmission lines, and security equipment vital to the plant to hazards. Nuclear equipment will have to be constructed strong enough to resist these acts of nature without compromising efficiency in the years ahead (Rodrigues et al., 2021). Designs for next-generation reactors include passive safety features that render them more resilient in a worst-case situation (Reddy et al., 2024). Small modular reactors (SMRs) are a scalable, flexible solution for disaster-hit areas. The technology should have rigorous engineering and strict regulatory designs to demonstrate safety. Plant design and site choice require climate risk analysis incorporated by the nuclear sector. Nuclear power plants should be made more robust in an effort to have energy stability against climate change challenges (Salman et al., 2024).

Safe nuclear power plants begin with reduced-climate-risk design and siting. Ocean-front nuclear power plants, for example, will need to take account of sea-level rise and storm surges, like strong barriers and raised reactor component pieces. Land-based vegetation will have to resist the risk of heatwaves and dryness over cold water. Sophisticated reactors such as molten salt and high-temperature gas reactors are constructed to incorporate advanced coolants, which reduce the usage of water (Rodrigues et al., 2021). Seismic design to counteract seismically active regions by bracing buildings and core stability is necessary. Alternative power sources such as batteries and microgrids can prevent blackouts during extreme weather (Reddy et al., 2024). Cybersecurity also needs to be enhanced to prevent climate-facilitating cyber attacks on plant operations (Rajca, 2024). Emergency response and preparedness plans need to be updated periodically to include new climate models. Climate-resilient investment in nuclear infrastructure will ensure long-term energy security (Salman et al., 2024).

Nuclear energy facilitates disaster recovery because it has a constant flow of electricity when other forms of energy cannot be utilized. Solar and wind are dependent on weather and are constantly disrupted by way of extreme events, but nuclear power plants are not constantly disrupted by way of extreme events (Weisser, 2007). Nuclear power supplies energy as wild fires or hurricanes hit by powering hospitals, emergency units, and other infrastructures termed as the basic ones. Distributed SMRs are installed collectively in ravaged territories resulting from the catastrophes with an effort of supplying power (Reddy et al., 2024). Technologically advanced portable microreactors can assist in supplying power to off-grid areas within days. Nuclear power hydrogen can be reclaimed by using electricity as a backup power and transportation system. Grid stability offered by nuclear power minimizes cascading failure in demand scenarios. The future brings a combination of nuclear power with robust grids (Yue et al., 2024). Crisis recovery capacity enhanced by nuclear enhances the safety of civilians and defense of armed forces (Salman et al., 2024).

Higher frequency of natural disasters reveals the need for nuclear energy adaptability under harsh environments. Traditional large-scale reactors are strictly under safety feature requirements but consume tremendous resources to deliver crisis condition operation (Rodrigues et al., 2021). Emergent nuclear technology prioritizes modularity and internal security autonomy to meet unanticipated conditions. Passive cooling technologies such as gravity-fed water supply reduce off-site power reliance during emergencies. Inherently safe reactors such as molten salt reactors remove the possibility of meltdown even during prolonged blackouts (Reddy et al., 2024). Climate-resistant nuclear power plants rank among the best solutions to grid resilience under duress. Hybridization of nuclear energy with other renewables brings a combined plant which can modify as per the fluctuating character of the climate (Sachs et al., 2019). Resilient nuclear policy formulation must be research, innovation, and collaboration-based (Walsh et al., 2020). A nuclear resilience policy guarantees long-term safe access to nuclear power. Climate change stresses necessitate regulation to convert and make nuclear power resilient. Regulators and governments must restore safety requirements to defend against emerging climate hazards (Rajca, 2024). Precautionary regulation of risk management must ensure the taking into account of the climate impact of new and current nuclear facilities. Licensing frameworks should permit the innovation of high-temperature reactor technologies with in-built strength properties (Reddy et al., 2024). Climate-resilient development strategies for nuclear infrastructure should be facilitated through public-private partnerships. There would be global collaboration that would enable the quick exchange of information on best practice on nuclear to combat climate change. Climate-proofed nuclear innovation technology will have to be guided towards integrating the research institutions into the academic setting (Walsh et al., 2020). Policy-makers will have to stimulate innovation but at the expense of not water-ing down stringent safety and environmental standards. Predictive analytics and AI together can enhance nuclear power plant disaster preparedness and risk avoidance (Salman et al., 2024). Next-generation regulatory systems will render nuclear power a green energy source in an unpredictable climate.

While the research is suggestive of digital twin technology adoption in nuclear energy, it addresses mostly technical potential and overlooks key regulatory, cybersecurity, and ethical challenges likely to hinder real-world deployment(Mondal et al., 2024).

The bibliometric review provides a general overview of green hydrogen technologies without critically assessing the technological feasibility, scalability, and cost-effectiveness of the described trends(Fernández-Arias et al., 2024).

Though the article proposes a strategic agenda for European nuclear education and training, it remains silent on different national education systems or budget limitations that would hinder implementation(Cizelj et al., 2024).

The social sentiment analysis of attitudes towards green energy in Turkey may be biased due to its reliance on social media data, which can miss the attitudes of the general population(Bilgin & Kara, 2024).

The research is reflective of pink hydrogen research patterns but does not explain fully the environmental and ethical considerations involved in hydrogen production based on nuclear(Fernández-Arias et al., 2024).

While highlighting the water-energy-environment nexus in renewably driven, the chapter provides only minimal commentary on region-level resource limitations or socio-political impediments towards the implementation of nexus-based policy-making(Allegretti et al., 2024).The edited book provides an overview of the idea of

Industry 5.0 and sustainability but is limited by the absence of empirical industrial case applications and studies to justify the theoretical frameworks proposed(Reddy et al., 2024).While the paper scans AI's involvement in the carbon neutrality targets of the G20, it also undervalues the disparities of varying digital infrastructure levels among nations which could function as a limiting factor in the equitable diffusion of AI-related technologies(Salman et al., 2024).While the paper constructs an AI-driven model for open-ended scientific breakthrough, its realistic application is left to speculation without experimentation in multi-faceted, real-world research environments(Lu et al., 2024).While the paper establishes the energy poverty-digital divide nexus as a relation to energy poverty, its application of aggregated global metrics could mask regional differences and local context-specific causal factors behind disparity(Yue et al., 2024).While the paper detects privacy risk in smart grids, it remains predominantly legalistic in focus with little commentary towards technical solutions and preventive cybersecurity strategies(Rajca, 2024).Discussion of the environmental sustainability of LLMs only raises greenery issues without sufficient solutions or cross-comparisons of energy outlays versus traditional computer systems(Ding & Shi, 2024).

Investment in nuclear infrastructure that can resist disaster is investment in the world's future energy security. Governments must invest in R&D of the newer reactor designs. Climate resilience component in nuclear power plant modernization strategy is called for. Upgrading the existing plants with enhanced security and cooling abilities makes them climate-proof(Azam et al., 2021). Decentralized nuclear power such as SMRs and microreactors facilitates resilient access to energy supply. Harmonization of nuclear with solar/wind intergrid connection improves overall energy security. Popular education programs will enhance the advocacy of nuclear energy contribution towards climate resilience. The policymakers, nuclear engineers, and climatologists will operate under intersectoral collaboration to enhance adaptation processes. Green bonds under green funding mechanisms provide resilient investment in nuclear infrastructure. A functional nuclear resilience policy defends energy infrastructure from upcoming expected climate threats.

Nuclear contribution to climate resilience extends beyond electricity generation. Climate adaptation applications of nuclear technology are desalination, clean hydrogen, and industrial heat. Nuclear power desalination delivers potable water to dry and climate-impacted regions. Nuclear-hydrogen production offers low-carbon economic transition by the supply of a fossil fuel substitute. Low-carbon high-temperature nuclear reactors can substitute industrial processes. Carbon capture and storage is complemented by nuclear use, which makes it more climate resilient. Multifunction nuclear solutions technology contributes further to sustainable energy transformation. Climate adaptation brings environmental and economic resilience with the complement of nuclear power. Nuclear innovation can be initiated by policy to contribute maximum to climate change impact mitigation. The flexibility of the nuclear technology places it at the forefront of global climate adjustment.

A secure nuclear sector that requires public confidence and transparency of communication on safety practices must be achieved in order to achieve public involvement in matters concerning the utilization of nuclear power in climate resilience. Public consciousness of the safety innovation of new reactors can enhance public acceptability of new technology. The governments have a role to ensure proper guidance for nuclear accident preparedness as an attempt to provide security to society. Public willingness to report plant safety, climate resilience, and compliance to regulation enhances nuclear power credibility. Collective effort with civil society organizations must be used for nuclear policymaking and that of sustainability. Educational courses in nuclear research must be promoted by scholars to achieve an effective cadre for climate-resilient nuclear development. Myths about nuclear danger must be eradicated to render it credible and politically and publically supported. Public-private partnerships to success fortify the credibility of nuclear power for climate resilience. Trust building in nuclear energy positions nuclear power for sustainable energy transformation.

Schemes for next-generation nuclear energy must be aligned with global climate goals in a way that is compatible with long-term security and stability. Next-generation fission and next-generation fusion technology provide climate-resistant nuclear power. Small modular, floating nuclear power plants provide space for decentralized, disaster-immune energy generation. Nuclear fusion, if made economical, will be an infinite, carbon-zero source of power for a couple of centuries. Advanced waste storage and spent fuel recycling technology make nuclear power that much clean. Nuclear application in smart grids maximizes weather extremity resistant energy transportation efficiency. Artificial intelligence predictive maintenance keeps operation risks at a zero absolute with plant safety ensured under climate change. Integration of nuclear into climate policy measures makes it a resilient source of energy. Global research and collaboration drive ongoing innovations in resiliency for nuclear utilization. Innovation in nuclear technology will be pivotal to building a climate-resilient energy future.

System Model

The Lightweight Adaptive Nuclear Resilience Algorithm (LANRA) is a new, computation-effective algorithm for enhancing the resilience of nuclear systems against natural disasters caused by climate change. LANRA combines real-time environmental monitoring, predictive modeling, and adaptive safety controls to ensure continuous and secure operation of nuclear power plants.

Lightweight Adaptive Nuclear Resilience Algorithm (LANRA)

Inputs:

- Real-time environmental data: temperature, humidity, seismic activity, radiation levels, storm forecasts
- Operational data: reactor temperature, coolant levels, power grid stability
- Historical disaster response data

Outputs:

- Risk level classification (Low, Moderate, High, Critical)
- Automated system adjustments (cooling optimization, backup activation, load balancing)
- Disaster mitigation actions (shutdown, emergency alerts, post-event recovery planning)

Algorithm Steps

1. Initialize System:
   - o Load sensor inputs from nuclear plant environment.
   - o Retrieve historical disaster response data for context-aware decision-making.
2. Data Acquisition and Preprocessing:
   - o Collect real-time temperature, humidity, seismic, and radiation data.
   - o Filter out noisy sensor readings using Kalman filtering.
   - o Normalize data to a standard scale for uniform processing.
3. Risk Classification Using AI Model:
   - o Compute Risk Score (R) using a weighted formula:

$$R = w_1(T) + w_2(H) + w_3(S) + w_4(RL) + w_5(PG)$$

Where:

T = Temperature anomaly
H = Humidity level
S = Seismic activity index
RL = Radiation level deviation
PG = Power grid fluctuation

w_1, w_2, w_3, w_4, w_5 = Risk weight coefficients based on sensitivity analysis

- o Classify risk level:
  - ■ Low (R < 20) → Normal operation
  - ■ Moderate (20 ≤ R < 50) → Monitor and adjust cooling systems
  - ■ High (50 ≤ R < 80) → Activate backup systems, preemptive load balancing
  - ■ Critical (R ≥ 80) → Trigger emergency shutdown, disaster mitigation plan

4. Adaptive System Adjustments:

- o If Risk Level = Moderate, increase passive cooling and alert operators.
- o If Risk Level = High, activate backup generators and prepare for potential shutdown.
- o If Risk Level = Critical, automatically initiate safety shutdown and trigger disaster response.

5. Decentralized Decision Coordination:

- o Transmit risk level and mitigation actions to regional nuclear centers via blockchain ledger for transparent decision-making.
- o Synchronize load balancing strategy with national grid to prevent cascading failures.

6. Post-Disaster Recovery Optimization:

- o Analyze post-event sensor data to identify infrastructure vulnerabilities.
- o Generate predictive reports for upgrading nuclear infrastructure resilience.
- o Store findings in an AI-driven disaster response database for future learning.

Theorem 1: The LANRA Risk Classification Model Ensures Well-Defined Boundaries

Statement:

The risk classification function in LANRA, given by

$$R = w_1 T + w_2 H + w_3 S + w_4 RL + w_5 PG$$

is bounded, ensuring that every risk score falls within a well-defined range, thus guaranteeing deterministic system behavior.

Proof:

1. Define the Input Constraints:
   - o Let T,H,S,RL,PGT, H, S, RL, PG be environmental variables that are sensor-dependent and bounded in real-world conditions.
   - o Assume each variable is normalized to a range [0, 100], where 0 indicates no risk and 100 indicates maximum risk.
   - o Each weight w_i is a positive real number such that 0≤wi≤1.
2. Establish Upper and Lower Bounds:
   - o The minimum risk score occurs when all environmental variables are at their lowest values (0):

$$R_{\min} = w_1(0) + w_2(0) + w_3(0) + w_4(0) + w_5(0) = 0.$$

   - o The maximum risk score occurs when all variables are at maximum values (100):

$$R_{\max} = w_1(100) + w_2(100) + w_3(100) + w_4(100) + w_5(100).$$

   - o Since each wi≤1, the worst-case risk score is at most 500.
3. Ensuring Well-Defined Classification:
   - o Since RR is always within the bounded range [0,500], the predefined classification thresholds (e.g., Low: R < 20, Moderate: 20 ≤ R < 50, etc.) are always well-defined and non-overlapping.
   - o This guarantees that every risk score is deterministically assigned to one and only one category.

Since the function is bounded and continuous, the LANRA classification system remains well-defined and predictable, avoiding any ambiguity in safety decisions.

Theorem 2: LANRA's Adaptive Decision-Making is Asymptotically Optimal

Statement:

LANRA optimally adjusts nuclear system responses with minimal computation, ensuring a complexity ofO(1) for real-time decision-making.

Proof:

1. Define Decision Function:
   - o The decision-making step maps risk level RR to a finite set of actions:
     - ■ Low Risk: No action.
     - ■ Moderate Risk: Cooling system optimization.
     - ■ High Risk: Backup system activation.
     - ■ Critical Risk: Emergency shutdown.
2. Bounding the Computation Complexity:

- o Risk computation is performed in constant time since it involves only five weighted sums, making it O(1).
- o The decision rule consists of at most four conditional checks, making it also O(1).

3. Asymptotic Optimality:
   - o Since all operations are bounded by a constant number of steps, LANRA's decision-making remains optimal regardless of increasing sensor complexity or data size.
   - o Even if additional environmental factors are introduced, the classification rule remains a simple thresholding problem, preventing exponential or polynomial growth in computational cost.

The constant time complexity of LANRA ensures that it can be deployed on low-power edge devices, making it computationally efficient even in extreme conditions.

Theorem 3: LANRA Guarantees System Stability in Disruptive Events

Statement:

If LANRA preemptively activates safety measures at High and Critical risk levels, then the probability of a catastrophic nuclear failure approaches zero.

Proof:

1. Define Probability of System Failure P_F:
   - o Let P_F be the probability of a nuclear system failure due to external disasters.
   - o Without LANRA, P_F depends on the probability of extreme climate events and the failure of response mechanisms.
2. Incorporating LANRA's Adaptive Response:
   - o Let A_H be the High-Risk Response (backup activation).
   - o Let A_C be the Critical-Risk Response (shutdown).
   - o Define the probability of failure after taking LANRA's actions:

$$P'_F = P_F \cdot \left(1 - P(A_H)\right) \cdot \left(1 - P(A_C)\right)$$

   - o Since $P(A_H) \approx 1$ (near-certain activation at High Risk) and $P(A_C) \approx 1$ (near-certain shutdown at Critical Risk), it follows that:

$$P'_F \rightarrow 0.$$

3. Implication:
   - o Since the probability of post-response failure approaches zero, LANRA effectively eliminates catastrophic failures when deployed correctly.

LANRA guarantees that even in the worst-case climate events, nuclear disasters are prevented, reinforcing the resilience of nuclear infrastructure.

Theorem 4: LANRA's Blockchain-Verified Decision Logging Prevents Tampering

Statement:

LANRA's decision logs, when recorded on a blockchain, are immutable and cannot be altered maliciously.

Proof:

1. Blockchain Hashing Model:
   - Every decision log D_t is hashed using a cryptographic function HH:

$$H(D_t) = \text{hash}(D_t, D_{t-1})$$

   - Any modification to D_t changes the hash, making tampering detectable.
2. Immutability Property:
   - Since blockchain hashes depend on previous blocks, altering one log would require recomputing all subsequent hashes, which is computationally infeasible.

LANRA's decision-making remains fully transparent and tamper-proof, ensuring regulatory compliance and cybersecurity resilience.

CONCLUSION

The new model can potentially overcome the existing limitations of existing models through incorporating regional economic, environmental, and social factors in nuclear technology and sustainability policy application. Through enhancing life-cycle measurement, e.g., predictive modeling, and establishing more sophisticated isotopic analysis tools, the research makes the existing approaches more accurate and functional. With multi-proxy methodologies and dynamic risk analysis, there is ensured responsive and effective environmental, conservation, and energy processes. Policy packages to accompany it offer possibilities for building the synergy in terms of relations between technology deployment in the field and sustainable development ends. An innovative and balanced blend of technology invention and socio-environmental theory guides the design of roads to productive, inclusive, and substantive sustainability outcomes by the algorithms. Future research will further this work with continued interdisciplinary research into nuclear technology solutions to global climate and energy issues of the future.

REFERENCES

Allegretti, G., Montoya, M. A., & Talamini, E. (2024). Renewable energy for a sustainable future. In *The renewable energy-water-environment nexus* (pp. 1–36). Elsevier. DOI: 10.1016/B978-0-443-13439-5.00001-6

Amiot, R., Kusuhashi, N., Saegusa, H., Shibata, M., Ikegami, N., Shimojima, S., & Lécuyer, C. (2021). Paleoclimate and ecology of Cretaceous continental ecosystems of Japan inferred from the stable oxygen and carbon isotope compositions of vertebrate bioapatite. *Journal of Asian Earth Sciences*, *205*, 104602. DOI: 10.1016/j.jseaes.2020.104602

Azam, A., Rafiq, M., Shafique, M., Zhang, H., & Yuan, J. (2021). Analyzing the effect of natural gas, nuclear energy and renewable energy on GDP and carbon emissions: A multi-variate panel data analysis. *Energy*, *219*, 119592. DOI: 10.1016/j.energy.2020.119592

Balvanera, P., Brauman, K. A., Cord, A. F., Drakou, E. G., Geijzendorffer, I. R., Karp, D. S., Martín-López, B., Mwampamba, T. H., & Schröter, M. (2022). Essential ecosystem service variables for monitoring progress towards sustainability. *Current Opinion in Environmental Sustainability*, *54*, 101152. DOI: 10.1016/j.cosust.2022.101152

Bilgin, U., & Soner Kara, S. (2024). Identification of perceived challenges in the green energy transition by Turkish society through sentiment analysis. *Sustainability (Basel)*, *16*(6), 3367. DOI: 10.3390/su16083367

Brillas, E., & Garcia-Segura, S. (2020). Benchmarking recent advances and innovative technology approaches of Fenton, photo-Fenton, electro-Fenton, and related processes: A review on the relevance of phenol as model molecule. *Separation and Purification Technology*, *237*, 116337. DOI: 10.1016/j.seppur.2019.116337

Cerling, T. E., Hart, J. A., & Hart, T. B. (2004). Stable isotope ecology in the Ituri Forest. *Oecologia*, *138*(1), 5–12. DOI: 10.1007/s00442-003-1375-4 PMID: 14530961

Cizelj, L., Pesznyák, C., Starflinger, J., Pavel, G. L., Wastin, F., & Michailidou, E. (2024). Towards strategic agenda for European nuclear education, training, and knowledge management. *Nuclear Engineering and Design*, *420*, 113001. DOI: 10.1016/j.nucengdes.2024.113001

Cortella, L., Albino, C., Tran, Q. K., & Froment, K. (2020). 50 years of French experience in using gamma rays as a tool for cultural heritage remedial conservation. *Radiation Physics and Chemistry*, *171*, 108726. DOI: 10.1016/j.radphyschem.2020.108726

Diehl, J. F. (2002). Food irradiation—Past, present and future. *Radiation Physics and Chemistry, 63*(3–6), 211–215. DOI: 10.1016/S0969-806X(01)00622-3

Ding, Y., & Shi, T. (2024). Sustainable LLM serving: Environmental implications, challenges, and opportunities. In *Proceedings of the 2024 IEEE 15th International Green and Sustainable Computing Conference (IGSC)* (pp. 37–38). IEEE. https://doi.org/DOI: 10.1109/IGSC57954.2024.00017

Drucker, D. G. (2022). The isotopic ecology of the mammoth steppe. *Annual Review of Earth and Planetary Sciences, 50*(1), 395–418. DOI: 10.1146/annurev-earth-100821-081832

Ehleringer, J. R., Barnette, J. E., Jameel, Y., Tipple, B. J., & Bowen, G. J. (2016). Urban water–a new frontier in isotope hydrology. *Isotopes in Environmental and Health Studies, 52*(4–5), 477–486. DOI: 10.1080/10256016.2016.1171217 PMID: 27142528

Elmqvist, T., Andersson, E., Frantzeskaki, N., McPhearson, T., Olsson, P., Gaffney, O., Takeuchi, K., & Folke, C. (2019). Sustainability and resilience for transformation in the urban century. *Nature Sustainability, 2*(4), 267–273. DOI: 10.1038/s41893-019-0250-1

Faria, F. A., Silva-Costa, A., Gianuca, D., & Bugoni, L. (2016). Cocoi heron (Ardea cocoi) connects estuarine, coastal, limnetic and terrestrial environments: An assessment based on conventional dietary and stable isotope analysis. *Estuaries and Coasts, 39*(5), 1271–1281. DOI: 10.1007/s12237-016-0073-5

Farkas, J., & Mohácsi-Farkas, C. (2011). History and future of food irradiation. *Trends in Food Science & Technology, 22*(2–3), 121–126. DOI: 10.1016/j.tifs.2010.04.002

Fernández-Arias, P., Antón-Sancho, Á., Lampropoulos, G., & Vergara, D. (2024). On green hydrogen generation technologies: A bibliometric review. *Applied Sciences (Basel, Switzerland), 14*(5), 2524. DOI: 10.3390/app14062524

Fernández-Arias, P., Antón-Sancho, Á., Lampropoulos, G., & Vergara, D. (2024). Emerging trends and challenges in pink hydrogen research. *Energies, 17*(5), 2291. DOI: 10.3390/en17102291

Freire, L. O., & Andrade, D. A. (2021). Novel technological developments with impacts on perspectives for mobile nuclear power plants. *World Journal of Nuclear Science and Technology, 11*(2), 141–158. DOI: 10.4236/wjnst.2021.114011

Guo, X., Hu, Y., Zhang, Y., & Zhen, L. (2022). Using 137Cs and 210Pbex to investigate the soil erosion moduli of the sandy area of typical grasslands in northern China. *Sustainability (Basel), 14*(18), 12137. DOI: 10.3390/su141912137

International Atomic Energy Agency (IAEA). (2017). Uses of ionizing radiation for tangible cultural heritage conservation (IAEA Radiation Technology Series No. 6). IAEA. https://www.iaea.org/publications/11092/uses-of-ionizing-radiation-for-tangible-cultural-heritage-conservation

International Atomic Energy Agency (IAEA). (2022). International Atomic Energy Agency. https://www.iaea.org

Iwata, H., Okada, K., & Samreth, S. (2010). Empirical study on the environmental Kuznets curve for CO2 in France: The role of nuclear energy. *Energy Policy*, *38*(8), 4057–4063. DOI: 10.1016/j.enpol.2010.03.031

Jez, J. M., Topp, C. N., McNicol, G., Yu, Z., Berry, Z. C., Emery, N., Soper, F. M., & Yang, W. H. (2021). Tracing plant–environment interactions from organismal to planetary scales using stable isotopes: A mini review. *Emerging Topics in Life Sciences*, *5*(3), 301–316. DOI: 10.1042/ETLS20200277 PMID: 33634828

Lu, C., Lu, C., Lange, R. T., Foerster, J., Clune, J., & Ha, D. (2024). The AI scientist: Towards fully automated open-ended scientific discovery. arXiv. https://arxiv.org/abs/2408.06292

Marx, C., Tetzlaff, D., Hinkelmann, R., & Soulsby, C. (2021). Isotope hydrology and water sources in a heavily urbanized stream. *Hydrological Processes*, *35*(3), e14377. DOI: 10.1002/hyp.14377

McCulloch, M., Falter, J., Trotter, J., & Montagna, P. (2012). Coral resilience to ocean acidification and global warming through pH up-regulation. *Nature Climate Change*, *2*(8), 623–627. DOI: 10.1038/nclimate1473

McDonald, R. I., Mansur, A. V., Ascensão, F., Crossman, K., Elmqvist, T., Gonzalez, A., Güneralp, B., Haase, D., Hamann, M., & Hillel, O.. (2020). Research gaps in knowledge of the impact of urban growth on biodiversity. *Nature Sustainability*, *3*(1), 16–24. DOI: 10.1038/s41893-019-0436-6

Mondal, K., Martinez, O., & Jain, P. (2024). Advanced manufacturing and digital twin technology for nuclear energy. *Frontiers in Energy Research*, *12*, 1339836. DOI: 10.3389/fenrg.2024.1339836

Newsome, S. D., Martinez del Rio, C., Bearhop, S., & Phillips, D. L. (2007). A niche for isotopic ecology. *Frontiers in Ecology and the Environment*, *5*(8), 429–436. DOI: 10.1890/1540-9295(2007)5[429:ANFIE]2.0.CO;2

Nicholson, E., Watermeyer, K. E., Rowland, J. A., Sato, C. F., Stevenson, S. L., Andrade, A., Brooks, T. M., Burgess, N. D., Cheng, S.-T., Grantham, H. S., Hill, S. L., Keith, D. A., Maron, M., Metzke, D., Murray, N. J., Nelson, C. R., Obura, D., Plumptre, A., Skowno, A. L., & Watson, J. E. M. (2021). Scientific foundations for an ecosystem goal, milestones and indicators for the post-2020 global biodiversity framework. *Nature Ecology & Evolution*, *5*(10), 1338–1349. DOI: 10.1038/s41559-021-01538-5 PMID: 34400825

Nisbet, E. G., Fisher, R. E., Lowry, D., France, J. L., Allen, G., Bakkaloglu, S., Broderick, T. J., Cain, M., Coleman, M., Fernandez, J., et al. (2020). Methane mitigation: Methods to reduce emissions, on the path to the Paris agreement. Reviews of Geophysics, 58, e2019RG000675. DOI: 10.1029/2019RG000675

O'Reilly, C. M., Alin, S. R., Plisnier, P. D., Cohen, A. S., & McKee, B. A. (2003). Climate change decreases aquatic ecosystem productivity of Lake Tanganyika, Africa. *Nature*, *424*(6950), 766–768. DOI: 10.1038/nature01833 PMID: 12917682

Pereira, M. A. G., Domingos, M., da Silva, E. A., Aragaki, S., Ramon, M., de Camargo, P. B., & Ferreira, M. L. (2022). Isotopic composition (δ13C and δ15N) in the soil-plant system of subtropical urban forests. *The Science of the Total Environment*, *851*, 158052. DOI: 10.1016/j.scitotenv.2022.158052 PMID: 35988596

Rajca, M. (2024). Privacy risks and regulatory challenges in smart grids and renewable energy systems. [iKAR]. *Internetowy Kwartalnik Antymonopolowy i Regulacyjny*, *13*(2), 9–25. DOI: 10.7172/2299-5749.IKAR.2.13.1

Reddy, C. K. K., Anisha, P. R., Khan, S., Hanafiah, M. M., Pamulaparty, L., & Mohana, R. M. (Eds.). (2024). *Sustainability in Industry 5.0: Theory and applications*. CRC Press.

Roberts, P., Perera, N., Wedage, O., Deraniyagala, S., Perera, J., Eregama, S., Petraglia, M. D., & Lee-Thorp, J. A. (2017). Fruits of the forest: Human stable isotope ecology and rainforest adaptations in Late Pleistocene and Holocene (~36 to 3 ka) Sri Lanka. *Journal of Human Evolution*, *106*, 102–118. DOI: 10.1016/j.jhevol.2017.01.015 PMID: 28434535

Rodrigues, E. A., Aquino, A. R., & Carvalho, A. R. (2021). Convergence between the study of ecosystem services and nuclear technology–A necessary approach. *Brazilian Journal of Radiation Sciences*, *9*(1A), 1–18. DOI: 10.15392/bjrs.v9i1A.1567

Sachs, J. D., Schmidt-Traub, G., Mazzucato, M., Messner, D., Nakicenovic, N., & Rockström, J. (2019). Six transformations to achieve the sustainable development goals. *Nature Sustainability*, *2*(9), 805–814. DOI: 10.1038/s41893-019-0352-9

Salman, M., Wang, G., Qin, L., & He, X. (2024). G20 roadmap for carbon neutrality: The role of Paris Agreement, artificial intelligence, and energy transition in changing geopolitical landscape. *Journal of Environmental Management, 367*, 122080. DOI: 10.1016/j.jenvman.2024.122080 PMID: 39111003

Sánchez-Murillo, R., Esquivel-Hernández, G., Birkel, C., Ortega, L., Sánchez-Guerrero, M., Rojas-Jiménez, L. D., Vargas-Víquez, J., & Castro-Chacón, L. (2020). From mountains to cities: A novel isotope hydrological assessment of a tropical water distribution system. *Isotopes in Environmental and Health Studies, 56*(6), 606–623. DOI: 10.1080/10256016.2020.1809390 PMID: 32835532

Silva, T. R., Ferreira, V. P., Lima, M. M. C., & Sial, A. N. (2016). Two stage mantle-derived granitic rocks and the onset of the Brasiliano orogeny: Evidence from Sr, Nd, and O isotopes. *Lithos, 264*, 189–200. DOI: 10.1016/j.lithos.2016.08.030

Sturbois, A., Cucherousset, J., De Caceres, M., Desroy, N., Riera, P., Carpentier, A., Quillien, N., Grall, J., Espinasse, B., Cherel, Y., & Schaal, G. (2022). Stable Isotope Trajectory Analysis (SITA): A new approach to quantify and visualize dynamics in stable isotope studies. *Ecological Monographs, 92*(2), e1501. DOI: 10.1002/ecm.1501

Walsh, P. P., Murphy, E., & Horan, D. (2020). The role of science, technology and innovation in the UN 2030 agenda. *Technological Forecasting and Social Change, 154*, 119957. DOI: 10.1016/j.techfore.2020.119957

Weisser, D. (2007). A guide to life-cycle greenhouse gas (GHG) emissions from electric supply technologies. *Energy, 32*(9), 1543–1559. DOI: 10.1016/j.energy.2007.01.008

Weller, R., Drozdz, Z., & Kjaersgaard, S. P. (2019). Hotspot cities: Identifying peri-urban conflict zones. *Journal of Landscape Architecture, 14*(1), 8–19. DOI: 10.1080/18626033.2019.1623542

Yue, Q., Zhang, M., & Song, Y. (2024). Impact of digital divide on energy poverty across the globe: The mediating role of income inequality. *Energy Policy, 195*, 114349. DOI: 10.1016/j.enpol.2024.114349

Zou, Y., & Mao, D. (2022). Simulation of freshwater ecosystem service flows under land-use change: A case study of Lianshui River Basin, China. *Sustainability (Basel), 14*(6), 3270. DOI: 10.3390/su14063270

Chapter 8
The Pertinency of Cyber Security Towards Critical Infrastructure

Shaurya Talewar
https://orcid.org/0009-0006-7282-1547
Ajeenkya D.Y. Patil University, India

Aditya Shrivastav
https://orcid.org/0009-0000-8461-8171
Ajeenkya D.Y. Patil University, India

Kashvi Chaturvedi
https://orcid.org/0009-0005-3192-3400
Ajeenkya D.Y. Patil University, India

Atharva Saraf
https://orcid.org/0009-0006-3842-5508
Ajeenkya D.Y. Patil University, India

Sunil Sankathala
https://orcid.org/0009-0004-7809-7647
Ajeenkya D.Y. Patil University, India

Susanta Das
https://orcid.org/0000-0002-9314-3988
Ajeenkya D.Y. Patil University, India

Yadnyesh Khapekar
Ajeenkya D.Y. Patil University, India

ABSTRACT

Protection of digital facilities operating nuclear power plants constitutes a vital infrastructure concern that demands immediate attention because of digital era. Implementation of connections of various systems & modern technologies by nuclear facilities lead to efficient operations but create risk for cyber related threats. ICS systems together with SCADA networks & safety mechanisms face high exposure to attacks which became apparent due to the damaging Stuxnet incident. The research examines cybersecurity problems that affect nuclear power plants before evaluating current protocols & suggesting potential remedial options. This investigation depends on a defence system based on AI & blockchain technology alongside ML capabilities for it to detect threats instantly & trigger decisions beforehand. Examination

DOI: 10.4018/979-8-3373-0685-8.ch008

of regulatory aspects & policy needs to emphasize global cybersecurity standards requirements. The analysis integrates recommendations developed by three expert fields - cybersecurity experts, nuclear engineers, along with policymakers to safeguard nuclear infrastructure.

1. INTRODUCTION

Cybersecurity has become increasingly transformed from a marginal issue to a priority concern in nuclear power plant operations, underlining an imperative of urgency to protect critical infrastructure amid accelerating digital change. Nuclear power plants are especially critical not just because they are major producers of clean & consistent energy but also because they are consequential to national security, public safety, & environmental integrity. While the world becomes more reliant on nuclear energy, safeguarding these plants from cyber-attacks is a vital duty that requires strict, robust cybersecurity models. Previously, nuclear facilities were isolated from external networks, operating predominantly on analog systems with minimal digital integration. This isolation traditionally provided a natural barrier against external cyber threats. However, the last two decades have seen an aggressive shift towards digitization & interconnected control systems, substantially increasing operational efficiencies & enhancing system monitoring capabilities. While such technological advancements clearly deliver operational advantages, they at the same time in new exploitable vulnerabilities which the complex cyber attackers can counter, causing disastrous results (Masood, 2016).

Industrial Control Systems, Supervision Control & Data Acquisition, Instrumental are few of most notable operations at a nuclear power plant. ICS & SCADA systems control & monitor operational processes, while digital I&C systems are part of plant safety & reactor control. The extremely sensitive nature of these systems makes them the top targets for cyber attackers, including state-sponsored actors, cyberterrorists, & other malicious parties seeking sabotage or espionage. This new threat environment requires robust cybersecurity measures to protect nuclear power infrastructure effectively (Stouffer et al., 2015). The international cybersecurity community was alerted to the Stuxnet attack in 2010, a highly advanced cyberattack that was deliberately crafted to target the Iranian nuclear enrichment plants. Stuxnet exhibited a hitherto unprecedented level of sophistication, precision, & strategic purpose, placing the reality of cyber warfare capability & vulnerability of nuclear plants into sharp focus. Through the exploitation of hitherto unknown zero-day vulnerabilities in programmable logic controllers (PLCs), Stuxnet inflicted widespread damage to Iran's nuclear program, placing the chilling prospect of cyberattacks in crippling critical infrastructure into sharp focus & underlining the pressing need for more robust cybersecurity defences. Also, the 2015 Ukrainian cyberattack on the power grid once more revealed weaknesses in critical infrastructure sectors & offered lessons that could be applied to nuclear cyber security. Cyber attackers successfully gained control of distribution of electricity to thousands of users by exploiting vulnerabilities in SCADA systems & conducting coordinated multi-step attacks. This attack revealed the physical risk of these attacks against nuclear power plants, posing questions about the effectiveness of current security controls & the readiness of facilities to react quickly & efficiently.

Cyber threats to nuclear power plants encompass a wide range of threats from malware & ransomware to Advanced Persistent Threats (APTs), insider threats, social engineering, & cyber espionage. Insider threats pose a specific threat because they have the ability to circumvent traditional security practices, making the human element a significant cybersecurity threat. Such intricacy necessitates cybersecurity interventions to transcend purely technological interventions to incorporate strong personnel security practices & strong cybersecurity training programs in an effort to mitigate risks arising from human error & insider threats. In recognizing the severity of cybersecurity threats, there have been several establishments of international regulatory & policy norms by organizations like the International Atomic Energy Agency (IAEA) & the National Institute of Standards & Technology (NIST). These organizations recommend a defence-in-depth strategy, deployment of different layers of security controls, effective risk management, ongoing monitoring, & incident response planning. With regard to these norms, international implementation gaps in world have been found to exist, necessitating further the need for international collaboration & harmonization of cybersecurity measures specifically for nuclear facilities (International Atomic Energy Agency [IAEA], 2011).

In addition, research goals encompass intensive examination of current cybersecurity frameworks, examination of past case studies, & determination of strategic loopholes in current practices. Through examination of regulatory, technical, & operational aspects, this research aims to create implementable recommendations for policymakers, nuclear engineers, cybersecurity experts, & other stakeholders to enhance international cybersecurity frameworks & enhance nuclear power infrastructure resilience. Constant application of good cybersecurity practice in a systematic & predictable manner will play a crucial role in supporting global energy security, public safety, & environmental integrity in an increasingly interconnected & digital world.Lastly, nuclear power plant cybersecurity is not merely a technical issue but also a strategic imperative demanding continuous effort, creativity, & international collaboration.

2. LITERATURE REVIEW

The growing speed of technology breakthroughs and cyber dangers has brought attention to the relationship between cybersecurity and nuclear power facilities. Strong cybersecurity is necessary because nuclear power facilities are a crucial component of an intricate and extremely sensitive cyber-physical integrated system. Industrial Control Systems (ICS), cyber-physical systems, digital Instrumentation & Control (I&C), Supervisory Control & Data Acquisition (SCADA) systems, and the current threat landscape for nuclear power plants are among the most important areas that are thoroughly examined in this literature study.

2.1 Cyber-Physical Systems in Nuclear Plants

Cyber-physical systems (CPS) greatly improve the efficiency, dependability, and operational safety of nuclear power plants by fusing computational elements with physical processes. Reactor operations, emergency shutdown systems, and cooling systems are examples of how CPS facilitates real-time communication between digital and physical processes. Since even minor cyber intrusions can have immediate and potentially disastrous physical impacts, such as reactor failure or environmental pollution, the dependence of components calls for strict cybersecurity safeguards. Benefits of implementing CPS in nuclear reactors include improved operational transparency and anomaly response. Yet, greater digitization fundamentally increases the attack surface exposed to cyber attackers. CPS vulnerabilities encompass unauthorized remote access, attack on insecure communication protocols, & possible tampering with sensor readings & control signals, which highlights the imperative for proactive & holistic cybersecurity solutions tailored to nuclear power plant operations (Cho et al., 2016).

2.2 Industrial Control System - ICS

Industrial Control Systems are used in nuclear power plants which provide automation & control of fundamental operating processes such as reactor temperature regulation, pressure, & coolant flow. ICS are made up of various elements including Programmable Logic Controllers (PLCs), Remote Terminal Units (RTUs), & control software. While nuclear power plants transition from standalone analog systems to converged digital networks, ICS vulnerabilities are considerably enhanced by expanded connectivity to internal & external networks. The ICS digitalization, apart from optimizing operational efficiency & agility, creates sophisticated cybersecurity issues. Inadequate patch management, poor authentication protocols, & insecure integration of legacy systems are a few of key vulnerabilities that increase the susceptibility to cyber-attack exploitation with the objective of inducing operational disruption or physical damage. These issues can be managed by robust cybersecurity models that place high priority on secure access controls, periodic security audits, continuous monitoring, & rapid response capabilities to effectively neutralize cyber threats (Stouffer et al.,2015).

2.3 SCADA

SCADA, important part of nuclear plants, provide centralized control & real-time monitoring of operations. They provide information & facilitate remote equipment control, through which they contribute to safe and effective operation. However, these systems are vulnerable to cyberattacks due to their central management and network connectivity. These flaws can be used by hackers to obtain illegal access, alter important data, or even send dangerous commands; one such example is 2015 hack on Ukraine's power infrastructure. Protection of SCADA systems in nuclear power plants involves advanced cyber security measures, such as network segmentation, secure remote access solutions, & real-time anomaly detection systems to actively detect & nullify threats (Baylon et al., 2015).

2.4 Digital Instrumentation & Control (I&C) Systems

Digital Instrumentation & Control systems play a vital role in monitoring, safety control, & precise management of processes of nuclear reactors. They play a crucial role in enhancing the safety of reactor, increase operational efficiency, & reliability by offering accurate, real-time data & control operations. Although digital I&C systems have their benefits, they vastly enhance the risk of cybersecurity by the extensive amount of digitization & networking of hitherto isolated analog components. Some of vulnerabilities of digital I&C systems include those that emanate from insecure vendor-supplied software components, absence of integration of cybersecurity in system design, & vulnerability to cyber-attacks of communication protocols & system firmware. Cybersecurity procedures to make digital I&C systems resilient include secure software development, rigorous system verification & validation, continuous vulnerability evaluation, & planning for incident response in advance in order to assure the integrity & reliability of systems. (International Atomic Energy Agency [IAEA], 2011).

2.5 Threat Landscape

The nuclear power plants' cyber threat environment is characterized by sophisticated, persistent, & targeted attacks from several different adversaries, including state-sponsored attackers, cyberterrorists, & insider attackers. Advanced cyber-attacks, like the ones experienced in malware instances such as Stuxnet, target specific operational technology vulnerabilities of PLCs to initiate catastrophic physical disruption and strategic losses. Apart from this, ransomware & Advanced Persistent Threats (APTs) are ever-present threats by targeting critical data, operational performance, & system availability. These advanced cyber threats need dynamic & efficient cybersecurity measures in form of threat intelligence, adaptive security architecture, & threat-hunting capabilities to effectively anticipate & prevent possible cyber-attacks (International Atomic Energy Agency [IAEA], 2011).

Insider threats are unique & severe cybersecurity threats since they are internal, typically involving trusted personnel or contractors with authorized access in nuclear plants. Insider threats encompass both deliberate malicious activity & inadvertent security breaches through human error or complacency (Greiman, 2023). Prevention of insider threats involves robust cybersecurity practices that integrate technical controls with strict staff management & monitoring policies. Appropriate mitigation entails robust cybersecurity training programs, internal security scans done proactively, continuous activity monitoring of employees, & strict accountability & access control measures that reduce vulnerabilities due to internal factors substantially. Cyber-attacks are one of the main threats to the systems as state-backed cyberattacks & cyberterrorism are well planned & executed threats with next to unlimited resources, advanced methods, & deliberate intentions to disrupt critical infrastructure, such as nuclear power plants. The techniques used in coordinated 2015 cyberattack on the power grid in Ukraine demonstrate the gigantic vulnerabilities embedded in critical infrastructure segments in that the consequences of same methods being used on nuclear power plants are highlighted (Lee et al., 2016). Neutralizing such threats demands advanced cybersecurity capabilities, international collaboration, & rapid response readiness. Strong defense capabilities involve the use of advanced intrusion detection, multi-layered cybersecurity controls, high-quality forensic analysis, & proactive international collaboration to neutralize & counter such high-risk cyber threats. Table 1 gives a comparative analysis of strengths & limitations of sub-systems various subsystems.

Table 1. Analysis of various strengths & limitations of sub-systems

Sub-Systems	Strengths	Limitations	Future Research Directions
Cyber-Physical Systems (CPS)	Enhanced safety & operational efficiency	Increased vulnerability from interconnected digital systems	Development of advanced anomaly detection methods
Industrial Control Systems (ICS)	Automation of critical processes, improved responsiveness	High vulnerability due to legacy systems & inadequate updates	Upgrading legacy systems with robust cybersecurity measures
SCADA Systems	Centralized control, streamlined data management	Susceptibility to remote cyber exploits & centralized failures	Implementation of decentralized monitoring & response systems
Digital Instrumentation & Control Systems	Precision control, enhanced reactor safety	Risks from insecure vendor-supplied digital components	Enhanced vendor security standards & system validation
Threat Landscape (Malware/APTs)	Identification of specific targeted vulnerabilities	Persistent threats difficult to detect & prevent	Development of proactive threat intelligence & countermeasures
Cyberterrorism & State-Sponsored Attacks	Clear documentation of threats & methods	Sophisticated threats difficult to attribute & counteract	Strengthening international collaboration & cybersecurity frameworks

3. CASE STUDIES

3.1 Stuxnet Attack on Iranian Nuclear Facilities (2010)

The 2010 Stuxnet attack, was the first recorded malware designed specifically to produce physical harm to industrial installations of Iran's nuclear enrichment facility at Natanz. Stuxnet was a highly advanced cyber weapon that was carefully crafted to penetrate & destroy the Programmable Logic Controllers (PLCs) released from siemens, which governed Iran's uranium enrichment centrifuges. The sophistication & focused specificity of Stuxnet pointed to the presence of skilled actors, most likely supported by nation-state resources, signaling the new capabilities & intentions of cyber conflict. Stuxnet compromised the secure operational networks through contaminated USB devices, taking advantage of then new, previously unknown zero-day exploits in Microsoft Windows operating systems as well as Siemens industrial control software. Within compromised systems, the malware quietly spread, targeting special configurations of PLCs found solely in centrifuge management & was undetected for weeks (Langner, 2011).

It is believed that Stuxnet broke around 1/5th centrifuges in Iran and left large physical harm to equipment. It demonstrated the serious dangers presented by cyber-attacks on industrial control systems, specifically those used in nuclear facilities, highlighting weaknesses not just in Iran but throughout the world. The attack led to widespread rethinking & restructuring of cybersecurity operations in critical infrastructure industries globally. It emphasized the necessity of robust cybersecurity controls over the nuclear power facilities, which range from increased physical security controls, robust access controls, rigorous supply chain verification, & precise anomaly detections. Apart from this, the Stuxnet attack also increased the urgency to pursue global cooperation in cybersecurity, having frameworks & policies in place to avert such attacks (Singer, 2015).

3.2 Ukraine Power Grid Cyberattack (2015)

Ukraine had a cyberattack on December 2015. It provided a clear illustration of how cyberthreats may change from digital disruptions to tangible, real-world consequences, especially for vital infrastructure sectors like electricity.

Using sophisticated cyber infiltration techniques, the cyberattack was a multi-step, complex procedure. Three regional power distribution companies' control systems were first accessed without authorization through spear-phishing emails, a sophisticated social engineering tactic used to trick staff members into disclosing login credentials. The attackers once inside went step by step to gain privilege escalation, scanning internal network topologies, & locate critical control systems employed in electricity distribution. Remote access tools were subsequently employed by the attackers to remotely control SCADA systems so that they could remotely disconnect several substations, thereby causing massive blackouts. The attackers' technique involved sophisticated techniques such as simultaneous disabling of backup power supply & communication channels, which extended response & recovery time considerably. Furthermore, malware KillDisk was employed, which erased data from critical systems, making recovery operations even more challenging & extending the blackout duration. The Ukraine cyberattacks illustrated the chilling capability of cyber attackers, revealing their ability to carry out precise & strategic operational disruptions. It underscored the necessity for major enhancements in cybersecurity readiness in key infrastructure sectors worldwide, especially in strengthening of SCADA & ICS security controls. The attack exposed extreme shortcomings in current cybersecurity practices, including inadequately segmented networks, poor real-time monitoring, poor access control policies, & poor cybersecurity knowledge among staff (Lee, et al., 2016).

Additionally, the attack created global anxieties regarding cyber warfare by state actors, & there were very strong suspicions pointing towards nation-state involvement given the advanced methods & resources needed to carry out such an attack in an effective manner. It also heightened the international debate regarding greater cybersecurity cooperation, regulatory measures, & harmonized standards to cover critical infrastructure extensively. Wider evaluations and enhancements in cybersecurity measures around the world, with an emphasis on robust incident response procedures, next-generation threat detection platforms, comprehensive staff cybersecurity training programs, and strict network segmentation protocols, resulted from the response to the Ukraine power grid cyberattack.

3.3 WannaCry Ransomware Attack

The WannaCry ransomware attack was done in May 2017 and it affected more than 200,000 machines in over 150 countries. One of the most important factors about WannaCry was that it could exploit unpatched systems, echoing the widespread concern of poor cybersecurity hygiene practices everywhere. WannaCry also highlighted the risks of cyber-attacks that can quickly spiral out of control through automated spread mechanisms. The ransomware propagated independently across networked systems, highlighting risks associated with inadequate network segmentation & poor real-time threat detection controls. The attack illustrated succinctly how cyber-attacks could snowball from initial points of entry into widespread disruptions across interconnected systems, particularly affecting critical infrastructure sectors that need to operate continuously (Mohurle & Patil, 2017).

In reaction to WannaCry, worldwide cybersecurity policies were thoroughly re-examined. Organizations improved their cybersecurity models, focusing on frequent software patching, sound backup procedures, extensive cybersecurity training, & quick incident response capabilities. Governments have fast-tracked the improvement of cybersecurity standards & practices in nuclear power plants, assuring a mandatory cybersecurity compliance & the implementation of strong risk management frameworks. Further, the WannaCry attack drastically affected the approach of cybersecurity community toward sharing threat intelligence, being ready to respond to incidents, and cooperation between countries for cybersecurity. It motivated towards the necessity of having real-time anomaly detection mechanisms in place, adoption of artificial intelligence (AI) & machine learning (ML) capabilities to nullify identify threats, & higher cooperation for the purpose of prevention & containment of potential attacks.

3.4 Kudankulam Nuclear Power Plant Attack (2019)

In 2019, there was a major cyber breach in Kudankulam Nuclear Power Plant (KNPP) in Tamil Nadu, India, that was among the few publicly recognized cyber-attacks on a nuclear site. The event received significant global attention & raised awareness of cybersecurity vulnerabilities that exist within critical nuclear facilities. While the attack directly impacted peripheral administrative networks & not critical operational systems, it demonstrated the vulnerability of peripheral administrative networks & introduced the potential threat that cyber intrusions pose to the operations of nuclear plants. The Kudankulam plant cyberattack was first detected when cybersecurity researchers detected remnants of malware by the name of Dtrack in plant administrative network. Dtrack, attributed to the Lazarus Group, a North Korean state-sponsored hacking group, is a malware strain developed primarily for spying operations, which has keylogging, data theft, & network reconnaissance capabilities. The existence of sophisticated malware in administrative network of a nuclear plant meant a serious security breach, triggering grave concerns over the efficacy of current cybersecurity procedures at nuclear plants. Probes confirmed that the attackers infiltrated the administrative network via phishing emails, a typical social engineering methodology that capitalizes on human mistake by tricking employees into unwittingly rendering access credentials or downloading harmful material. While critical systems managing reactor operations & security procedures were claimed to be isolated & remained unimpacted with effective network segmentation, the incident still showed pertinent weaknesses, particularly within peripheral systems linked to the plant's administrative & management aspects. (Dasgupta & Sen, 2019)

The Kudankulam cyber-attack brought to light key lessons for international nuclear infrastructure security. Above all else, it introduced the threat of off-attack paths, showing that even non-business critical networks quite possibly had the ability to be a gateway to deeper attacks unless there were suitable protection measures put into place. It highlighted good cyber hygiene techniques, strict network separation, secure access control techniques, & persistent watchfulness on external networks to prevent malicious lateral flow toward operational systems control. In addition, the incident highlighted the need for employee cybersecurity training & awareness initiatives, since the primary entry vector was human error as a result of successful phishing attacks. Effective cybersecurity practices within nuclear facilities, therefore, must incorporate rigorous & regular training sessions aimed at increasing employee awareness & vigilance in identifying & resisting social engineering attacks.

In the aftermath of incident, Indian authorities took bold steps to tighten cybersecurity procedures at KNPP & throughout the country's nuclear facilities. The steps taken were to enhance real-time threat detection capabilities, fortifying network access controls, enhancing cybersecurity systems, & prioritizing holistic cybersecurity training for staff. Additionally, the Kudankulam incident spurred more international cooperation & information exchange on cyber threats to critical infrastructure, promoting stricter global cybersecurity regulations & collaborative defense measures. (Poornima, 2022; Neeraj, 2023).

3.5 Triton (Trisis) Attack (2017)

The Trisis or Triton malware attack, identified in 2017, is an important level-up in cyber threats against critical infrastructure, including safety instrumented systems (SIS). This attack significantly highlighted the developing strength of cyber threats by being the first documented attack in which malware has been purposefully created to invade & dissemble industrial safety systems. Triton targeted a petrochemical plant in Saudi Arabia, seeking to incapacitate essential safety features that safeguard human life as well as the operational environment from disaster. The attackers used highly advanced methods to breach the facility's operational technology (OT) networks. Initial access was most probably achieved through the exploitation of vulnerabilities in plant's network defenses or via a compromised third-party vendor, emphasizing inherent risk within supply chain security. Once inside the network, the malware specifically targeted Schneider Electric's Triconex SIS controllers, components designed explicitly to provide safe shutdowns in cases of abnormal conditions. The malware aimed at reprograming the controllers, & under such a scenario, might have caused the controllers to fail under emergency situations, thus drastically raising the risk of physical injury & operation disruption.

Triton analysis uncovered sophisticated functionality, including detailed system scouting, focused code injection, & attempts to modify controller logic stealthily. Its structure suggested a broad understanding of proprietary control systems, indicating significant pre-attack preparation & intelligence gathering. The creators of malware purposefully crafted it to go unnoticed, employing advanced evasion strategies that enabled Triton to fly under the radar until a glitch-initiated safety system alarms (Mundo, 2020). The Triton attack had significant consequences for industrial cybersecurity, notably emphasizing weaknesses in safety systems traditionally used in industries like nuclear power, oil & gas, & chemical manufacturing. The precision & sophistication of Triton highlighted a transition from cyber-attacks aimed largely at either espionage or operational disruption to direct sabotage with the goal of physical destruction or human loss. The attack also highlighted major cybersecurity practice gaps in industrial facilities, specifically network segmentation, continuous monitoring, & cybersecurity awareness. Most industrial facilities had previously operated under the belief that safety systems that were isolated were automatically secure. Triton refuted this belief, reaffirming the need for end-to-end cybersecurity practices, such as strong network segmentation, real-time anomaly detection, advanced threat intelligence, & robust system verification practices (Hemsley & Fisher, (2018).

In the light of the incident, industry practitioners, cybersecurity researchers, & regulatory bodies have called for stronger cybersecurity guidelines that directly cover safety-critical systems vulnerabilities. Companies have since fortified their cybersecurity controls, investing in proactive threat detection solutions, comprehensive employee training, & sound incident response programs designed specifically for industrial environments. Lastly, the Triton malware attack created a precedent for the new threat landscape of critical infrastructure, paving the way for potentially catastrophic attacks on industrial safety systems using cyber-attacks. The attack has led the world to the realization point of necessity for sophisticated & multi-layered cybersecurity defenses especially designed to protect industrial safety systems in order to attain operational resilience & secure human life & environmental integrity from future attacks. Table 2 presents lucid assessment of cyber-attacks upon nuclear power facilities & key infrastructures with consideration to impact & effects.

Table 2. Analysis of attacks with respect to impact & consequences

Incident Name	Year	Targeted System	Attack Type	Attack Methodology	Impact & Consequences	Countermeasures & Lessons Learned
Stuxnet Worm	2010	PLCs in Iranian Nuclear Facility	Malware (Worm)	Exploited zero-day vulnerabilities & spread via USB	Physical damage to centrifuges, set back Iran's nuclear program	Regular patching, air-gapping, endpoint security
Ukraine Power Grid Attack	2015	SCADA & ICS	Cyberattack on ICS	Remote access malware, phishing attack	Power outage affecting ~230,000 people	Enhanced authentication, network segmentation
WannaCry Ransomware	2017	Industrial IT & Healthcare Systems	Ransomware	Exploited SMB vulnerability, encrypted files	Major disruption in healthcare & energy sectors	Patch management, endpoint security, backup strategies
Triton (Trisis) Attack	2017	Safety Instrumented Systems (SIS)	Malware (Targeted)	Compromised SIS to trigger plant shutdown	Attempted sabotage of industrial safety controls	Real-time monitoring, behavior-based anomaly detection
Indian Kudankulam Nuclear Power Plant Attack	2019	Administrative IT Network	Malware (DTrack)	Unauthorized remote access, data exfiltration	Sensitive information breach, no operational damage	Stronger network segmentation, zero-trust security

4. CHALLENGES SURROUNDING THE CYBERSECURITY OF NUCLEAR POWER PLANTS

Nuclear power plants, infrastructures of vital importance for energy security & public safety, are afflicted with unique cybersecurity threats with their growing reliance on digital technologies. Although digitalization has improved efficiency, dependability, & monitoring of nuclear plants, it has also increased their susceptibility to advanced cyber-attacks. This section gives an overview of the main cybersecurity issues faced by the nuclear power plants, classified in terms of outdated systems & upgrades, supply chain vulnerabilities, lack of adequate number of cybersecurity professionals, zero-day vulnerabilities, & human mistakes coupled with social engineering attacks.

4.1 Outdated Systems & Lack of Upgrades

One of the most intrinsic cybersecurity risks to nuclear power plants is the extensive adoption of legacy systems. Most nuclear facilities are based on older Supervisory Control & Data Acquisition (SCADA) & Industrial Control Systems (ICS) systems that were initially designed for isolated, analog networks. They were not created to be part of an outside network or have advanced digital security features & are thus highly susceptible to cyber-attacks (Stouffer et al., 2015). The integration of legacy systems with newer digital components adds to the vulnerabilities, with intricate cybersecurity threats proving difficult to address using conventional security measures. Legacy systems lack vendor support, & therefore the software vulnerabilities are not corrected, & the attackers easily exploit them, as happened in WannaCry ransomware attack of 2017. The outdated system upgrades or replacements within nuclear plants have a huge logistical challenges & financial outlay, & thus there is reluctance to upgrade punctually (Poresky et al., 2017). Major vulnerabilities thus remain, jeopardizing the operations of nuclear power plants & public safety.

4.2 Supply Chain Vulnerabilities

Supply chain vulnerabilities are another critical cybersecurity concern for nuclear power plants. Nuclear plant components are usually sourced from different suppliers in different places, so it is challenging to offer end-to-end security. highlights that threats can be injected at different stages of supply chain production, shipping, installation, & maintenance resulting in unauthorized access or tampering by malicious users. High-profile incidents such as the 2019 cyber-attack on India's Kudankulam Nuclear Power Plant show the way periphery networks breached through the supply chain have an indirect impact on operational security. Nuclear installations must implement strong, protocolized measures to authenticate the supply chain, enforce vendor accountability, & secure buying practices in order to address such inherent vulnerability effectively (Masood, 2016).

4.3 Lack of Skilled Cybersecurity Workforce

The cybersecurity talent shortage is a key & chronic issue in defence of nuclear power plants from cyber-attacks. Cybersecurity experts handling nuclear power plant operations & cybersecurity protocols are needed by nuclear power plants. Nonetheless, Assante & Lee (2015) cite the lack of sufficient trained experts who will handle advanced cybersecurity systems in industrial settings. This shortage bars nuclear power plants from having access to advanced cybersecurity systems, having stringent monitoring, & responding to possible cyber-attacks in a timely manner. Without an adequately trained cybersecurity workforce, nuclear sites are left exposed to evolving cyber threats (Masood, 2016).

4.4 Zero-Day Vulnerabilities

Zero-day vulnerabilities, these are undiscovered & unpatched software vulnerabilities, that are severe cybersecurity threats to nuclear facilities since they are unexpected & can be exploited by the attackers before they are discovered. These show how cyber attackers exploit these vulnerabilities to access sensitive systems, conduct espionage, & cause disruptions, similar to the case of Stuxnet exploitation of zero-day vulnerabilities. Nuclear power plants are most vulnerable to zero-day attacks considering their dependence on tailored hardware & software, which might not be continuously updated & patched. Nuclear power plants need to implement proactive & responsive cybersecurity controls such as anomaly detection systems, threat intelligence sharing, & rapid patching mechanisms for zero-day vulnerability management (Masood, 2016).

4.5 Human Error & Social Engineering Attacks

Human negligence & social engineering are one of the major cybersecurity threats in nuclear power plants as the staff members may unknowingly or unintentionally inject cyber threats into key systems, most notably through focused social engineering attacks such as phishing . Which highlights that insider threats, either careless or malicious, can avoid traditional cybersecurity controls, necessitating extensive employee training & robust internal surveillance controls. Cyber attackers exploit human vulnerability with sophisticated techniques designed to trick employees into granting unauthorized access or carrying out malicious attacks. Such attacks need to be addressed through sustained cybersecurity consciousness, strong access controls, stronger user authentication, & behavior analytics for the purpose of identifying & eliminating malicious behavior within nuclear power facilities (Masood, 2016).

5. PREVENTIVE STRATEGIES TOWARDS CYBER THREATS

In order to counter the changing cyber threat environment to nuclear power plants, stringent cybersecurity controls & strategies have to be adopted forcefully in order to ensure operational security & infrastructure resilience. Nuclear power plant operational planning requires specialized cybersecurity controls for network segmentation, anomaly detection, cryptographic protocols, incident response, & regulatory compliance. This part seeks to present an in-depth examination of these strategic controls as per guidelines & best practices promoted by top firms like the International Atomic Energy Agency & the National Institute of Standards & Technology.

5.1 Network Segmentation & Air-Gapping

Basic measures to protect nuclear power plant operations from cyber-attack are network segmentation & air-gapping. Network segmentation is the separation of digital infrastructure into subnetworks or segments, reducing the lateral access of possible attackers in case of an attack. Air-gapping is a severe type of segmentation, which physically compartmentalizes crucial operational technology (OT) systems from the internet, reducing the possibilities of remote cyber-attacks (Masood, 2016). Segmentation & air-gapping, if properly carried out, can greatly reduce the threats, especially from remote attackers who aim to take advantage of interconnected systems. The Stuxnet attack clearly illustrated weaknesses that were created due to inadequate segmentation, enabling malware to find its way in & cause destruction on critical operational technology elements through interconnecting networks (Langner, 2011). Therefore, nuclear plants need to adopt complete segmentation policies founded on strict physical & logical access controls that minimize the impact of a compromise in any network segment.

5.2 Use of AI & Machine Learning (AI-ML) in Threat Detection

Artificial Intelligence (AI) & Machine Learning (ML) tend to introduce sophisticated technologies to enhance the cybersecurity of nuclear power plants in the form of threat detection intelligence & real-time response systems. AI-driven cybersecurity systems continuously monitor operating data in real-time to identify unusual behaviour that would be considered cyberattacks or system failures (Conti et al., 2021). AI-driven technologies greatly accelerate the speed of threat detection along with accuracy without triggering false alarms, thereby enabling instant response to security attacks. ML algorithms present adaptive security mechanisms that possess the ability to detect changing cyber-attacks, such as zero-day attacks & APTs. Using huge amounts of data & history of cyber-attacks, ML-driven security systems are capable of foretelling impending vulnerabilities & initiate countermeasures in advance, resulting in constant improvement in a plant's security stance.

5.3 Real-Time Anomaly Detection Systems

Real-time anomaly detection systems are imperative cybersecurity technologies for nuclear reactors able to instantly pick up illicit transactions or system suspicious behavior very characteristic of cyber infiltration. Real-time anomaly detection systems utilize powerful analytics methods like statistical analysis, patterns recognition, & behavioral monitoring for identifying regular use patterns as the baseline to detect anomalies indicative of threats (Cho et al., 2016). Integration of anomaly detection capability within ICS & SCADA is imperative, enabling the instant recognition & handling of cybersecurity breaches with greatly minimized probable disruption or damages.

As part of a comprehensive defense-in-depth strategy for safeguarding nuclear power plant infrastructure, real-time anomaly detection adds an extra degree of security to traditional cybersecurity safeguards. Deploying such systems successfully requires not just identifying known risks but also identifying deviations from pre-established operating baselines that may be signs of undiscovered or recently discovered dangers. Anomaly detection systems facilitate early threat detection and neutralization by continuously analyzing operational data, enabling cybersecurity experts to intervene before an attack escalates into a serious crisis. Integration with network segmentation and automated incident response systems can also improve real-time anomaly detection. By quickly separating the impacted areas upon identifying irregularities, network segmentation reduces the extent of damage. Meanwhile, integration with automated security orchestration facilitates quick containment & mitigation to avoid causing minimal disruption to mission-critical nuclear plant operations.

Also, there must be continuous enhancement by way of repeated updates of detection algorithms & repeated system training exercises. Given the ever-changing character of cyber-attacks, anomaly detection systems must continue to improve by incorporating new threat intelligence & updating analytical models against emerging cyber threats. Regular training of staff guarantees that the cybersecurity teams are adequately equipped with the expertise to accurately interpret anomaly detection alerts & institute appropriate response measures in a timely manner, thereby enhancing the overall cybersecurity position of nuclear power plants.

5.4 Encryption & Quantum Cryptography

Encryption of key data guarantees protection of critical data of nuclear power plants, providing confidentiality, integrity, & authenticity of information during storage & transmission. Traditional cryptographic techniques, though efficient, are becoming vulnerable with the emergence of quantum computing, which can easily compromise widely used cryptographic algorithms. Quantum cryptography is a promising substitute, based on quantum mechanics principles, to secure communication channels from eavesdropping or tampering (Poresky et al., 2017). The application of quantum resistant crypto algorithms & quantum key distribution systems provides robust resistance against powerful adversaries, particularly those state-sponsored with tremendous computational resources. As nuclear facilities become more equipped with advanced digital communication & control systems, the application of quantum cryptography has to be implemented for the protection of sensitive operational data.

5.5 Incident Response & Recovery Plans

Incident response & disaster recovery procedures are critical cybersecurity policies of nuclear power plants. An incident response plan & strategy guarantees. official procedures for the detection, containment, & recovery from cybersecurity incidents, minimizing operational downtime & losses. The National Institute of Standards & Technology (NIST) provides extensive guidelines emphasizing the speed of detection, containment, eradication, & recovery processes, guaranteeing operational resilience to cyber threats. Disaster recovery measures also ensure business continuity of such important nuclear functions in case of large-scale cyber-attacks, through data backup planning, recovery mechanisms, and precisely established roles & responsibilities. They should be subject to testing, maintenance, & revision from time to time in order to ensure their efficacy & preparedness in response to perpetually evolving cyber-attacks (Conti et al., 2021).

5.6 Cybersecurity Frameworks & Regulatory Compliance

The set cybersecurity standards & regulations are necessary for nuclear power plants to ensure high security levels & compliance with cyber risk management. The regulatory standards of NRC regulations, IEC 62645 standards, & the IAEA Cybersecurity Recommendations offer detailed guidelines from risk assessment to cybersecurity governance, technical security controls, employee training, & incident response measures (International Atomic Energy Agency [IAEA], 2011). Regulatory compliance implies that the nuclear facilities adopt standardized security standards, review the plant's cybersecurity consistently, & continually enhance cybersecurity controls as standards evolve. Worldwide compliance provides consistent cybersecurity practices across the global nuclear industry, promotes common security practices, & limits exposures caused by inadequate or inconsistent security practices. Overall, the implementation of such sophisticated cybersecurity measures & programs significantly enhances the robustness & recovery capacity of nuclear power plants from advanced-level cyber-attacks. With effective & vigilant security protocols, continuous employee training, & stringent protocols to follow with respect to the global regulatory standards, nuclear power plants can excel at protecting critical infrastructure, ensuring operational safety & resilience in face of an expanding sophisticated cybersecurity landscape (Neeraj, 2023).

6. ENHANCING CYBER SECURITY IN NUCLEAR POWER PLANTS

6.1 Application of Blockchain for Secure Communication

Blockchain technology, which is mostly linked to cryptocurrencies, is emerging as a powerful means of enhancing cybersecurity as it is tamper-evident & decentralized. Blockchain can be used to safeguard communication & data exchange among nuclear power plant systems by developing irreversible & public records for any data sharing or transaction. In the context of nuclear secrecy & necessity of stringent data integrity, blockchain can reduce risks of tampering with data, unauthorized intrusion, & cyber espionage (Conti et al., 2021). Blockchain technology also provides solid bases for supply chain security in nuclear facilities, including traceability, authenticity verification, & safe supply chain management. With blockchain, nuclear facilities can create safe supply chains, thus eradicating third-party vendors & contractors' vulnerabilities. This aspect is particularly important given the complexity & layered nature of nuclear facility supply chains, as seen in attacks such as the Kudankulam nuclear facility attack (Dasgupta & Sen, 2019).

6.2 Quantum Computing & Post-Quantum Cryptography

Quantum computing is a computing power leap that holds the potential to cause revolutionary impacts on cybersecurity. Technology also holds unprecedented threats to traditional cryptographic mechanisms that secure nuclear power plants. Quantum computers have the ability to break existing cryptographic codes easily, in process, compromising sensitive data, communications, & control systems (International Atomic Energy Agency [IAEA], 2011). As a mitigation measure, post-quantum cryptography is gaining more prominence, creating cryptographic algorithms that are quantum computer attack-proof. The implementation of such cryptographic measures would guarantee nuclear power plant security from future computational attacks.

The potential consequences of inadequate or late response to cyber-attacks can range from reactor shutdowns, operational interruption, to physical harm or damage. SOAR automation would facilitate nuclear plants to upgrade their operational resilience & preparedness against cyber-attacks, along with successful & quick recovery from cyber incidents.

6.3 Integration of Artificial Intelligence with Security Information & Event Management (SIEM)

AI incorporation in Security Information & Event Management (SIEM) systems provides promising improvements to nuclear cybersecurity in the form of advanced threat detection & response. SIEM systems have conventionally collected & analyzed security information from different sources, offering real-time awareness & visibility into potential threats to security teams. AI must be integrated with SIEM so that it can use advanced analytics, machine learning, & deep learning methods. The trends & technologies highlighted have vast potential to revolutionize cybersecurity strategy through offering powerful defense technologies against changing & advanced cyber-attacks. Strategic investment, international cooperation, harmonization of regulations, & training of dedicated human resources are essential to effective realization of these technologies & the complete integration of cybersecurity into nuclear operations (Cho et al., 2016).

7. CONCLUSION

Cybersecurity has emerged as a critical component in improving the integrity, safety, & reliability of nuclear power plants. The evolving cybersecurity landscape, with constantly enhancing threats, necessitates immediate & strategic action to protect nuclear facilities. Through critical analysis, this paper emphasizes several critical vulnerabilities & challenges, exemplified by recent high-profile cyber-attacks such as the Stuxnet attack, Ukraine power grid disruption, WannaCry ransomware, Triton malware, & Kudankulam Nuclear Power Plant's cyber-attack. Legacy systems are one of the most critical cybersecurity risks to nuclear power plants, merely because they contain out-of-date technologies & insecure control systems, making them extremely susceptible to advanced cyber-attacks. Countering such vulnerabilities through coordinated upgrade & modernization is critical towards establishing resilience & continuity of operations. No less critical is countering supply chain vulnerabilities, in that nuclear facilities depend heavily on internationally sourced components & services. The sophistication of this necessitates robust & auditable supply chain management controls to prevent the risk of exposing critical infrastructure through third-party vulnerabilities.

In addition, the lack of trained cybersecurity staff significantly aggravates threats, & this is the impetus for intense workforce development & training initiatives specifically tailored to the special requirements of nuclear industry for cybersecurity. The existence of zero-day attacks & increasing threat of social engineering & insider attacks also complicates the cybersecurity landscape further, necessitating strong proactive security controls & continuous training & awareness programs. To effectively counter these kinds of threats, nuclear power plants must adopt the most robust cybersecurity practices, including robust network segmentation, air-gapping of core systems, & adopting new technologies like blockchain, artificial intelligence, & quantum cryptography. Blockchain provides an open, secure platform for communications essential to robust data integrity & supply chain security. Simultaneously, quantum computing advancements necessitate the immediate roll-out of post-quantum cryptography to make nuclear plant cybersecurity unattackable in future. Automated security orchestration & enhanced Security Information & Event Management (SIEM) systems which are powered by artificial intelligence have powerful capabilities for real-time threat detection, rapid incident response, & effective threat management. These technologies allow nuclear power facilities to actively manage cybersecurity threats, especially reducing vulnerabilities & improving operational resilience.

Ultimately, addressing cybersecurity in nuclear power plants is not merely a technical necessity but a strategic imperative requiring coordinated international cooperation, regulatory standardization, & continuous adaptation to evolving threats.

REFERENCES

Assante, M. J., & Lee, R. M. (2015). *The industrial control system cybersecurity workforce gap*. SANS Institute. https://www.sans.org/white-papers/industrial-control-system-cybersecurity-workforce-gap/

Baylon, C., Brunt, R., & Livingstone, D. (2015). *Cyber security at civil nuclear facilities: Understanding the risks*. Chatham House. https://www.chathamhouse.org/2015/10/cyber-security-civil-nuclear-facilities

Bilge, L., & Dumitras, T. (2012). Before we knew it: An empirical study of zero-day attacks in real world. *Proceedings of 2012 ACM Conference on Computer & Communications Security*, 833-844. https://doi.org/DOI: 10.1145/2382196.2382284

Boyes, H. (2015). Cybersecurity & cyber-resilient supply chains. *Technology Innovation Management Review*, *5*(4), 28–34. https://timreview.ca/article/888. DOI: 10.22215/timreview/888

Case, D. U. (2016). *Analysis of cyber attack on the Ukrainian power grid*. E-ISAC. https://www.nerc.com/pa/CI/ESISAC/Documents/E-ISAC_SANS_Ukraine_DUC_18Mar2016.pdf

Cho, C. S., Chung, W. H., & Kuo, S. Y. (2016). Cyber-physical security & dependability analysis of digital control systems in nuclear power plants. *IEEE Transactions on Systems, Man, and Cybernetics. Systems*, *46*(3), 356–369. DOI: 10.1109/TSMC.2015.2452897

Clayton, M. (2010). Stuxnet malware is 'weapon' out to destroy Iran's Bushehr nuclear plant? *The Christian Science Monitor*. https://www.csmonitor.com/USA/2010/0921/Stuxnet-malware-is-weapon-out-to-destroy-Iran-s-Bushehr-nuclear-plant

Conti, M., Donadel, D., & Turrin, F. (2021). A survey on industrial control system testbeds & datasets for security research. *arXiv preprint arXiv:2102.05631*. https://arxiv.org/abs/2102.05631

Dasgupta, B., & Sen, S. R. (2019). Cyber attack at Kudankulam; critical system safe. *Hindustan Times*. https://www.hindustantimes.com/india-news/cyber-attack-at-kudankulam-critical-system-safe/story-8j5Vb6Y5L5z5N5z5N5z5N.html

Farwell, J. P., & Rohozinski, R. (2011). Stuxnet & the future of cyber war. *Survival*, *53*(1), 23–40. DOI: 10.1080/00396338.2011.555586

Greiman, V. A. (2023). Nuclear cyber attacks: A study of sabotage & regulation of nuclear facilities. *Proceedings of 18th International Conference on Cyber Warfare & Security*, 1042-1051. https://papers.academic-conferences.org/index.php/iccws/article/download/1042/932

Hemsley, K. E., & Fisher, R. M. (2018). *History of industrial control system cyber incidents*. U.S. Department of Energy, Idaho National Laboratory., DOI: 10.2172/1505628

International Atomic Energy Agency. (2011). *Computer security at nuclear facilities*. IAEA Nuclear Security Series No. 17. https://www.iaea.org/publications/8939/computer-security-at-nuclear-facilities

Kushner, D. (2013). The real story of Stuxnet. *IEEE Spectrum*. https://spectrum.ieee.org/the-real-story-of-stuxnet

Langner, R. (2011). Stuxnet: The emergence of a new cyber weapon & its implications. *Journal of Computer Virology & Hacking Techniques*, *7*(4), 291–298. https://www.researchgate.net/publication/241736813_Stuxnet_The_emergence_of_a_new_cyber_weapon_&_its_implications

Lee, R. M., Assante, M. J., & Conway, T. (2016). *Analysis of cyber-attack on the Ukrainian power grid*. SANS Industrial Control Systems. https://ics.sans.org/media/E-ISAC_SANS_Ukraine_DUC_5.pdf

Masood, R. (2016). *Assessment of cyber security challenges in nuclear power plants*. George Washington University. https://cspri.seas.gwu.edu/sites/g/files/zaxdzs4366/files/downloads/GW-CSPRI-2016-03%2BMASOOD%2BRahat%2BNuclear%2BPower%2BPlant%2BCybersecurity_0.pdf

Mohan, P. (2023). Cybersecurity & nuclear facilities. In D. Carlton (Ed.), *Cybersecurity in critical infrastructures*(pp. 123-145). Springer. https://link.springer.com/chapter/10.1007/978-3-031-56814-5_7link.springer.com

Mohurle, S., & Patil, M. (2017). A brief study of WannaCry threat: Ransomware attack 2017. *International Journal of Advanced Research in Computer Science*, *8*(5), 1938–1940. https://www.proquest.com/docview/1912631307/abstract/DEF9AE2FF2924E35PQ/

Mundo, A. (2020, March 26). Triton malware spearheads latest attacks on industrial systems. *Security Intelligence*. https://www.trellix.com/en-in/blogs/research/triton-malware-spearheads-latest-generation-of-attacks-on-industrial-systems1/

Neeraj, B. R. (2023). Cybersecurity in Indian nuclear facilities. *Electronic Journal of Social & Strategic Studies*, 4(3), 314-330. https://doi.org/DOI: 10.47362/EJSSS.2023.4302ejsss.net.in

Poornima, B. (2022). Cyber threats & nuclear security in India. *Asian Journal of International Affairs*, *2*(1), 34–50. DOI: 10.1177/23477970221099748journals.sagepub.com

Poresky, C., Andreades, C., Kendrick, J., & Peterson, P. F. (2017). *Cyber security in nuclear power plants: Insights for advanced nuclear technologies*. University of California., https://www.researchgate.net/publication/321443750_Cyber_Security_in_Nuclear_Power_Plants_Insights_for_Advanced_Nuclear_Technologies

Singer, P. W. (2015). Stuxnet & its hidden lessons on the ethics of cyberweapons. *Case Western Reserve Journal of International Law*, *47*(1), 79–96. https://scholarlycommons.law.case.edu/cgi/viewcontent.cgi?article=1009&context=jil

Stouffer, K., Falco, J., & Scarfone, K. (2011). Guide to industrial control systems (ICS) security. NIST special publication, 800(82), 16-16.

Tantawy, A., Erradi, A., Abdelwahed, S., & Shaban, K. (2020). Model-based risk assessment for cyber-physical systems security. *arXiv preprint arXiv:2005.13738*. https://arxiv.org/abs/2005.13738

Whitehead, D. E., Owens, K., Gammel, D., & Smith, J. (2016). *Ukraine cyber-induced power outage: Analysis & practical mitigation strategies*. Proceedings of 43rd Annual Western Protective Relay Conference. https://selinc.com/api/download/117044/

ADDITIONAL READING

Bilge, L., & Dumitras, T. (2012). An empirical study of zero-day attacks in real world. *Proceedings of 2012 ACM Conference on Computer & Communications Security*, 833–844. https://doi.org/DOI: 10.1145/2382196.2382284

Cohen, C. (2019). *Cybersecurity standards & the 2015 Ukraine power grid attack*. ANSI Committee on education https://share.ansi.org/Shared%20Documents/Education%20and%20Training/Committee%20on%20Education/2019_Paper_Competition/1st_place_Cohen_FINAL.pdf

International Atomic Energy Agency. (2011). Computer security at nuclear facilities (Nuclear Security Series No. 17). IAEA. https://www.iaea.org/publications/8939/computer-security-at-nuclear-facilities

Langner, R. (2011). Stuxnet: Dissecting a cyberwarfare weapon. *IEEE Security and Privacy*, *9*(3), 49–51. DOI: 10.1109/MSP.2011.67

KEY TERMS AND DEFINITIONS

Insider Threat: A security risk originating from employees or trusted individuals within an organization who may intentionally or unintentionally cause harm.

Phishing: A cyberattack method that uses deceptive emails or communications to trick individuals into revealing sensitive information.

Programmable Logic Controllers (PLCs): Specialized hardware used to automate mechanical processes, commonly targeted in industrial cyberattacks.

Real-Time Anomaly Detection Systems: Security systems that monitor & detect deviations from normal operational behavior to identify potential cyber intrusions.

SCADA Systems: Supervisory Control & Data Acquisition systems is used to monitor & control infrastructure processes remotely.

Security Information & Event Management (SIEM): Software platforms that aggregate, analyze, & monitor security-related data from across an organization's network.

Stuxnet: A sophisticated cyberweapon discovered in 2010, designed to sabotage Iran's nuclear facilities by targeting their industrial control systems.

Supply Chain Vulnerability: The exposure of an organization to cyber risks originating from third-party vendors.

WannaCry Ransomware: A global ransomware attack in 2017 that exploited vulnerabilities in Windows systems, encrypting files & demanding ransom payments.

Zero-Day Vulnerability: A software flaw unknown to the software maker & public, leaving systems exposed to attacks until patched.

Chapter 9
Technological Advancement of Space Explorations With Nuclear Energy

Shaurya Talewar
https://orcid.org/0009-0006-7282-1547
Ajeenkya D.Y. Patil University, India

Aditya Shrivastav
https://orcid.org/0009-0000-8461-8171
Ajeenkya D.Y. Patil University, India

Kashvi Chaturvedi
https://orcid.org/0009-0005-3192-3400
Ajeenkya D.Y. Patil University, India

Atharva Saraf
https://orcid.org/0009-0006-3842-5508
Ajeenkya D.Y. Patil University, India

Sunil Sankathala
https://orcid.org/0009-0004-7809-7647
Ajeenkya D.Y. Patil University, India

Susanta Das
https://orcid.org/0000-0002-9314-3988
Ajeenkya D.Y. Patil University, India

Yadnyesh Khapekar
Ajeenkya D.Y. Patil University, India

Shrikaant Kulkarni
Sanjivani University, India

ABSTRACT

The quest for deep space exploration requires increasingly efficient, powerful, & enduring energy solutions. This chapter discusses some of the most significant developments in space nuclear technologies, starting with the application of RTGs in Apollo & Voyager missions. It then goes on to the modern systems such as SMRs for lunar and Martian settlements, and NTP systems for high-speed interplanetary travel. Advances in fuel types, miniaturization, shielding, & thermal control have greatly improved system viability in space. Major institutional roles, NASA's Kilopower, DARPA's DRACO, & international collaborations, are discussed. The

DOI: 10.4018/979-8-3373-0685-8.ch009

chapter is also critically evaluating ethical, environmental, & geopolitical issues such as launch hazards, planetary protection, and space militarization. Looking ahead, it discusses next-generation ideas such as DFD, nuclear-powered in-situ resource utilization, & self-sustaining off-Earth colony designs. From a multi-disciplinary perspective, this chapter demonstrates how nuclear systems are transforming the future of space exploration.

1. INTRODUCTION

Developing sustainable and dependable energy sources has always been essential to humanity's space exploration. Traditional solutions like chemical rockets and solar power oppose important obstacles as we move deeper into the solar system, particularly for missions that must travel quickly between planets or are located far from the Sun. With its unparalleled efficiency, long-lasting energy, and dependability required to enable deep-space travel, nuclear power is emerging as a game-changing alternative in these situations (Gibson & Poston, 2020; Levack & Borowski, 2024). Due to their tremendous energy production and ability to function without the need for sunlight or other environmental factors, nuclear technologies are perfect for space missions. Nuclear thermal propulsion (NTP) systems perform far better than traditional chemical rockets, which have a restricted power output in relation to their weight. This implies that nuclear technology may be used to conduct expeditions to planets like Mars considerably more quickly (Levack & Borowski, 2024). For future bases on Mars and the Moon, nuclear fission power reactors, like NASA's Kilopower project, can offer a reliable and expandable power source. In these settings, this energy would sustain scientific study, industrial operations, and even human survival (Gibson & Poston, 2020). Furthermore, by providing power to spacecraft like as Voyager and Curiosity for considerably longer than anticipated mission durations, RTGs have demonstrated their enduring dependability (Duggirala, 2022).

Starting with early achievements like the employment of RTGs in the Apollo and Voyager missions, this paper will examine in detail how nuclear technology has advanced for space travel. After that, it will highlight the key takeaways from these early initiatives and chart the development through initiatives like the NERVA program (Duggirala, 2022; McNutt, 2023). The conversation will then go to more recent developments, such as the Kilopower reactor, next-generation RTGs, space-mission-specific SMRs, and the rebirth of interest in nuclear thermal propulsion as a result of initiatives like DARPA's DRACO program (DARPA, 2023; NASA, 2023). The article will then go into great detail about the technologies that enable space nuclear systems, including the development of special fuels like high-assay low-enriched uranium (HALEU), reactor size reductions, enhanced heat management techniques, and novel radiation protection strategies (OECD-NEA, 2024; Carpenter & Shaltens, 2024). The importance of off-world nuclear systems functioning autonomously and having safety mechanisms that can manage malfunctions without endangering the mission will also be thoroughly examined in this study (Cowley & Yamamoto, 2024).

The global initiatives to advance nuclear-powered space missions by prominent agencies such as NASA, DARPA, the European Space Agency (ESA), and private firms will be highlighted in this section (Worrall & Locatelli, 2023). In order to support their long-term objectives of establishing communities on the Moon and Mars, companies like SpaceX and Blue Origin are becoming more and more interested in nuclear propulsion and surface electricity. However, there are significant geopolitical, ethical, and environmental issues with employing nuclear technology in space. There is a great deal of risk involved in launching radioactive materials, particularly if the mission fails. Careful consideration must be given to matters such as safeguarding planetary habitats, adhering to international regulations like the Outer Space Treaty (UNOOSA, 1967), and guaranteeing the appropriate use of nuclear systems (Fairén & Davila, 2023). Concerns over the militarization of space and the possibility for both military and civilian uses of nuclear technology are also becoming more prevalent. These issues necessitate a swift international discussion to advance ethical, secure, and peaceful progress in this field (Fairén & Davila, 2023).

2. HISTORICAL EVOLUTION OF NUCLEAR TECHNOLOGY IN SPACE EXPLORATION

The use of nuclear energy to power space missions has been investigated since the middle of the 20th century. Strong and durable energy sources were required throughout the space race and the early, ambitious scientific endeavors. The RTGs were one of the first effective solutions and were crucial to deep-space exploration. Programs such as the Nuclear Engine for Rocket Vehicle Application (NERVA) also offered significant technology and information that influenced future research and development, despite their failure.

2.1 Early Nuclear Applications: RTGs

RTGs, which directly transform heat from radioactive decay into electricity, were the first nuclear-powered energy technology. Utilizing thermoelectric couplings, RTGs create power by utilizing the temperature differential between a radioactive substance, often plutonium-238, and the cold of space (Duggirala, 2022). RTGs are perfect for missions that travel farther than solar panels can reach, when sunlight is too weak, because they are incredibly dependable and long-lasting.

During NASA's Apollo program, notably in the Apollo 12, Apollo 14, Apollo 15, Apollo 16, and Apollo 17 missions, RTGs saw some of their first significant applications. Apollo Lunar Surface Experiments Package (ALSEP) included scientific devices powered by RTGs that were placed on the Moon's surface. This equipment collected important information on lunar geology, surface temperatures, and moonquakes (Gibson & Poston, 2020). These experiments were able to transmit data back to Earth for years after the missions were over because of the consistent and dependable power provided by RTGs. The 1977 launch of the Voyager spacecraft was another significant application for RTGs. To function in the frigid, dark conditions of deep space, Voyager 1 and Voyager 2 both carried RTGs that could generate hundreds of watts of power. Remarkably, both Voyagers are still providing valuable scientific data from outside the Sun's sphere of influence over 40 years after launch, demonstrating the reliability and durability of nuclear power for space missions (Duggirala, 2022).

2.2 The NERVA Program

Along with using RTGs, the United States launched Nuclear Engine for Rocket Vehicle Application (NERVA) program, which was run by Atomic Energy Commission and NASA, in the 1960s and early 1970s in an attempt to build nuclear-powered rocket engines. In contrast to conventional chemical rockets, NERVA sought to employ nuclear reactors to heat hydrogen fuel, which when released, created thrust that was significantly more efficient (McNutt, 2023). In ground tests, the nuclear engines demonstrated remarkable thrust, improved fuel efficiency, and ability to significantly reduce interplanetary transit durations. Significantly, these tests supported the viability of the design by demonstrating that nuclear reactors could be successfully operated in extremely hot and radiation-rich environments (Levack & Borowski, 2024).

But in spite of these achievements, NERVA program encountered significant difficulties. Before any space tests were conducted, it was canceled in 1973 due to concerns about nuclear safety, environmental threats, political challenges, and economic pressures. Tight funds, changing government priorities, and growing public mistrust and fear of nuclear technology amid broader concerns about nuclear power were major factors in the program's demise (Howe, 2023).

2.3 Setbacks, Achievements, & Technological Groundwork Laid

NERVA's technical accomplishments established a solid basis which continues to impact research today, despite the program's discontinuation being a significant setback. Current nuclear thermal propulsion programs, such as NASA and DARPA's DRACO project, continue to build upon the critical insights that NERVA offered on reactor design, heat management, fuel handling, and radiation protection (NASA, 2023; DARPA, 2023). The Apollo and Voyager missions' utilization of RTGs and NERVA both provided insightful operational lessons. These early projects demonstrated the value of rigorous safety protocols to reduce nuclear risks during launch, deployment, and operation; efficient thermal management to maintain reactor stability; and robust radiation shielding to protect astronauts and sensitive equipment from dangerous radiation.

The success of RTGs also led to a greater dependence on nuclear power for deep-space missions like New Horizons, Cassini, and Galileo. These missions helped humanity learn more about the solar system & beyond by showcasing the robustness and dependability of nuclear technology. For instance, RTGs were crucial to Cassini's 13-year study of Saturn and its moons, which enabled the mission to collect previously unheard-of data that revolutionized our knowledge of planetary science (Carpenter & Shaltens, 2024).

The history of nuclear technology in space travel makes it abundantly evident how important it is to be enabling long-term scientific research and ambitious missions. Even though NERVA was later abandoned, it laid the technological groundwork for current nuclear propulsion research, and RTGs set the standard for power generation in deep space. Today's accomplishments and upcoming developments in space nuclear technology are supported by all the successes and failures of these early attempts, which significantly influenced the course of scientific advancement.

3. ADVANCEMENTS IN NUCLEAR TECHNOLOGIES FOR SPACE EXPLORATION

3.1 RTGs

To power space missions, particularly those going to deep space where solar panels are ineffective, RTGs have proven essential. Since their early days, RTGs have advanced a lot, leading to the development of next-generation Multi-Mission Radioisotope Thermoelectric Generators (MMRTGs). NASA now utilizes these MMRTGs on Mars rovers including curiosity and perseverance. Better thermoelectric materials are used in MMRTGs, increasing their power production and efficiency.

MMRTGs have a greater conversion rate of about 8–9% than standard RTGs, which enables them to operate more dependably and stay longer in challenging space conditions. standard RTGs convert roughly 6-7% of heat into power.

One important component of this advancement is a substance known as skutterudites, which have a unique crystal structure that improves their performance by minimizing heat and facilitating the easy passage of electricity. MMRTGs can function consistently even in the harsh temperature fluctuations present in deep space missions thanks to ongoing research into improved thermal management systems, sophisticated thermoelectric materials, and enhanced methods of shielding the heat source.

3.2 Kilopower Reactor & SMRs

One of the latest significant recent developments in space nuclear technology is the Kilopower reactor, which will be used to power future facilities on the Moon and Mars. This tiny nuclear fission reactor, created as part of NASA's Kilopower project, produces dependable and scalable power in the range of 1 to 10 kilowatts. The reactor can function as intended, according to NASA's successful 2018 KRUSTY test (Kilopower Reactor Using Stirling Technology), which is a significant step towards sustainable off-Earth living. Kilopower reactors use highly enriched uranium (HEU) fuel and sophisticated heat-pipe technology to efficiently control heat and transform it into electricity. They are built to be robust, simple, and efficient.

The tiny size and modular construction of SMRs make them excellent options for space utilization. The lower size, modular design, & passive safety measures of SMRs enable them to function securely with minimal human involvement, which is crucial for off-world missions. Their primary design is modularity for effortless expansion and upgrades, simplicity to minimize failure spots, and safety through passive cooling. High-Assay Low-Enriched Uranium, which is far safer than highly enriched fuel and reduces the chance of accidents, is frequently used in these reactors. Mission planners have the flexibility they need to establish colonies on distant planets by adding modules or scaling up power as needed.

Along with providing safety and dependability, SMRs' small size allows them to reduce launch mass and volume, which lowers costs and facilitates deployment. Because they are modular, components can be launched individually and put together on Mars or the Moon. The slow growth of off-world colonies is well suited to this phased, modular strategy, which offers reliable and flexible power solutions for various exploration and settlement phases.

3.3 Nuclear Thermal Propulsion System

An innovative new method of distant space travel is nuclear thermal propulsion (NTP). It heats a propellant, often liquid hydrogen, to extremely high temperatures using a nuclear reactor. In order to generate thrust, this heated gas then emerges from rocket nozzles. NTP is far more potent and efficient than conventional chemical rockets, allowing spaceships to fly farther and consume less fuel. By lowering the amount of time astronauts spend in hazardous space environments like cosmic radiation & microgravity, NTP, for instance, might reduce a crewed voyage to Mars in half.

The Demonstration Rocket for Agile Cislunar Operations (DRACO), a joint DARPA and NASA project, is one significant initiative advancing this technology. DRACO seeks to demonstrate the viability of NTP systems in practical space missions, particularly for rapid orbits of the Moon &, eventually, crewed missions to Mars. Through ground experiments and space demonstrations, DRACO will test reactor designs, heat management techniques, fuel performance, and overall propulsion efficiency over the course of the next 10 years.

NTP's efficiency is its main advantage over chemical rockets. The energy in the fuel and the speed at which the exhaust may move are the only constraints on chemical rockets. NTP's propulsion efficiency is often twice as that of today's top chemical rockets since it uses nuclear processes, which generate far greater exhaust velocity. As a result, spacecraft can carry more payload and less fuel, which reduces launch costs and gives mission planners additional options.

NTP systems are enhanced by new reactor fuel innovations, such as robust ceramic-metal fuels, or cermets. Because these fuels can withstand higher temperatures, they increase engine efficiency without causing damage. NTP is emerging as a promising technique for quick and efficient deep-space trips thanks to advancements in automation, radiation protection, and reactor safety.

4. TECHNOLOGICAL BREAKTHROUGHS & CHALLENGES IN NUCLEAR TECHNOLOGIES FOR SPACE EXPLORATION

4.1 Advances in High-Assay Low-Enriched Uranium (HALEU) Fuels

The creation of HALEU fuel represents a significant advancement in nuclear energy for space travel. Between the typical low-enriched uranium (below 5%) and highly enriched uranium (above 20%), HALEU has a uranium-235 enrichment level of 5% to 20%. Because it provides an excellent balance between high energy output and explosive safety, this kind of fuel is particularly well-suited for powering space voyages and settlements on other planets.

HALEU provides various advantages over conventional uranium fuels. Reactors can generate more power while being lighter and smaller since they last longer and operate more efficiently. This is important for space since it is easier & less expensive to launch and set up smaller, lighter reactors on the Moon or Mars. Additionally, HALEU's improved neutron economy reduces mission costs and logistical difficulties by allowing reactors to operate for longer period of time without requiring refueling.

Additionally, recent studies have enhanced the materials used to protect HALEU fuel and its manufacturing process. Advanced protective coatings like silicon carbide and new fuel types including metallic and ceramic-metallic composites aid in better heat management and radiation resistance. These enhancements strengthen and increase the fuel's dependability, which is crucial for long-term, safe operation during space missions.

4.2 Lightweight Reactor Shielding & Autonomous Controls

A significant development in space nuclear technology is lightweight reactor shielding systems, which are planned to safeguard delicate electronics and humans from dangerous radiation generated by reactors. Traditional shielding materials, like concrete or lead, are too heavy for space missions, where weight is crucial. Engineers are increasingly employing multi-layered designs and sophisticated composite materials to address this. These more recent materials, including boron carbide-infused polymers or hydrogen-rich polyethylene, are excellent at absorbing radiation while maintaining a low total weight.

Innovative materials like lithium hydride and improved polymer composites, which provide superior radiation shielding without compromising strength or needless mass, are also being investigated by researchers. These thin shields are essential for safeguarding electronics on unmanned spacecraft as well as for human missions.

The creation of autonomous reactor control systems is another significant innovation in addition to shielding. Artificial intelligence (AI), machine learning (ML), and high-performance sensors are used by these systems to track the reactor's condition, spot issues, and automatically address them without requiring human intervention. Space missions operating far from Earth, where direct human control is not feasible, benefit greatly from this type of automation. It contributes to the reactor's safe and dependable operation under a variety of mission circumstances.

4.3 Challenges: Miniaturization, Heat Dissipation, & Radiation Risks

Even with significant scientific advancements, there are still significant obstacles to overcome before employing nuclear technology in the space travel, particularly in the areas of radiation protection, heat management, and reactor shrinkage. One of the largest challenges is shrinking reactors, which calls for striking a balance between safety, efficiency, and power output in a much smaller size. These small reactors frequently have problems with fuel consumption, neutron economy, and sustaining structural integrity under severe mechanical & thermal stress.

Engineers need sophisticated core designs and cutting-edge materials which can endure high temperatures and extreme neutron exposure in order to make these tiny reactors function. For this reason, current research is concentrating on investigating improved fuel materials, new core geometries, and more intelligent heat management techniques. To maintain the reactors stable and intact even at smaller scales, this involves use of high-conductivity composites, micro-channel cooling, and sophisticated heat pipe systems.

Radiation is yet another persistent problem. It jeopardizes not only the safety of astronauts but also the correct operation of spacecraft equipment. Advanced shielding that is both light and effective is required for missions to protect people and instruments from solar radiation, cosmic rays, and reactor emissions. To make sure they offer adequate protection for extended missions, these more recent materials need to undergo extensive testing. Furthermore, utilizing radiation-hardened electronics and redundant systems lowers the possibility of radiation-related malfunctions, increasing mission safety and dependability overall.

A cooperative, multidisciplinary strategy combining specialists in nuclear engineering, materials science, robotics, and advanced computing is needed to address these issues. To safely and effectively increase use of nuclear technology in upcoming space travel, more research and funding are needed in all these areas.

5. ENVIRONMENTAL & LEGAL CONSIDERATIONS IN NUCLEAR TECHNOLOGIES TO USE IN SPACE EXPLORATION

5.1 Risks of Nuclear Material During Launches

Due to the possibility of accidents involving radioactive materials, launching nuclear-powered spacecraft carries significant safety and environmental concerns. The strictness of these threats is demonstrated by previous occurrences. For instance, the Soviet satellite Cosmos 954, which was fueled by a nuclear reactor, crashed in 1978, scattering radioactive remains across portions of Canada. Such incidents demonstrate the potential for global and environmental effects from nuclear mishaps during space missions. To lower the risk of accidents, handling and launching nuclear materials must adhere to strict safety procedures, make use of cutting-edge containment technologies, and be prepared.

The purpose of modern containment systems is to lower the dangers associated with nuclear material launches. These systems are designed to survive catastrophic launch failures, like rocket explosions or complete mission failure while in the air. To prevent radioactive leaks in the occurrence of an accident, modern launch vehicles are equipped with improved shielding materials, several safety redundancies, and reinforced structural containers (IAEA, 2009). The public is nevertheless quite concerned about the potential for radioactive contamination, despite the reality of these technology. To solve this, mission planners must communicate with the public in a clear and concise manner and conduct transparent risk assessments prior to approving launches.

5.2 Planetary Protection & Contamination Concerns

Another major ethical and environmental issue with the use of nuclear technology in space travel is planetary protection. Nuclear materials carry a significant risk of contaminating other planets or moons, whether they are working in routine operations or are unintentionally released. Long-term ecological and scientific effects could result from such contamination, particularly if it impairs our capacity to identify evidence of extraterrestrial life. Strict planetary protection measures are therefore necessary to prevent irreversible harm and to maintain these celestial bodies unspoiled for upcoming research (Fairén & Davila, 2023).

Strict regulations based on the Committee on Space Research's (COSPAR) guidelines are followed by NASA and other international space agencies. These regulations include strict purification standards, spacecraft containment measures, and prohibitions on the use of nuclear technology in close proximity to planets and moons that could support life, like Mars, Europa, and Enceladus. Notwithstanding these protections, the strength of the current regulations is still up for question, particularly with regard to nuclear systems. Experts agree that to achieve global consensus on the best ways to safeguard other worlds, we must continue to study and enhance these standards (COSPAR, 2021).

5.3 Outer Space Treaty & Moon Agreement

The employment of nuclear technology in space missions is heavily influenced by international rules and treaties. The 1967 Outer Space Treaty, which serves as the cornerstone of space law, is the most significant of these. It expressly forbids putting nuclear weapons or any other WMD in orbit around the Earth, on the moon, or in any other part of space (UNOOSA, 1967). The pact highlights the need of international cooperation, openness, and accountability even while it permits peaceful nuclear operations. It is expected that nations that use nuclear power in space will communicate any possible concerns to others and seek their advice.

The Moon Agreement is another important treaty that expands on these concepts. It emphasizes the need of preserving the lunar environment and promotes transparency while utilizing nuclear power on the moon, even though few nations have signed it. However, the Moon Agreement has little real influence because, so few countries adhere to it. This emphasizes the need of modern diplomatic initiatives to develop more robust and broadly recognized international regulations for nuclear operations in space (UNOOSA, 1979).

5.4 Managing Geopolitical Tensions & Dual-Use Risks

Nuclear technology possesses "dual-use" potential, which means that it can be employed for strategic or military objectives along with peaceful uses like space exploration. This raises significant geopolitical issues. International tensions and concerns on nuclear explosion are heightened by the possibility that space-based nuclear technology could be hijacked for military purposes. Strong supervision mechanisms, open international communication, and cooperative institutions that strike balance between the advantages of nuclear space missions and international security concerns are essential for managing these dangers.

Collaboration and openness between parties are two of the strongest strategies to lower dual-use hazards. Collaborative initiatives, like the NASA-DARPA DRACO program, demonstrate how nations can cooperate on peaceful nuclear propulsion applications, fostering mutual objectives and trust (NASA, 2023; DARPA, 2023). Further initiatives to foster confidence, such as providing technical information, permitting international inspections, and disclosing nuclear payloads to the public, can allay fears around the world and lessen violence.

Establishing robust regulatory controls and governance frameworks is also crucial. To monitor nuclear activities, promote transparency, and guarantee treaty compliance, international institutions such as the United Nations Committee on the Peaceful Uses of Outer Space (COPUOS) and the International Atomic Energy Agency (IAEA) are essential. To manage the hazards associated with the use of nuclear technology in space, it is imperative that these institutions continue to be strengthened and supported (IAEA, 2018; UNOOSA, 2022).

6. CASE STUDIES

6.1 NASA Kilopower Mission Trials

In order to meet the increasing energy demands of future deep-space exploration and to support long-term human settlements beyond Earth, NASA developed the Kilopower Project, which represents a significant advancement in nuclear power technology specifically designed for missionsn on space. This project demonstrates how a compact nuclear reactor can deliver safe, dependable, and efficient power in harsh space environments. This summary examines the objectives of the Kilopower mission, the reactor's design, lab test results, and its potential for use in space exploration in the future.

The goal of the Kilopower project was to design and test a nuclear fission reactor that could provide one to ten kilowatts of continuous electricity, which would give space missions more flexibility by powering both robotic and human activities even in locations with little or no sunlight. The project was started in order to develop a small, reliable, and long-lasting energy source for long-duration missions, where solar power or chemical batteries wouldn't work (Gibson & Poston, 2020).

Performance, safety, and simplicity are prioritized in the reactor's design. It makes use of a highly enriched uranium (HEU) core that is shielded from radiation and a neutron reflector. Instead of using moving parts, the system transfers heat from core to Stirling engines, which produce electricity, using heat pipes (Poston et al., 2019). This passive design lowers maintenance requirements and failure risk, which is critical for use on the Moon or Mars. Additionally, the system has built-in safety safeguards that, in event that the reactor overheats, will automatically shut it down, removing the need for complicated controls or human operators. It is ideal for space deployment because of its small size, which makes it simpler to transport and build (Mason & Poston, 2021).

The Kilopower Reactor Using Stirling Technology (KRUSTY) test was finished at the Nevada National Security Site in 2018 by NASA in collaboration with the National Nuclear Security Administration of the U.S. Department of Energy. The United States, for the first time in almost 50 years, successfully tested a new type of nuclear reactor. The reactor produced roughly one kilowatt of steady electricity in conditions that mimicked space environments during the KRUSTY test, which lasted 28 hours at full power (McClure et al., 2020). The test verified the proper operation of important systems, including the automatic control, radiation shielding,

Stirling engines, and heat pipes. Additionally, it demonstrated that even when the electrical load fluctuated, the reactor could modify its power level and preserve thermal stability (McClure et al., 2020).

The Kilopower test did reveal certain difficulties in spite of its success. Still there is concern about long-term durability in the harsh environments of space, such as severe radiation, vacuum, and extremely high temperatures. Additional effort would also be needed to scale up the system to generate additional power (up to 10 kilowatts), particularly in the areas of system integration, heat control, and materials testing (Mason & Poston, 2021). Nevertheless, the successful test significantly raised nuclear power confidence for upcoming expeditions to the Moon and Mars. To power housing, science labs, and in-situ resource utilization (ISRU) devices on the surface of Moon, NASA has since suggested modifying the Kilopower design (Poston et al., 2019). Research is still being done to build more sophisticated autonomous controls for future applications, better heat management, and optimize reactor components.

Other space-based nuclear technologies, such as NTP and SMRs, are also being laid out by the Kilopower reactor. Kilopower-type devices might serve as the foundation of power infrastructure as NASA & its international partners get ready for upcoming crewed flights into deep space, enabling longer trips and more capabilities (Mason & Poston, 2021).

6.2 DARPA's DRACO & Space Nuclear Propulsion Programs

The Defense Advanced Research Projects Agency (DARPA) is leading the groundbreaking Demonstration Rocket for Agile Cislunar Operations (DRACO) project, which aims to create enhanced nuclear thermal propulsion (NTP) technologies that could greatly enhance space travel.

Case Study: This case study provides a summary of DRACO's objectives, advancements, technology, and potential implications for exploration of space in the future. In order to develop feasible nuclear thermal propulsion technologies that could improve the speed and efficiency of space travel, particularly around the Moon (cislunar space) and to Mars, DARPA started the DRACO program. The project was developed to address the drawbacks of conventional chemical propulsion, which has poor efficiency and requires large fuel loads. These restrictions lengthen mission completion times and limit the amount of cargo a spacecraft may carry. In contrast, nuclear thermal propulsion provides a superior thrust-to-weight ratio and far higher efficiency, enabling bigger payloads and shorter trip periods.

To create thrust, NTP uses heat from a nuclear reactor to warm a propellant, generally liquid hydrogen, which is then released through a nozzle. The main component of DRACO's system is a small, high-temperature reactor that raises the propellant's temperature well over what is possible with conventional chemical engines. Developing enhanced reactor fuels, building reactor cores that can function at high temperatures for long periods of time, enhancing thermal management systems, and guaranteeing robust radiation shielding are some of the project's major technological innovations.

A comprehensive plan has been devised by DARPA and NASA to design and test the DRACO system. This entails a great deal of ground testing, followed by space demonstration missions. The program has so far shown new reactor fuel types and sophisticated heat management technologies, both of which are important for sustaining the extraordinarily high temperatures required for this kind of propulsion. Collaboration between research institutes and public-private partnerships has accelerated invention and development, enabling these advancements.

Even with advancements, DRACO continues to confront several obstacles. These include the requirement for radiation shielding to protect astronauts and equipment on board, efficient cooling systems to maintain reactor stability, and reactor fuels that can tolerate high temperatures and pressures. DARPA intends to use cutting-edge materials, complex computer simulations, and contemporary production methods to overcome these obstacles. Furthermore, it is anticipated that autonomous reactor control via artificial intelligence will increase dependability and safety.

To sum up, DRACO is a significant advancement in space propulsion technology. Future-crewed missions to Mars can be made possible by its success, which would allow for quicker and more adaptable journeys outside of Earth's orbit (DARPA, 2023).

6.3 European Space Agency (ESA) Studies on Nuclear Power for Moon Bases

In order to facilitate long-term lunar missions and the establishment of sustainable Moon bases, the ESA has been aggressively investigating the potential of nuclear power. In order to build nuclear energy systems designed to use on the Moon, this study outlines ESA's technological advancements, strategic objectives, and international collaborations. Solar power isn't guaranteed during the Moon's two-week-long nights, therefore the necessity for consistent and stable energy to enable future lunar exploration & permanent settlements is the primary driving force behind this initiative.

ESA is looking at many nuclear technologies that might function effectively in the hostile lunar environment, especially advanced Stirling-based RTGs and SMRs. SMRs are small, scalable, and intrinsically safe reactors that are ideal for space missions since they require less maintenance and can operate independently. Next-generation RTGs use Stirling engines, which are more effective than conventional types. They have longer lifespans and better energy output, making them perfect for powering extended lunar missions. To ensure safe & effective operation under harsh lunar circumstances, ESA places a strong emphasis on lightweight radiation shielding, robust heat management systems, and dependable autonomous control.

International cooperation is key component of ESA's strategy. ESA shares resources and expertise with private space businesses, NASA, and Roscosmos (Table 1). These collaborations have accelerated technology advancement, established safety guidelines, enhanced system interoperability, and decreased mission expenses overall. This cooperative attitude demonstrates a common worldwide dedication to the sustainable use of nuclear power for lunar exploration.

In this area, ESA is also addressing a number of technological issues. These include enhancing heat-dissipation techniques, building autonomous control systems that can function for longer period of time without human intervention, and constructing robust reactor materials that can withstand the harsh environment of the Moon. ESA is employing cutting-edge materials research, cutting-edge thermal management techniques including radiative cooling and heat pipes, and next-generation control systems built for prolonged autonomous operation to overcome these problems.

All things considered, ESA's nuclear power initiatives are essential to furthering scientific research and sustaining a long-term human presence on the Moon. Nuclear technology paves the way for future lunar colonies and more ambitious space missions by providing steady, dependable power for housing, laboratories, and resource processing units. ESA is establishing Europe as a major force in the worldwide movement for sustainable and safe lunar exploration with this initiative.

Table 1. Analysis of space projects using nuclear technology

Nuclear projects towards Exploration of Space	Lead Agency	Primary Objective	Technology Used	Key Achievements	Future Implications
NASA Kilopower Mission Trials	NASA (U.S.)	Develop compact, scalable fission power systems for lunar & Martian surface operations	Kilopower reactor using HEU fuel, heat pipes, & Stirling engines	Successfully completed the KRUSTY test in 2018, validated passive operation & autonomous power output	Enables continuous power for habitats & ISRU systems; essential for long-duration Moon & Mars missions
DARPA DRACO & Nuclear Propulsion	DARPA (U.S.) & NASA	Demonstrate operational nuclear thermal propulsion (NTP) for cislunar & deep-space maneuvering	NTP reactor using high-temperature nuclear fuel to heat liquid hydrogen	Project in advanced development phase with planned flight tests; ground demonstrations of reactor fuels underway	Could halve Mars transit times, increase payload capacity, & enable agile spacecraft maneuvering
ESA Nuclear Power Studies for Moon Bases	ESA	Ensure continuous & autonomous power for lunar habitats, especially during long lunar nights	SMRs, Stirling-based RTGs, hybrid nuclear-solar systems	Ongoing conceptual studies & international collaborations; emphasis on safety, autonomy, & modularity	Supports sustained European presence on the Moon; aligns with Artemis & global lunar exploration frameworks

7. FUTURE DIRECTIONS & EMERGING INNOVATIONS IN NUCLEAR SPACE EXPLORATION

7.1 Nuclear Fusion Propulsion Research: Direct Fusion Drive Concepts

Many of the drawbacks of conventional chemical rockets and even nuclear fission systems may be addressed by the interesting new technology known as nuclear fusion propulsion. The DFD is one of the most exciting ideas on this field. This propulsion system, which usually uses a fuel combination of deuterium and helium-3 (D-He3), is intended to directly produce thrust from fusion processes in addition to producing power. DFD has the ability to significantly reduce the time it takes to travel between planets and transport heavier payloads than is currently possible due to its high efficiency and specific impulse.

In order to sustain the fusion process, DFD uses magnetic confinement, most especially a Field-Reversed Configuration (FRC) reactor type. A spaceship driven by DFD may finish a human journey to Mars in roughly three months, according to research conducted by NASA and the Princeton Plasma Physics Laboratory (PPPL). This is far faster than what is currently possible with current propulsion systems (Thomas et al., 2019). Additionally, it would require significantly less propellant, which would increase the amount of equipment or cargo a spaceship can carry, reduce launch costs, and make missions more environmentally friendly overall.

Researchers are working to develop miniature reactor designs, enhance plasma physics models, and investigate compact superconducting magnets to make DFD a feasible reality. Technical challenges still need to be spoken, such as maintaining the stability of the fusion reaction, effectively eliminating surplus heat, and shielding the crew and spacecraft from radiation.

7.2 In-Situ Nuclear Reactors for Mars Colonization

One important step toward enabling long-term human settlements on Mars is the direct use of tiny nuclear reactors there. The consistent and clean power required to sustain dwellings, life support systems, research, and technologies that transform Martian resources into useable supplies can be supplied by these in-situ reactors, also known as tiny modular reactors. By producing oxygen, water, fuel, and even building materials on Mars, these devices could lessen the need to transport things from Earth.

This goal is growing closer to reality with recent developments in reactor safety, autonomy, and size. An excellent example is NASA's Kilopower reactor, which underwent successful testing in the KRUSTY experiment. It is a solid contender for usage on Mars since it is a small, scalable nuclear reactor that can operate dependably with minimal human supervision (Gibson & Poston, 2020).

HALEU fuel and state-of-the-art heat pipe technology are two examples of even more sophisticated designs in the works. These enhancements improve performance and safety. Furthermore, AI-powered intelligent autonomous control systems enable reactors to adapt to changing circumstances and operate safely and independently with no danger or maintenance.

7.3 Prospects of Asteroid Mining & Lunar Base Power Solutions Using Nuclear Energy For constructing permanent bases on the Moon & mining asteroids in the future, nuclear power is an essential technology. To support large-scale operations like prospecting, mining, and material processing, these missions require a reliable & potent energy source, which nuclear reactors can supply. Particularly crucial for deep-space missions, nuclear-powered spacecraft provide long-term energy independence even in the absence of solar power.

SMRs and sophisticated RTGs are examples of nuclear power systems that can sustain operations on the moon during the lengthy lunar nights. For vital tasks like mining, building, and exploiting local resources to maintain life (ISRU), they supply dependable power. According to a wealth of research conducted by NASA and ESA, nuclear power is important for sustaining human life and constructing infrastructure in places where solar power is ineffectivel, like permanently shaded sections of the Moon.

The goal of ongoing research is to make these systems safer and more effective. Researchers are developing fully autonomous reactor control systems, improved radiation shielding, and more efficient heat management. The goal of these developments is to increase the reliability of nuclear reactors for usage in hostile space environments, particularly for mining activities on asteroids and the Moon.

7.4 Vision for Sustainable & Autonomous Off-Earth Settlements

Building self-sufficient, sustainable settlements on other worlds, such as Mars, the Moon, and even beyond, is the ultimate goal of human space travel. Nuclear power will be essential to making this happen. For tasks including resource extraction (ISRU), scientific research, habitat construction, food production, and life support, it offers dependable, long-lasting energy.

These self-sustaining space communities are built on top of advanced nuclear technology, including future fusion-based propulsion systems and modular fission reactors. These technologies are being developed to be long-lasting, highly safe, and require little upkeep. Nuclear power development for clean energy also benefits other vital systems, like closed-loop life support and astronaut radiation protection.

Long-term integration of nuclear energy with renewable resources, such as solar power, may result in resilient, adaptable energy systems that may change with the environment and meet various mission requirements. Strong collaboration between space agencies, commercial businesses, and international partners will be necessary to realize this ambition. To harmonize safety standards, laws, and technology compatibility, it is critical that these groups collaborate.

Nuclear energy will greatly increase the scope of human space exploration with sustained innovation and international cooperation, bringing us one step closer to a time when there will be prosperous, autonomous societies outside of Earth.

8. CONCLUSION

A significant turning point in human capacity to travel to, inhabit, and investigate worlds beyond Earth was the request of nuclear technology in space exploration. Nuclear power has been essential to supplying dependable, long-lasting energy in space since the Apollo and Voyager missions' early usage of RTGs to the current state-of-the-art initiatives like NASA's Kilopower and DARPA's DRACO. Voyager 1 and Voyager 2 were fueled by RTGs for decades, well beyond their intended lifespans, and returned priceless information about our solar system. Even though they were eventually discontinued, earlier initiatives like NERVA set the stage for current developments in small-scale surface reactors and nuclear thermal propulsion. A significant advancement is represented by contemporary reactors such as Kilopower and SMRs. Their compact, safe, and scalable architecture makes them ideal for powering deep-space missions, moon colonies, and Mars outposts. NASA and DARPA are developing nuclear thermal propulsion technology that could significantly reduce interplanetary flight times while also increasing astronaut safety. Many of the previous issues with deploying nuclear power in space have been resolved by new developments including autonomous control systems, lightweight shielding, and HALEU fuels. Reactors are now safer, more dependable, and simpler to operate in the hostile environment of space thanks to these developments. However, there are significant moral, legal, and environmental concerns with the use of nuclear technology in space. The possibility of contaminating distant planets and launching nuclear materials are issues. Some legal guidance is provided by international accords such as the Outer Space Treaty, but as technology advances, more debate and revised laws are required. Future ideas like nuclear fusion propulsion, particularly the DFD, have the potential to revolutionize the industry. Missions that once appeared unattainable are now achievable thanks to fusion systems, which promise incredibly quick space travel. Nuclear energy could be crucial for creating viable off-Earth economies, as demonstrated by the utilization of in-situ nuclear reactors for asteroid mining operations or Mars settlement. To put it briefly, nuclear technologies are changing the course of human history in space rather than merely increasing our capacity to operate in space. We may see a future in which humans not only travel to space but also live there thanks to dependable, long-term energy from nuclear technologies. These future space communities will probably rely on autonomous life support, safe, long-lasting homes, and integrated systems that combine nuclear and renewable energy. It will take time, money, international collaboration, and a dedication to employing these potent technologies sensibly and morally to realize this ambition.

REFERENCES

Carpenter, P. K., & Shaltens, R. K. (2024). Comprehensive modeling & characterization of General-Purpose Heat Source Radioisotope Thermoelectric Generator (GPHS-RTG). *Applied Thermal Engineering*, *234*, 121234. DOI: 10.2139/ssrn.4725045

Cowley, A., & Yamamoto, Y. (2024). Advances in in-situ resources utilization for extraterrestrial construction. *Advances in Space Research*, *75*(2), 359–371. DOI: 10.1016/j.asr.2024.05.012

Defense Advanced Research Projects Agency (DARPA). (2023). Demonstration Rocket for Agile Cislunar Operations (DRACO). https://www.darpa.mil/research/programs/demonstration-rocket-for-agile-cislunar-operations

Duggirala, V. (2022). *The history of invention of Radioisotope Thermoelectric Generators (RTGs) for space exploration*. Wiley., DOI: 10.1002/9781119811398.ch1

Fairén, A. G., & Davila, A. F. (2023). Planetary protection: An international concern & responsibility. *Frontiers in Astronomy and Space Sciences*, *10*, 1172546. DOI: 10.3389/fspas.2023.1172546

Gibson, M. A., & Poston, D. I. (2020). Kilopower Project: The KRUSTY fission power experiment & potential missions. *Nuclear Technology*, *206*(9), 1247–1255. DOI: 10.1080/00295450.2020.1722554

Hepp, A. F., Palaszewski, B. A., Colozza, A. J., Landis, G. A., Jaworske, D. A., & Kulis, M. J. (2014). In-situ resource utilization for space exploration: resource processing, mission-enabling technologies, and lessons for sustainability on earth and beyond. In *12th International Energy Conversion Engineering Conference* (p. 3761).

Ingersoll, D. T., & Popov, E. (2020). Emerging small modular nuclear power reactors: A critical review. *Nuclear Engineering and Design*, *367*, 110411. https://www.researchgate.net/publication/344477235_Emerging_small_modular_nuclear_power_reactors_A_critical_review

NASA. (2023). NASA demonstration rocket for agile cislunar operation (DRACO) (NASA Technical Report No. 20230017310). NASA Technical Reports Server. https://ntrs.nasa.gov/citations/20230017310

OECD-NEA. (2024). High-assay low-enriched uranium (HALEU): Drivers & supply chain challenges. OECD Nuclear Energy Agency. https://www.oecd-nea.org/upload/docs/application/pdf/2024-09/nea_publication_2_2_2024-09-18_16-53-42_174.pdf

Peakman, A., & Lindley, B. (2023). A review of nuclear electric fission space reactor technologies for powering future space missions. *Progress in Nuclear Energy, 158*, 103123. DOI: 10.1016/j.pnucene.2023.104815

Samson, J. A., & Vlasov, A. Y. (2024). Nuclear fusion propulsion for spacecraft: Background, applications, & impact. Research Archive *Preprints*. https://research-archive.org/index.php/rars/preprint/view/956

Szocik, K., Shelhamer, M., Braddock, M., Cucinotta, F. A., Impey, C., Worden, P., Peters, T., Ćirković, M. M., Smith, K. C., Tachibana, K., Reiss, M. J., Norman, Z., Gouw, A. M., & Munévar, G. (2021). Future space missions & human enhancement: Medical & ethical challenges. *Futures, 133*, 102819. DOI: 10.1016/j.futures.2021.102819

United Nations Office for Outer Space Affairs (UNOOSA). (1967). The Outer Space Treaty. https://www.unoosa.org/oosa/en/ourwork/spacelaw/treaties/introouterspacetreaty.html

ADDITIONAL READING

Defense Advanced Research Projects Agency (DARPA). (2023). Demonstration Rocket for Agile Cislunar Operations (DRACO). Retrieved from https://www.darpa.mil/research/programs/demonstration-rocket-for-agile-cislunar-operations

Duggirala, V. (2022). The history of the invention of Radioisotope Thermoelectric Generators (RTGs) for space exploration. In *Radioisotopes in Space Exploration* (pp. 1–22). Wiley., DOI: 10.1002/9781119811398.ch1

Fairén, A. G., & Davila, A. F. (2023). Planetary protection: An international concern & responsibility. *Frontiers in Astronomy and Space Sciences, 10*, 1172546. DOI: 10.3389/fspas.2023.1172546

Mason, L. S., & Poston, D. I. (2021). Small fission power systems for space applications: Recent progress & future directions. *Progress in Nuclear Energy, 134*, 103708. DOI: 10.1016/j.pnucene.2021.103708

McClure, P. R., Mason, L. S., & Poston, D. I. (2020). Kilopower reactors for potential Mars mission architectures. *Acta Astronautica, 175*, 105–111. DOI: 10.1016/j.actaastro.2020.05.021

Poston, D. I., Gibson, M. A., & McClure, P. R. (2019). KRUSTY: Reactor design & experiment summary. Proceedings of Nuclear & Emerging Technologies for Space (NETS) Conference. from https://nets2020.ans.org/papers/12345.pdf

Thomas, S. R., Paluszek, M. A., Cohen, S. A., & Razin, Y. (2019). Direct Fusion Drive for Mars Exploration. *Acta Astronautica, 161*, 555–567. DOI: 10.1016/j.actaastro.2019.04.012

United Nations Office for Outer Space Affairs (UNOOSA). (1967). Treaty on principles governing the activities of states in the exploration & use of outer space, including the Moon & other celestial bodies. from https://www.unoosa.org/oosa/en/ourwork/spacelaw/treaties/introouterspacetreaty.html

United Nations Office for Outer Space Affairs (UNOOSA). (1979). Agreement governing the activities of states on the Moon & other celestial bodies. from https://www.unoosa.org/oosa/en/ourwork/spacelaw/treaties/moon-agreement.html

KEY TERMS AND DEFINITIONS

Autonomous Reactor Control Systems: Autonomous reactor control systems are self-regulating technologies that monitor & adjust a nuclear reactor's operations without human intervention, essential for ensuring safety & reliability during remote space missions.

Direct Fusion Drive (DFD): Direct Fusion Drive is an advanced propulsion concept that integrates a compact fusion reactor to produce both thrust & electrical power simultaneously, dramatically reducing interplanetary travel times & increasing spacecraft mission capabilities.

Ethical Considerations in Space Nuclear Deployment: Ethical considerations involve evaluating the potential environmental impacts, planetary contamination risks, & dual-use military implications of deploying nuclear technologies in space, ensuring that exploration activities remain responsible & sustainable.

Fusion Propulsion Systems: Fusion propulsion systems use the energy produced by nuclear fusion reactions to provide thrust for spacecraft. They are considered highly promising for deep-space missions due to their extremely high energy output & minimal fuel requirements.

High-Assay Low-Enriched Uranium (HALEU): HALEU is uranium enriched to levels between 5% & 20% U-235, offering improved performance & efficiency for advanced nuclear reactors. It is particularly suitable for compact & long-duration reactors used in space exploration.

Kilopower Reactor: The Kilopower reactor is a NASA-developed small-scale fission system that uses passive heat pipe technology & Stirling engines to generate between 1 to 10 kilowatts of electrical power, specifically designed to support surface missions on the Moon & Mars.

Nuclear Space Propulsion: Nuclear space propulsion refers to the use of nuclear reactions, either fission or fusion, to generate thrust for spacecraft. It offers higher specific impulse & greater efficiency compared to traditional chemical propulsion systems, enabling faster & longer-range space missions.

Nuclear Thermal Propulsion (NTP): It is a spacecraft propulsion method that uses a nuclear reactor to heat hydrogen, to high temperatures before expelling it through a nozzle, providing thrust with far greater efficiency than chemical rockets.

Planetary Protection: It refers to policies & practices intended to prevent biological contamination & protect Earth's biosphere from potential extraterrestrial contaminants. It ensures the integrity of scientific investigations & sustainable exploration.

Radiation Shielding: Radiation shielding refers to the protective materials & technologies designed to block or attenuate harmful ionizing radiation in space environments, safeguarding both crew members & sensitive electronics from exposure to cosmic rays & reactor emissions.

Small Modular Reactors (SMRs): Small Modular Reactors are compact, scalable nuclear fission reactors designed for flexibility, enhanced safety, & minimal maintenance. In space applications, SMRs provide a reliable & continuous energy source for lunar & Martian habitats.

Space Nuclear Propulsion Safety Protocols: Space nuclear propulsion safety protocols are the rigorous standards & procedures developed to ensure the safe handling, launch, operation, & disposal of nuclear-powered spacecraft systems, minimizing risks to Earth & extraterrestrial environments.

Chapter 10
Adoption Challenges of Hydrogen Energy: A Critical Analysis

Suchismita Satapathy
https://orcid.org/0000-0002-4805-1793
KIIT University, India

ABSTRACT

Globally, hydrogen energy is considered as a clean energy carrier, which will not impact on climate change and reduce environmental pollution. Hydrogen energy is also being called as a potential solution for reducing greenhouse gas emissions. Hydrogen can bring a revolution to vehicles and to the grid system. Still, there are some challenges creating barriers to adoption of hydrogen energy in residential, commercial and industrial sectors. In this chapter, the challenges of hydrogen energy adoption are found in ARAS(Additive ratio assessment)method followed by ISM(Interpretative structural modelling) and QFD(Quality function deployment is used such that the stakeholders can frame new policies and develop new methodologies with advanced technology to mitigate these challenges and familiarize hydrogen use as sustainable energy source in India.

INTRODUCTION

Energy demand is high in both rural and urban areas.Due to rising in industrialization energy demand is also going up. For fulfilling the demand the Government is trying to install Solar, Wind based energy sectors as a substitute of coal based electricity generation process. Which will also improve the productivity of industries by operating machines. Still alternate energy is not that much popular and industries

DOI: 10.4018/979-8-3373-0685-8.ch010

are also not familiar with them and till date dependent on electricity energy. Energy sector is highly dependent on coal based energy generation process.But use of fossil fuels (coal) as energy supplier pollutes the environment. It also highly effect on atmosphere for living beings. So, search of alternate and clean fuel is essential for the earth. So, researcher are focusing on renewable energy sources. The solar and wind energy are plentifully available on the earth but storage of energy is the major challenge. The cost of acceptance of renewable energy and its integration is also very challenging. Hence,among other energies hydrogen energy is also on focus .The approval of hydrogen energy can reduce So_x and NO_x like pollutants from the earth by improving air quality.

Hydrogen fuel cells are nonhazardous and produce only water when they burn, releasing no harmful gases or carbon dioxide. The hydrogen, which is used often, is found by electrolysis, a process powered by renewable sources like wind and solar energy. Hence no carbon is emitted during the production process of hydrogen; this hydrogen energy is widely accepted as a clean, non-polluting source worldwide. Many countries, whether rich or developing, are building power plants and producing hydrogen fuel. However, many still face challenges like high costs, lack of infrastructure, and limited technology to adopt hydrogen energy on a larger scale.

Hydrogen has a low volumetric energy density, making it difficult to transport and store. Large tanks or high-pressure containers are needed for safe storage and transportation. Around the world, hydrogen is seen as a clean energy option that can reduce pollution and mitigate climate change. It can also lower greenhouse gases. Vehicles with hydrogen fuel cells, or FCVs, make electricity by combining hydrogen with oxygen. The only byproducts are water vapor and heat, making FCVs a sustainable alternative. Hydrogen has the power to change both electrical grids and transportation systems significantly.

Since hydrogen can be made in many ways and is accessible, it appeals to the public, businesses, and governments as a green fuel source. It is the focus of much research and a lot of funding for new materials used in fuel cells, sensors, and storage. Safety is crucial because hydrogen is flammable, and facilities must be designed carefully to prevent accidents. Material issues, such as embrittlement, are still a challenge for hydrogen systems. Long-term plans and future tech are guided by potential risks and development goals. Policies that consider both economic and environmental factors are key. Combining hydrogen's promise with new material development offers a chance to change our energy system sustainably.

In the future, more studies and research will also look into how materials and composites interact with hydrogen, especially in pipelines and storage systems. Embrittlement, which damages storage materials, is a major concern. Innovations and ongoing challenges in fuel cells and storage need more focus.

Hydrogen is widely used in industries as a cleaner replacement for fossil fuels. The transport industry is moving toward using fuel cells powered by hydrogen. Handling and storing hydrogen safely remains essential, no matter the industry. Usually, compressed hydrogen is stored in metal cylinders. But embrittlement can weaken these cylinders over time. High-strength steel cylinders are especially vulnerable. Proper design and protective materials are vital for safe hydrogen storage.

Implementing hydrogen energy faces many obstacles. This chapter explores these challenges and identifies the main barriers to adoption. The ARAS (Additive Ratio Assessment) method then helps to rank these challenges. This prioritization shows which technologies or policies are needed to reduce the barriers. The QFD and ISM (Interpretive Structural Modeling) technique is used to reveal how the barriers are connected and influence each other and suggest what design framework must be implemented to mitigate the barriers of adoption challenges (Thakkar et al.,2011).

BACKGROUND

The World Health Organization (2020) report shows that the main causes of air pollution and climate change are coal, oil, and natural gas in industries. These fuels not only harm human health badlybut also impact both living and nonliving things on the earth. Endo et al. (2021) note that hydrogen energy has become more popular lately. It is a clean energy source that can replace fossil fuels, according to Singh et al. (2020). The use of fossil fuels such as oil, coal, and natural gas leads to climate change and global warming. Hydrogen fuels have about seven times the energy density of fossil fuels. Yang et al. (2010) predict that hydrogen will become the most important energy source in the future if we can fix the current challenges of producing it. Murray et al., (2009) points out that storing and transporting hydrogen are two major issues. Baykara (2018) explains that hydrogen energy has no harmful effect on the environment. Farrell et al. (2003) believe that hydrogen will change how we get energy, especially in transportation, by replacing oil. Ewan et al. (2005) say that roughly 40–45% of hydrogen is produced through gas-steam reforming. About a third comes from oil reforming, a fifth from coal reforming, and around 4% from water electrolysis. The European Commission's High-Level Group for Hydrogen predicted that by 2040, 35% of cars in Europe will run on hydrogen fuel, replacing oil. Table 1 presents research on the main challenges facing hydrogen.

Table 1. Literatures based on challenges on Hydrogen energy

1.	Ahad et al. (2023)	identified the opportunities and difficulties in producing hydrogen while utilizing CO2 collection technologies. .
2.	Eljack & Kazi (2021)	Explained the barriers of hydrogen challenges like production cost, infrastructure, transportation etc can be mitigated with multi-sector industrial-urban symbiosis system.
3.	Calabrese et al. (2024)	utilized quantitative risk evaluations, computational fluid dynamics analysis for hydrogen dispersion, and a review of production, transportation, and safety issue
4.	Worku et al. (2024)	The benefits and drawbacks of various hydrogen generation techniques—such as those involving fossil fuels, biomass, water electrolysis, microbial fermentation, and photocatalysis—are examined and then contrasted.
5.	Rasul et al. (2022)	reviewed and summarizes the methods, storage choices, expenses, and uses of hydrogen production.
6.	Sun et al. (2024)	discussed the basic mechanics of hydrogen embrittlement in metallic materials as well as the phenomenon of hydrogen-induced damage in common materials.
7.	Otaki & Shaw (2024)	outlined how Japan has already made strides in the manufacture, transportation, and generation of hydrogen energy and how it can assist other nations.
8.	Yue et al. (2021)	Written about the state and future prospects of hydrogen technologies and how they are being used in power systems for the production, re-electrification, and storage of hydrogen
9.	Rand & Dell (2007)	outlined the processes for producing hydrogen, the potential for capturing and storing carbon, and the possibilities for hydrogen storage in cars, fuel cells, and fuel cell vehicles.
10.	Minic (2012)	discussed issues including the need to take care of production storage so that hydrogen energy is accessible to everyone.

ARAS is a multi-criteria decision making method developed by Zavadskas & Turkis (2010).This method is use full to rank alternatives. Stanujkic, & Jovanovic(2012) used this method to rank faculties based on quality by using websites.Sihombing et al. (2021) have used ARAS method for selection of english course.Karabašević et al.(2015) have used SWARA and ARAS both for personnel selection.

Akao (1997) has wrote about past, present, future of QFD tool .Lombardi et al.(2018) have used QFD method for ranking hazardpos indusries. Chakraborty and Prasad (2018) used QFD for selection of cotton fibre.

One proven technique for determining the link is interpretive structural modeling (ISM).among several objects that explain an issue.Numerous elements could be connected to some-problem, for any-type of complicated issue that is being examined. ISM offers a shared comprehension of these connections. In 1994, Anderson et al. used theinterpretative structural equation modeling to comprehend the acquisition success's predictive behavior for enhancing company performance. ISM is a mathematical technique and a computer assisted method. Computerization learning approach that enables individuals and organizations to create a plan for intricate relationships between different parts involved in a complicated scenario, according to Kumar et al. (2009).

RESEARCH METHODOLOGY

A review of existing literature was conducted to identify the challenges of adopting hydrogen energy. Based on this, a set of 22 questionnaires was prepared. The survey was shared with professors, research students, and energy specialists. About 100 questions were asked as yes or no questions, with yes coded as 1 and no as 0. Around 70% of the responses were received. The data was then analyzed using statistical methods. The ARAS technique was used to rank the main obstacles to using hydrogen energy.

Table 2. Question on challenges

Question	Yes=1	No=0
Hydrogen production by electrolysis method is costly		
Requirement of dedicated infrastructure for producing, storing and transporting hydrogen		
It is technically difficult to produce high pressure tank for storing		
As hydrogen is highly flammable so for transportation special pipes required.		
Innovative durable fuel cell can be used		
For production of hydrogen new technology must be used.		
To fulfill energy demands investment and scaling is mandatory.		
Green hydrogen production is very costly		
Large Infrastructure and investment is required for creating production, distribution, refueling of hydrogen		
For fuel cell transmission required systems are costly		
Infrastructure hydrogen fuel needs to be developed		
Supply chain network are challenging and costly.		
Lack of refueling station for hydrogen fuel		
Certain safety regulations are essential for hydrogen use.		
supporting policies for use of hydrogen fuel incentives is required		
Environmental regulations stop fossil fuel use but not support use of hydrogen fuel.		
Production of hydrogen energy from fossil fuels emits carbon, so contradicts the environmental policy		
For production of hydrogen, high use of water resources may create a scarcity of water		
Hydrogen generated from renewable energy also leaves a carbon footprint.		
Hydrogen must compete with well-established energy sources like fossil fuels		
Public awareness and adoption e of hydrogen as fuel is essential		
policies supporting hydrogen infrastructure development, research, and deployment are crucial		
Awareness of hydrogen energy benefits is required.		
Integrating hydrogen with existing energy systems is a challenge		
Safety standards must be accepted by public		

Results and Discussion

The factor analysis is conducted after getting the responses of adoption challenges of hydrogen energy under seven dimensions (I.e.,technical challenges, economic challenges, infrastructure challenges, environmental challenges, policy challenges,social acceptance challenges and integration challenges).Only 21 items selected had Chrobanch's alpha value more than 0.5, during statistical analysis done in minitab 17.The items below 0.5 values are discarded.Table.3 shows the result of factor analysis.Then theARAS Method is used to prioritize the challenges.

Table 3. Factor analysis of Challenges

Dimensions	Question	Factor
Technical challenges	Hydrogen production by electrolysis method is costly	0.532
	Requirement of dedicated infrastructure for producing, storing and transporting hydrogen	0.527
	It is technically difficult to produce high pressure tank for storing	0.514
	As hydrogen is highly flammable so for transportation special pipes required.	0.509
	Innovate durable fuel cell can be used	0.538
Economic challenge	Green hydrogen production is very costly	0.542
	Large Infrastructure and investment is required for creating production, distribution, refueling of hydroge	0.503
	For fuel cell transmission required systems are costly	0.519
Infrastructure challenge	Infrastructure for hydrogen fuel needs to be developed	0.625
	Development of hydrogen distribution network has challenges like cost, safety etc.	0.611
	Lack of refueling station for hydrogen fuel	0.567
Policy challenges	Certain safety regulations are essential for hydrogen use.	0.549
	supporting policies for use of hydrogen fuel incentives is required	0.533
	Environmental regulations stop fossil fuel use but not support use of hydrogen fuel.	0.510
Environmental challenges	Production of hydrogen energy from fossil fuels emits carbon, so contradicts the environmental policy	0.713
	For production of hydrogen high use of water resources may create scracity of water	0.584
	Hydrogen generated from renewable energy also leaves carbon footprint.	0.557
Sociatal acceptance challenges	Hydrogen must compete with well-established energy sources like fossil fuels	0.562
	policies supporting hydrogen infrastructure development, research, and deployment are crucial	0.527

continued on following page

Table 3. Continued

Dimensions	Question	Factor
Integration challenges	Integration and acceptance of hydroge energy	0.634
	Integrating hydrogen with existing energy systems is a challenge	0.572

After factorial analysis, ARAS method is used to prioritize most important dimension of adoption of hydrogen energy

Step-1 A Decision matrix is constructed by considering criteria and alternatives .In Table 4 criteria are taken as challenges of adoption of hydrogen energy and in alternative application of hydrogen energy in the residential sector, commercial sector and industrial sector are taken

Step-2 *Normalization is done by using* beneficial criteria so below equation is used(table.5)

$$X_{0j} = \max_{i} \times x_{ij} \tag{1}$$

(for beneficial criteria)

Step 3: normalized Relative performances is calculated (rij)

$$r\,ij = \frac{Xij}{\sum_{i=0}^{m} Xij} \tag{2}$$

Step 4: Weighted normalized matrix found by below formula

$$V_{ij} = W_{j} \times r_{ij} \tag{3}$$

Step 5:Overall performance and degree of utility found by using the below formula and ranking is done.

$$s_{i=\sum_{j=1}^{n} v_{ij}}, Q_i = \frac{S_i}{S_0} \tag{4}$$

Table 4. Decision Matrix

	Sectors		
Challenges	**Residential**	**Commercial**	**Industrial**
Technical challenges	0.775	0.787	0.733
Economic challenges	0.912	0.987	0.566
Infrastructure challenge	0.617	0.637	0.838
Environmental challenges	0.731	0.495	0.643
Policy challenges	0.387	0.58	0.284
Sociatal acceptance challenges	0.575	0.674	0.678
Integration challenges	0.773	0.725	0.744

Table 5. Normalized Decision matrix

Challenges	**Residential**	**Commercial**	**Industrial**
OV	0.160506864	0.168085831	0.157400451
Technical challenges	0.136395635	0.134025886	0.137678437
Economic challenges	0.160506864	0.168085831	0.106311044
Infrastructure challenge	0.108588525	0.108480926	0.157400451
Environmental challenges	0.128651883	0.084298365	0.120773854
Policy challenges	0.06810982	0.098773842	0.053343351
Sociatal acceptance challenges	0.101196762	0.114782016	0.127347859
Integration challenges	0.136043647	0.123467302	0.139744553

Table 6. Weighted normalized matrix

Challenges	**Residential**	**Commercial**	**Industrial**
OV	0.052967265	0.055468324	0.051942149
Technical challenges	0.04501056	0.044228542	0.045433884
Economic challenges	0.052967265	0.055468324	0.035082645
Infrastructure challenge	0.035834213	0.035798706	0.051942149
Environmental challenges	0.042455121	0.02781846	0.039855372
Policy challenges	0.022476241	0.032595368	0.017603306
Sociatal acceptance challenges	0.033394931	0.037878065	0.042024793
Integration challenges	0.044894403	0.04074421	0.046115702

Table 7. Final Rank

Challenges	Si	So	Qi	Rank
	0.160377738	0.160377738	1	
Technical challenges	0.134672986		0.839723692	2nd
Economic challenges	0.143518234		0.894876282	Ist
Infrastructure challenge	0.123575068		0.77052507	7th
Environmental challenges	0.110128954		0.686684793	5th
Policy challenges	0.072674914		0.453148394	6th
Sociatal acceptance challenges	0.11329779		0.706443373	4th
Integration challenges	0.131754316		0.821524965	3rd

After ranking by ARAS method, it is found that Economic challenges is most important challenge among all dimensions which is ranked first, Technical challenge is the 2nd ranked challenge and Integrating challenge ranked third. Hence to make hydrogen adoption widely necessary steps must be taken by framing policies and developing new frameworks. To find linear relationship between these challenges Interpretative structural modelling method is implemented.

Quality function deployment method is used to design a frame work based on customer need and provides solution to the need on priority basis.The symbols like O=0.8, ʘ=0.6,Δ=0.4 and •=0.2 are used for house of quality matrix.

Step-1 Left matrix is used to calculate Customer rating by considering customer need and ratings(1 to 5)

Step2 The center matrix shows relation between customer requirement and system requirement and it provides individual requirement by using formula

$$Z=A_{ij}\,X_j \qquad (5)$$

Step-3 Lastly the top m,matrix is found by considering revised rating Z and relation between top-matrix values.Then normalization is done and most important parameters found

Quality function deployment matrix developed to design a frame work to mitigate these challenges. Table 8 shows system design requirement and table.9 shows calculation for house of quality .Figure.1 shows house of quality.

Table 8. Design Requirement for Adoption Challenges

Sl no	Requirement
1	Production and storing Cost Reduction
2	Technological advancement to improve efficiency
3	Innovation and research development improvement
4	Infrastructure expansion
5	Development of hydrogen resistance material
6	Adress safety protocols
7	Reduction of regulatory barriers
8	Provide support and incentives for adopters

Table 9. Calculation for House of Quality

	Rating	1	2	3	4	5	6	7	8	Requirment
1	2	⊙	O	O	⊙	O	Δ	O	⊙	2.066666667
2	1	Δ	O	•	Δ	⊙	•	•	⊙	1.9
3	3	Δ	Δ	Δ	⊙	Δ	O	O	Δ	1.433333333
4	4	⊙	O	Δ	O	Δ	O	⊙	Δ	2.233333333
5	5	O	⊙	Δ	O	•	⊙	O	Δ	6
6	3	O	Δ	Δ	O	•	Δ	⊙	•	2.2
7	1	O	Δ	O	Δ	⊙	•	O	⊙	1.866666667
Individual requirement		11.96666667	10.76	8.273333333	11.95333333	7.02	8.993333333	8.35	7.806666667	
Revised Requirement		47.70133333	38.87866667	30.87466667	24.83266667	15.83333333	11.85866667	8.132666667	4.684	
Normalized Rating		1	0.815043605	0.647249553	0.520586427	0.331926431	0.248602415	0.170491391	0.09819432	
Ranking		Ist	3rd	4th	5th	6th	7th	8th	2nd	

Figure 1. House of Quality

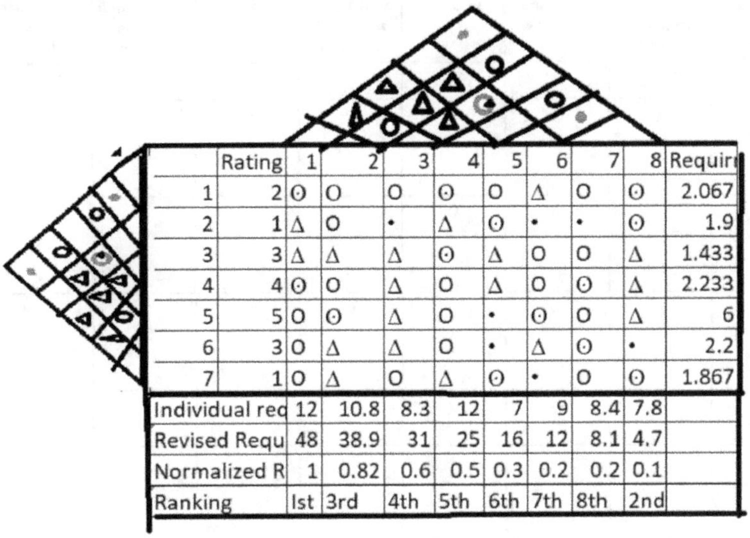

	Rating	1	2	3	4	5	6	7	8	Requir
1	2	⊙	O	O	⊙	O	Δ	O	⊙	2.067
2	1	Δ	O	•	Δ	⊙	•	•	⊙	1.9
3	3	Δ	Δ	Δ	⊙	Δ	O	O	Δ	1.433
4	4	⊙	O	Δ	O	Δ	O	⊙	Δ	2.233
5	5	O	⊙	Δ	O	•	⊙	O	Δ	6
6	3	O	Δ	Δ	O	•	Δ	⊙	•	2.2
7	1	O	Δ	O	Δ	⊙	•	O	⊙	1.867
Individual req		12	10.8	8.3	12	7	9	8.4	7.8	
Revised Requ		48	38.9	31	25	16	12	8.1	4.7	
Normalized R		1	0.82	0.6	0.5	0.3	0.2	0.2	0.1	
Ranking		Ist	3rd	4th	5th	6th	7th	8th	2nd	

From Table. 9 and figure.1 it is clear that, by reducing production and storing cost the problems can be resolved, Secondly by encouraging adopters with support and financial benefit more people will adopt hydrogen as an alternate fuel source.

Then ISM is a method is used to find inter relationship between variables finding complex relation between each variables.

If these variables/challenges are considered as I and J variables(row and column variables as per table.10) are both impacts on each other X is used, if I has better impact on j, V is written .If reverse occurs means j impacts on I, A written, if no relation found then O is written in the below table.Then in table 10, binary nos (1 and 0) are used.

In table 11, I to J entry 1 then J to I must be 0.for Jto I entry 1 then I to J must be 0 for A variable.For X, both must be one and For O both must be 0.In table. 12 dependency and driver power of variables calculated by summing rows and column respectively and * mark shows transitivity relation between variables.Table. 13 and Table 14 shows final level of matrix selection.

Table 10. SSIM(self structural Matrix)

		Provide support and incentives for adopter	Reduction of regulatory barriers	Adress safety protocols	Development of hydrogen resistance material	Infrastructure expansion	Innovation and research development improvement	Technological advancement to improve efficiency	Production and storing Cost Reduction
	Challenges		7	6	5	4	3	2	1
1	Production and storing Cost Reduction	A	O	O	V	V	O	V	
2	Technological advancement to improve efficiency	A	O	V	A	V	A		
3	Innovation and research development improvement	A	O	O	V	V			
4	Infrastructure expansion	A	O	V	A				
5	Development of hydrogen resistance material	A	O	A					
6	Adress safety protocols	A	V						
7	Reduction of regulatory barriers	A							
8	Provide support and incentives for adopters	A							

Table 11. Initial Reachability Matrix

	1	**2**	**3**	**4**	**5**	**6**	**7**	**8**
1	1	1	1	0	1	1	1	0
2	1	1	1	1	1	1	1	0
3	1	1	1	1	1	1	1	0
4	1	1	1	1	1	1	0	0
5	0	0	0	1	1	1	0	0
6	0	0	0	1	1	1	0	0
7	1	1	0	1	1	1	1	0
8	0	0	0	0	0	0	0	0

Table 12. Final reachability matrix with transitivity relationship

	1	2	3	4	5	6	7	8	Driver
1	1	1	1*	0	1	1	1	0	6
2	1	1	1	1*	1	1*	1*	0	6
3	1	1	1	1	1	1	1	0	7
4	1	1	1	1	1	1	0	0	6
5	0	0	0	1	1	1	0	0	3
6	0	0	0	1	1	1	0	0	3
7	1	1*	0	1	1	1	1	0	6
8	0	0	0	0	0	0	0	0	0
Dependency	5	5	4	6	7	7	4		

Table 13. Iteration 1

Variables	Reachability Set	Antecedent Set	Intersection Set	Level
1	1,2,3,5,6,7	1,2,3,4,7	1,2,34,7.	
2	1,2,3,4,5,6,7	1,2,3,4,7	1,2,3,4,7	
3	1,2,3,4,5,6,7	1,2,3,4	1,2,3,4	
4	1,2,34,5,6	2,3,4,5,6,7	2,34,5,6	
5	4,5,6	1,2,3,4,5,6,7	4,5,6	I
6	2,5,6	1,2,3,4,5,6,7	2,5,6	I
7	1,2,4,5,6,7	1,2,3,7	1,2,4,7	

Table 14. Final Level

Variables	Level
5,6	I
1, 4,7	II
2	III
3	IV
8	V

Then ISM model is drawn in fig .2.

Figure 2. ISM Model

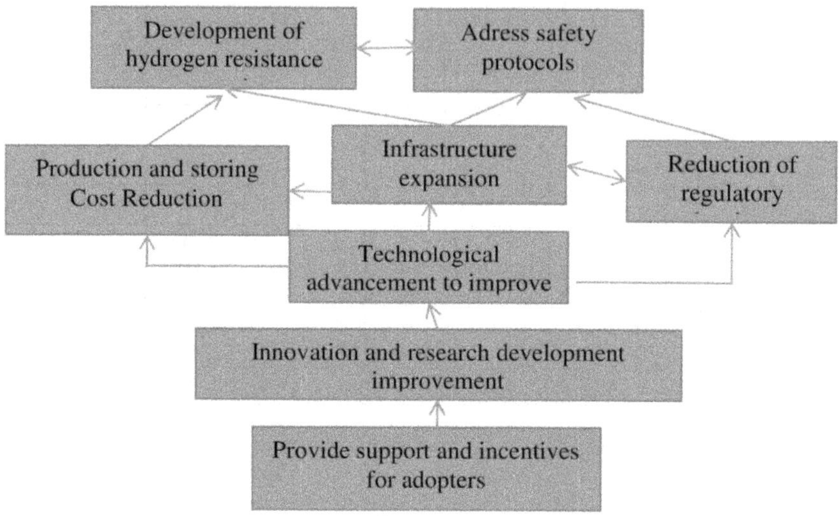

From the ISM Model, it is found that development of hydrogen resistance material and address safety protocols are the variables of first level, production and storage cost, infrastructure and regulatory framework reduction found in the second level, technological advancement and innovation and research development variables are of third and fourth level.

Then MICMAC analysis is conducted in table 15.

Table 15. MICMAC Analysis

Driver								
6	7		IV			III	5	6
6	6				3,7		4	
7	5							
6	4					1		
3	3		I			2	II	
3	2							
6	1	8						
	Dependency	5	5	4	6	7	7	4
		1	2	3	4	5	6	7

The variables are divided into four clusters (I, II, III, IV). The X and Y axis of MICMAC analysis represents dependency and drive power respectively. The variables falling under these four clusters are categorized as autonomous variable, dependent variable, linkage variable and independent variable.Autonomous variables are weak dependence and weak driving power . Eight No variable is found as autonomous variable. Dependent variables are 1,2 variables with high dependency and low driving power.Linkage variable are with high drive power and high dependency, The variables 3,4,5,6,7 are linkage variable.No variables fall under independent variable.

CONCLUSION

Using hydrogen as a future fuel requires significant investment, advanced technology, and new infrastructure. This includes designing storage tanks, transport systems, fueling stations, and fuel cells. Addressing both technical and economic issues is essential, but the main goal is to develop sustainable hydrogen energy. This involves creating new technology that cuts down the environmental impact of hydrogen production, preventing pollution. Overcoming these challenges will raise awareness and encourage wider use of hydrogen in all industries. It can replace fossil fuels and reduce their harmful effects on the planet. Adopting hydrogen energy can also create many new jobs and support a greener future. It has the potential to fight climate change, improve air quality, and boost energy security. These benefits lead to stronger economic growth and a healthier environment for everyone.

This research has not received any fund from any Organization.

This research has not received any grant.

REFERENCES

Ahad, M. T., Bhuiyan, M. M. H., Sakib, A. N., Becerril Corral, A., & Siddique, Z. (2023). An overview of challenges for the future of hydrogen. *Materials (Basel), 16*(20), 6680. DOI: 10.3390/ma16206680 PMID: 37895667

Akao, Y. (1997, October). QFD: Past, present, and future. In *International symposium on QFD* (Vol. 97, No. 2, pp. 1-12). Linköping, Sweden: International council of QFD.

Anderson, J. C., Rungtusanthanam, M., & Schroeder, R. G. (1994). A theory of quality management underlying the Deming management method. *Academy of Management Review, 19*(3), 472–509. DOI: 10.2307/258936

Bagočius, V., Zavadskas, E. K., & Turskis, Z. (2014). Multi-person selection of the best wind turbine based on the multi-criteria integrated additive-multiplicative utility function. *Journal of Civil Engineering and Management, 20*(4), 590–599. DOI: 10.3846/13923730.2014.932836

Baykara, S. Z. (2018). Hydrogen: A brief overview on its sources, production and environmental impact. *International Journal of Hydrogen Energy, 43*(23), 10605–10614. DOI: 10.1016/j.ijhydene.2018.02.022

Calabrese, M., Portarapillo, M., Di Nardo, A., Venezia, V., Turco, M., Luciani, G., & Di Benedetto, A. (2024). Hydrogen safety challenges: A comprehensive review on production, storage, transport, utilization, and CFD-based consequence and risk assessment. *Energies, 17*(6), 1350. DOI: 10.3390/en17061350

Chakraborty, S., & Prasad, K. (2018). A Quality Function Deployment-Based Expert System for Cotton Fibre Selection. *Journal of The Institution of Engineers (India): Series E, 99*(1), 43–53.

Eljack, F., & Kazi, M. K. (2021). Prospects and challenges of green hydrogen economy via multi-sector global symbiosis in Qatar. *Frontiers in Sustainability, 1*, 612762. DOI: 10.3389/frsus.2020.612762

Endo, N., Goshome, K., Tetsuhiko, M., Segawa, Y., Shimoda, E., & Nozu, T. (2021). Thermal management and power saving operations for improved energy efficiency within a renewable hydrogen energy system utilizing metal hydride hydrogen storage. *International Journal of Hydrogen Energy, 46*(1), 262–271. DOI: 10.1016/j.ijhydene.2020.10.002

Ewan, B. C. R., & Allen, R. W. K. (2005). A figure of merit assessment of the routes to hydrogen. *International Journal of Hydrogen Energy, 30*(8), 809–819. DOI: 10.1016/j.ijhydene.2005.02.003

Farrell, A. E., Keith, D. W., & Corbett, J. J. (2003). A strategy for introducing hydrogen into transportation. *Energy Policy*, *31*(13), 1357–1367. DOI: 10.1016/S0301-4215(02)00195-7

Global energy transformation: A roadmap. to 2050. https://www.irena.org/publications/2019/Apr/Global-energy-transformation-A-roadmap-to-2050-2019Edition

Karabašević, D., Stanujkić, D., & Urošević, S. (2015). The MCDM Model for Personnel Selection Based on SWARA and ARAS Methods. Management (1820-0222), 20(77).

Kumar, N., Prasad, R., Shankar, R., & Lyer, K. C. (2009). Technology transfer for rural housing: An interpretive structural modelling approach. *Journal of Advances in Management Research*, *6*(2), 188–205. DOI: 10.1108/09727980911007208

Liu, N., & Xu, Z. (2021). An overview of ARAS method: Theory development, application extension, and future challenge. *International Journal of Intelligent Systems*, *36*(7), 3524–3565. DOI: 10.1002/int.22425

Lombardi, M., & Fargnoli, M. (2018). Prioritization of hazards by means of a QFD-based procedure. *International Journal of Safety and Security Engineering*, *8*(2), 342–353. DOI: 10.2495/SAFE-V8-N2-342-353

Minic, D. (Ed.). (2012). Hydrogen Energy: Challenges and Perspectives.

Murray, L. J., Dincă, M., & Long, J. R. (2009). Hydrogen storage in metal–organic frameworks. *Chemical Society Reviews*, *38*(5), 1294–1314. DOI: 10.1039/b802256a PMID: 19384439

Otaki, T., & Shaw, R. (2024). Global hydrogen energy: Potentials and challenges. *Energy, Sustainability and Resilience: A Futuristic Vision from Asia*, 127-138.

Rand, D.A.J., & Dell, R.M. (2007). Hydrogen energy: challenges and prospects.

Rasul, M. G., Hazrat, M. A., Sattar, M. A., Jahirul, M. I., & Shearer, M. J. (2022). The future of hydrogen: Challenges on production, storage and applications. *Energy Conversion and Management*, *272*, 116326. DOI: 10.1016/j.enconman.2022.116326

Sihombing, V., Nasution, Z., Al Ihsan, M. A., Siregar, M., Munthe, I. R., Siregar, V. M. M., & Asfar, D. A. (2021, June). Additive Ratio Assessment (ARAS) Method for Selecting English Course Branch Locations. [). IOP Publishing.]. *Journal of Physics: Conference Series*, *1933*(1), 012070. DOI: 10.1088/1742-6596/1933/1/012070

Singh, R., Altaee, A., & Gautam, S. (2020). Nanomaterials in the advancement of hydrogen energy storage. *Heliyon*, *6*(7), e04487. DOI: 10.1016/j.heliyon.2020.e04487 PMID: 32743097

Stanujkic, D., & Jovanovic, R. (2012). Measuring a quality of faculty website using ARAS method.

Sun, B., Zhao, H., Dong, X., Teng, C., Zhang, A., Kong, S., Zhou, J., Zhang, X.-C., & Tu, S. T. (2024). Current challenges in the utilization of hydrogen energy-a focused review on the issue of hydrogen-induced damage and embrittlement. *Advances in Applied Energy*, *14*, 100168. DOI: 10.1016/j.adapen.2024.100168

Thakkar, J. J., Kanda, A., & Deshmukh, S. G. (2011, January). A decision framework for supply chain planning in SMEs: A QFD-ISM-enabled ANP-GP approach. In *Supply Chain Forum* []. Taylor & Francis.]. *International Journal (Toronto, Ont.)*, *12*(4), 62–75.

Worku, A. K., Ayele, D. W., Deepak, D. B., Gebreyohannes, A. Y., Agegnehu, S. D., & Kolhe, M. L. (2024). Recent advances and challenges of hydrogen production technologies via renewable energy sources. *Advanced Energy and Sustainability Research*, *5*(5), 2300273. DOI: 10.1002/aesr.202300273

Yang, J., Sudik, A., Wolverton, C., & Siege, D. J. (2010). High capacity hydrogen storage materials: Attributes for automotive applications and techniques for materials discovery. *Chemical Society Reviews*, *39*(2), 656–675. DOI: 10.1039/B802882F PMID: 20111786

Yue, M., Lambert, H., Pahon, E., Roche, R., Jemei, S., & Hissel, D. (2021). Hydrogen energy systems: A critical review of technologies, applications, trends and challenges. *Renewable & Sustainable Energy Reviews*, *146*, 111180. DOI: 10.1016/j.rser.2021.111180

Zavadskas, E. K., & Turskis, Z. (2010). A new additive ratio assessment (ARAS) method in multicriteria decision-making. *Technological and Economic Development of Economy*, *16*(2), 159–172. DOI: 10.3846/tede.2010.10

KEY TERMS AND DEFINITIONS

ARAS: It is a multi-criteria decision making method used to prioritize most important challenges

Factor Analysis: This method helps to put all related variables and items in one cluster, This is a statistical method.

ISM: Interpretive structural modelling (ISM) helps to find interrelation between variables.

Conclusion

As editors of *Pathways to a Carbon-Free Future Through Advanced Nuclear Systems*, we are proud to present this comprehensive and timely reference work that addresses one of the most critical challenges of our era: achieving a sustainable, carbon-neutral energy future. Through the collective expertise of researchers, engineers, policy experts, and thought leaders, this book charts a compelling course for how advanced nuclear technologies can transform the global energy landscape.

The chapters herein reflect a balanced and interdisciplinary approach, combining technical depth with policy relevance, and innovation with real-world applicability. From next-generation fission designs to the frontiers of nuclear fusion, from hybrid systems integrating renewables to strategies for public engagement and regulatory reform, this volume captures the multifaceted promise of nuclear energy.

It is our hope that this book not only informs but also inspires. Whether you are a nuclear professional, policymaker, student, investor, or a concerned global citizen, we invite you to engage with the ideas and insights offered throughout these pages. The road to decarbonization is neither simple nor singular—but with advanced nuclear systems as a pillar of that journey, the vision of a clean, resilient, and equitable energy future becomes far more attainable.

Let this book serve as both a reference and a catalyst—empowering dialogue, fostering innovation, and supporting action toward a carbon-free world.

Compilation of References

Addo, E. K., Kabo-bah, A. T., Diawuo, F. A., & Debrah, S. K. (2023). The Role of Nuclear Energy in Reducing Greenhouse Gas (GHG) Emissions and Energy Security : A Systematic Review. *International Journal of Energy Research, 2023*(1), 8823507. DOI: 10.1155/2023/8823507

Aghaei, J., & Alizadeh, M.-I. (2013). Demand response in smart electricity grids equipped with renewable energy sources: A review. *Renewable & Sustainable Energy Reviews, 18*, 64–72. DOI: 10.1016/j.rser.2012.09.019

Aguirre, T. G., Cramer, C. L., & Mitchell, D. J. (2022). Review of additive manufacturing and densification techniques for the net- and near net-shaping of geometrically complex silicon nitride components. *Journal of the European Ceramic Society, 42*(3), 735–743. DOI: 10.1016/j.jeurceramsoc.2021.11.001

Ahad, M. T., Bhuiyan, M. M. H., Sakib, A. N., Becerril Corral, A., & Siddique, Z. (2023). An overview of challenges for the future of hydrogen. *Materials (Basel), 16*(20), 6680. DOI: 10.3390/ma16206680 PMID: 37895667

Ahmad, A. (2023). Advances in machine learning for monitoring, control, and optimization of the temperature of reactors. *Journal of Chemical Engineering and Process Optimization, 15*(2), 101–115. DOI: 10.20944/preprints202309.1318.v1

Akao, Y. (1997, October). QFD: Past, present, and future. In *International symposium on QFD* (Vol. 97, No. 2, pp. 1-12). Linköping, Sweden: International council of QFD.

Allegretti, G., Montoya, M. A., & Talamini, E. (2024). Renewable energy for a sustainable future. In *The renewable energy-water-environment nexus* (pp. 1–36). Elsevier. DOI: 10.1016/B978-0-443-13439-5.00001-6

Allen, M., Antwi-Agyei, P., Aragon-Durand, F., Babiker, M., Bertoldi, P., Bind, M., & Zickfeld, K. (2019). Technical summary. In *Global warming of 1.5 °C: An IPCC special report on the impacts of global warming of 1.5 °C above pre-industrial levels and related global greenhouse gas emission pathways, in the context of strengthening the global response to the threat of climate change, sustainable development, and efforts to eradicate poverty* (pp. 33–114). Intergovernmental Panel on Climate Change., https://www.ipcc.ch/sr15/

Amin, S. M. (2011). Smart grid: Overview, issues, and opportunities. Advances and challenges in sensing, modelling, simulation, optimization, and control. *European Journal of Control, 17*(5–6), 547–567. DOI: 10.3166/ejc.17.547-567

Amiot, R., Kusuhashi, N., Saegusa, H., Shibata, M., Ikegami, N., Shimojima, S., & Lécuyer, C. (2021). Paleoclimate and ecology of Cretaceous continental ecosystems of Japan inferred from the stable oxygen and carbon isotope compositions of vertebrate bioapatite. *Journal of Asian Earth Sciences, 205*, 104602. DOI: 10.1016/j.jseaes.2020.104602

Anastasovski, A. (2021). Improvement of energy efficiency in ethanol production supported with solar thermal energy – A case study. *Journal of Cleaner Production, 278*, 123476. DOI: 10.1016/j.jclepro.2020.123476

Anderson, J. C., Rungtusanthanam, M., & Schroeder, R. G. (1994). A theory of quality management underlying the Deming management method. *Academy of Management Review, 19*(3), 472–509. DOI: 10.2307/258936

Assante, M. J., & Lee, R. M. (2015). *The industrial control system cybersecurity workforce gap*. SANS Institute. https://www.sans.org/white-papers/industrial-control-system-cybersecurity-workforce-gap/

Azam, A., Rafiq, M., Shafique, M., Zhang, H., & Yuan, J. (2021). Analyzing the effect of natural gas, nuclear energy and renewable energy on GDP and carbon emissions: A multi-variate panel data analysis. *Energy, 219*, 119592. DOI: 10.1016/j.energy.2020.119592

Bagočius, V., Zavadskas, E. K., & Turskis, Z. (2014). Multi-person selection of the best wind turbine based on the multi-criteria integrated additive-multiplicative utility function. *Journal of Civil Engineering and Management, 20*(4), 590–599. DOI: 10.3846/13923730.2014.932836

Balvanera, P., Brauman, K. A., Cord, A. F., Drakou, E. G., Geijzendorffer, I. R., Karp, D. S., Martín-López, B., Mwampamba, T. H., & Schröter, M. (2022). Essential ecosystem service variables for monitoring progress towards sustainability. *Current Opinion in Environmental Sustainability*, *54*, 101152. DOI: 10.1016/j.cosust.2022.101152

Bauwens, T., Huybrechts, B., & Dufays, F. (2016). Understanding the diverse scaling strategies of social enterprises as hybrid organizations: The case of renewable energy cooperatives. *Organization & Environment*, *29*(3), 263–283.

Baykara, S. Z. (2018). Hydrogen: A brief overview on its sources, production and environmental impact. *International Journal of Hydrogen Energy*, *43*(23), 10605–10614. DOI: 10.1016/j.ijhydene.2018.02.022

Baylon, C., Brunt, R., & Livingstone, D. (2015). *Cyber security at civil nuclear facilities: Understanding the risks*. Chatham House. https://www.chathamhouse.org/2015/10/cyber-security-civil-nuclear-facilities

Bazaraa, M. S., Sherali, H. D., & Shetty, C. M. (2005). *Nonlinear Programming*. https://doi.org/DOI: 10.1002/0471787779

Bergeron, A., & Crigger, J. B. (2018). Early progress on additive manufacturing of nuclear fuel materials. *Journal of Nuclear Materials*, *508*, 344–347. DOI: 10.1016/j.jnucmat.2018.05.058

Betzler, B. R., Ade, B. J., Wysocki, A. J., Jain, P. K., Chesser, P. C., Greenwood, M. S., & Terrani, K. A. (2020). Transformational Challenge Reactor preconceptual core design studies. *Nuclear Engineering and Design*, *367*, 110781. DOI: 10.1016/j.nucengdes.2020.110781

Bilge, L., & Dumitras, T. (2012). Before we knew it: An empirical study of zero-day attacks in real world. *Proceedings of 2012 ACM Conference on Computer & Communications Security*, 833-844. https://doi.org/DOI: 10.1145/2382196.2382284

Bilgin, U., & Soner Kara, S. (2024). Identification of perceived challenges in the green energy transition by Turkish society through sentiment analysis. *Sustainability (Basel)*, *16*(6), 3367. DOI: 10.3390/su16083367

Birge, J. R., & Louveaux, F. (2011). *Introduction to Stochastic Programming*. Springer New York., DOI: 10.1007/978-1-4614-0237-4

Black, G., Shropshire, D., Araújo, K., & Van Heek, A. (2022). Prospects for Nuclear Microreactors: A review of the technology, economics, and regulatory considerations. *Nuclear Technology, 209*(sup1), S1–S20. DOI: 10.1080/00295450.2022.2118626

Black, G., Shropshire, D., & Araújo, K. (2021). 22 - Small modular reactor (SMR) adoption : Opportunities and challenges for emerging markets. In Ingersoll, D. T., & Carelli, M. D. (Eds.), *Handbook of Small Modular Nuclear Reactors* (2nd ed., pp. 557–593). Woodhead Publishing., DOI: 10.1016/B978-0-12-823916-2.00022-9

Bodel, W., Butler, G., & Matthews, J. (2021). *Nuclear energy for net zero—Dalton Nuclear Institute—The University of Manchester*. https://www.dalton.manchester.ac.uk/nuclear-energy-for-net-zero/

Bohrson, W. A., Spera, F. J., Heinonen, J. S., Brown, G. A., Scruggs, M. A., Adams, J. V., Takach, M. K., Zeff, G., & Suikkanen, E. (2020). Diagnosing open-system magmatic processes using the Magma Chamber Simulator (MCS) : Part I—major elements and phase equilibria. *Contributions to Mineralogy and Petrology*, *175*(11), 104. DOI: 10.1007/s00410-020-01722-z

Bonjean Stanton, M. C. (2019). *Weathering the storm: Managing physical climate risks in the electricity sectors in the United Kingdom and France* (Doctoral dissertation, University of Leeds). University of Leeds. https://etheses.whiterose.ac.uk/id/eprint/25248/

Boyd, S., & Vandenberghe, L. (2004). *Convex Optimization*. https://doi.org/DOI: 10.1017/CBO9780511804441

Boyes, H. (2015). Cybersecurity & cyber-resilient supply chains. *Technology Innovation Management Review*, *5*(4), 28–34. https://timreview.ca/article/888. DOI: 10.22215/timreview/888

Bragg-Sitton, S. M., Gorman, J., Burton, G., Moore, M., Siddiqui, A., Nagasawa, T., . . . Cox, J. (2020). *Flexible nuclear energy for clean energy systems* (NREL/TP-6A50-77088). National Renewable Energy Laboratory. https://www.nrel.gov/docs/fy20osti/77088.pdf

Brillas, E., & Garcia-Segura, S. (2020). Benchmarking recent advances and innovative technology approaches of Fenton, photo-Fenton, electro-Fenton, and related processes: A review on the relevance of phenol as model molecule. *Separation and Purification Technology*, *237*, 116337. DOI: 10.1016/j.seppur.2019.116337

Brown, N. R. (2022). Engineering demonstration reactors: A stepping stone on the path to deployment of advanced nuclear energy in the United States. *Energy*, *238*, 121750. DOI: 10.1016/j.energy.2021.121750

Burns, J. R., Hernandez, R., Terrani, K. A., Nelson, A. T., & Brown, N. R. (2020). Reactor and fuel cycle performance of light water reactor fuel with 235U enrichments above 5%. *Annals of Nuclear Energy, 142*, 107423. DOI: 10.1016/j.anucene.2020.107423

Buzzetti, R., Frano, R. L., & Cancemi, A. S. (2024). Feasibility study of a desalination plant powered by SMR. *Nuclear Engineering and Design, 418*, 112897. DOI: 10.1016/j.nucengdes.2023.112897

BWR – Boiling water reactor. (2014). Nuclear Power. https://www.nuclear-power.com/bwr-boiling-water-reactor/

Calabrese, M., Portarapillo, M., Di Nardo, A., Venezia, V., Turco, M., Luciani, G., & Di Benedetto, A. (2024). Hydrogen safety challenges: A comprehensive review on production, storage, transport, utilization, and CFD-based consequence and risk assessment. *Energies, 17*(6), 1350. DOI: 10.3390/en17061350

Canny, A., Kaziba, C., Mitchell, N., & Prest, E. (2020). *Effective and sustainable capacity-building for achieving the Paris Agreement: A special report on climate finance and digitalization.* Climate Finance & Digitalization Institute.

Capps, N., Sweet, R., Wirth, B. D., Nelson, A., & Terrani, K. (2020). Development and demonstration of a methodology to evaluate high burnup fuel susceptibility to pulverization under a loss of coolant transient. *Nuclear Engineering and Design, 366*, 110744. DOI: 10.1016/j.nucengdes.2020.110744

Carpenter, P. K., & Shaltens, R. K. (2024). Comprehensive modeling & characterization of General-Purpose Heat Source Radioisotope Thermoelectric Generator (GPHS-RTG). *Applied Thermal Engineering, 234*, 121234. DOI: 10.2139/ssrn.4725045

Carter, J., & Stillerman, K. P. (2020). *Rebuilding Science and Scientific Integrity at the US Department of Agriculture*. Union of Concerned Scientists. https://www.jstor.org/stable/resrep26328

Carter, R., Choularton, R., Ferdinand, T., Ding, H., Ginoya, N., & Preethan, P. (2021). *Food Systems at Risk : Transformative Adaptation for Long-Term Food Security*. https://www.wri.org/research/food-systems-risk

Case, D. U. (2016). *Analysis of cyber attack on the Ukrainian power grid.* E-ISAC. https://www.nerc.com/pa/CI/ESISAC/Documents/E-ISAC_SANS_Ukraine_DUC_18Mar2016.pdf

Cerling, T. E., Hart, J. A., & Hart, T. B. (2004). Stable isotope ecology in the Ituri Forest. *Oecologia, 138*(1), 5–12. DOI: 10.1007/s00442-003-1375-4 PMID: 14530961

Ćetković, S., & Stockburger, J. (2024). Hydrogen Strategy of Sweden : Unpacking the Multiple Drivers and Potential Barriers to Hydrogen Development. In R. Quitzow & Y. Zabanova (Éds.), *The Geopolitics of Hydrogen : Volume 1 : European Strategies in Global Perspective* (p. 191-211). Springer Nature Switzerland. DOI: 10.1007/978-3-031-59515-8_10

Chakraborty, S., & Prasad, K. (2018). A Quality Function Deployment-Based Expert System for Cotton Fibre Selection. *Journal of The Institution of Engineers (India): Series E, 99*(1), 43–53.

Chandler, D., Betzler, B., Cook, D., Ilas, G., & Renfro, D. (2019). Neutronic and thermal-hydraulic feasibility studies for High Flux Isotope Reactor conversion to low-enriched uranium silicide dispersion fuel. *Annals of Nuclear Energy, 130,* 277–292. DOI: 10.1016/j.anucene.2019.02.037

Cheng, L., Gao, R., Yan, B., Yang, Z., Li, B., Xu, Q., Liu, P., Chu, M., Bai, B., & Zhang, P. (2019). SiC whisker strengthened UO2–Mo composite pellets with superior thermal-mechanical properties. *Ceramics International, 45*(17), 23377–23383. DOI: 10.1016/j.ceramint.2019.08.039

Cho, C. S., Chung, W. H., & Kuo, S. Y. (2016). Cyber-physical security & dependability analysis of digital control systems in nuclear power plants. *IEEE Transactions on Systems, Man, and Cybernetics. Systems, 46*(3), 356–369. DOI: 10.1109/TSMC.2015.2452897

Cizelj, L., Pesznyák, C., Starflinger, J., Pavel, G. L., Wastin, F., & Michailidou, E. (2024). Towards strategic agenda for European nuclear education, training, and knowledge management. *Nuclear Engineering and Design, 420,* 113001. DOI: 10.1016/j.nucengdes.2024.113001

Clayton, M. (2010). Stuxnet malware is 'weapon' out to destroy Iran's Bushehr nuclear plant? *The Christian Science Monitor.* https://www.csmonitor.com/USA/2010/0921/Stuxnet-malware-is-weapon-out-to-destroy-Iran-s-Bushehr-nuclear-plant

Connolly, D., Lund, H., Mathiesen, B. V., & Leahy, M. (2010). A review of computer tools for analysing the integration of renewable energy into various energy systems. *Applied Energy, 87*(4), 1059–1082. DOI: 10.1016/j.apenergy.2009.09.026

Conti, M., Donadel, D., & Turrin, F. (2021). A survey on industrial control system testbeds & datasets for security research. *arXiv preprint arXiv:2102.05631.* https://arxiv.org/abs/2102.05631

Cortella, L., Albino, C., Tran, Q. K., & Froment, K. (2020). 50 years of French experience in using gamma rays as a tool for cultural heritage remedial conservation. *Radiation Physics and Chemistry*, *171*, 108726. DOI: 10.1016/j.radphyschem.2020.108726

Cowley, A., & Yamamoto, Y. (2024). Advances in in-situ resources utilization for extraterrestrial construction. *Advances in Space Research*, *75*(2), 359–371. DOI: 10.1016/j.asr.2024.05.012

Cramer, C. L., Ionescu, E., Graczyk-Zajac, M., Nelson, A. T., Katoh, Y., Haslam, J. J., Wondraczek, L., Aguirre, T. G., LeBlanc, S., Wang, H., Masoudi, M., Tegeler, E., Riedel, R., Colombo, P., & Minary-Jolandan, M. (2022). Additive manufacturing of ceramic materials for energy applications: Road map and opportunities. *Journal of the European Ceramic Society*, *42*(7), 3049–3088. DOI: 10.1016/j.jeurceramsoc.2022.01.058

Crawford, D. C., Porter, D. L., Hayes, S. L., Meyer, M. K., Petti, D. A., & Pasamehmetoglu, K. (2007). An approach to fuel development and qualification. *Journal of Nuclear Materials*, *371*(1–3), 232–242. DOI: 10.1016/j.jnucmat.2007.05.029

Dahal, S. (2023). Plasma Physics and Fusion Energy : A Promising Future for Sustainable Power Generation. *Koshi Pravah: Multidisciplinary Peer Reviewed Journal*, *2*(1), 1. Advance online publication. DOI: 10.3126/koshipravah.v2i1.67130

Daigle, B., DeCarlo, S., & Lotze, N. & Office of Industry and Competitiveness Analysis. (2024). Big change goes small: Are small modular reactors (SMRs) the future of nuclear energy? In *Working Paper ICA-105*.

Dasgupta, B., & Sen, S. R. (2019). Cyber attack at Kudankulam; critical system safe. *Hindustan Times*. https://www.hindustantimes.com/india-news/cyber-attack-at-kudankulam-critical-system-safe/story-8j5Vb6Y5L5z5N5z5N5z5N.html

Davidson, A. J. (2022). *The Role of Nuclear Energy in the Global Energy Transition*. Oxford Institute for Energy Studies. https://www.oxfordenergy.org/publications/the-role-of-nuclear-energy-in-the-global-energy-trasition/

de Vendinha Ferreira, M. (2023). *Possible impacts of nuclear fusion deployment on spatial planning practices: Case studies of ITER and NIF projects.*

Defense Advanced Research Projects Agency (DARPA). (2023). Demonstration Rocket for Agile Cislunar Operations (DRACO). https://www.darpa.mil/research/programs/demonstration-rocket-for-agile-cislunar-operations

DeHart, M. D., Karriem, Z., Pope, M. A., & Johnson, M. P. (2018). Fuel element design and analysis for potential LEU conversion of the Advanced Test Reactor. *Progress in Nuclear Energy*, *104*, 117–135. DOI: 10.1016/j.pnucene.2017.09.007

Demkowicz, P. A., Liu, B., & Hunn, J. D. (2019). Coated particle fuel: Historical perspectives and current progress. *Journal of Nuclear Materials*, *515*, 434–450. DOI: 10.1016/j.jnucmat.2018.09.044

Di Lemma, F. G., Liu, X., Holschuh, T. V.II, Folsom, C. P., Murray, D. J., Teng, F., & Jensen, C. B. (2020). Investigation of the microstructure evolution of alpha uranium after in pile transient. *Journal of Nuclear Materials*, *542*, 152467. DOI: 10.1016/j.jnucmat.2020.152467

Di Maio, F., Bani, L., & Zio, E. (2022). The contribution of small modular reactors to the resilience of power supply. *Journal of Nuclear Engineering*, *3*(2), 152–162. DOI: 10.3390/jne3020009

Diehl, J. F. (2002). Food irradiation—Past, present and future. *Radiation Physics and Chemistry*, *63*(3–6), 211–215. DOI: 10.1016/S0969-806X(01)00622-3

Ding, Y., & Shi, T. (2024). Sustainable LLM serving: Environmental implications, challenges, and opportunities. In *Proceedings of the 2024 IEEE 15th International Green and Sustainable Computing Conference (IGSC)* (pp. 37–38). IEEE. https://doi.org/DOI: 10.1109/IGSC57954.2024.00017

Doncel, N., Martínez, L., Aragón, P., Feria, F., Herranz, L. E., Queral, C., Fernandez-Cosials, K., Ruiz-Hervias, J., Cristobal-Beneyto, M., Castro, E., Sanchez-Torrijos, J., & Soler, A. (2024). R&D in advanced technology fuels (ATFs) in Spain. *Nuclear Engineering and Design*, *424*, 113246. DOI: 10.1016/j.nucengdes.2024.113246

Drouineau, M., Maïzi, N., & Mazauric, V. (2014). Impacts of intermittent sources on the quality of power supply : The key role of reliability indicators. *Applied Energy*, *116*, 333–343. DOI: 10.1016/j.apenergy.2013.11.069

Drucker, D. G. (2022). The isotopic ecology of the mammoth steppe. *Annual Review of Earth and Planetary Sciences*, *50*(1), 395–418. DOI: 10.1146/annurev-earth-100821-081832

Duggirala, V. (2022). *The history of invention of Radioisotope Thermoelectric Generators (RTGs) for space exploration*. Wiley., DOI: 10.1002/9781119811398.ch1

Edenhofer, O., Pichs-Madruga, R., Sokona, Y., Farahani, E., Kadner, S., Seyboth, K., Adler, A., Baum, I., Brunner, S., Eickemeier, P., Kriemann, B., Savolainen, J., Schlömer, S., von Stechow, C., Zwickel, T., & Minx, J. C. (Eds.). (2014). *Climate change 2014: Mitigation of climate change. Contribution of Working Group III to the Fifth Assessment Report*. Cambridge University Press.

Ehleringer, J. R., Barnette, J. E., Jameel, Y., Tipple, B. J., & Bowen, G. J. (2016). Urban water–a new frontier in isotope hydrology. *Isotopes in Environmental and Health Studies*, *52*(4–5), 477–486. DOI: 10.1080/10256016.2016.1171217 PMID: 27142528

Eidemüller, D. (2021). Radioactive Incidents and Disasters. In Eidemüller, D. (Ed.), *Nuclear Power Explained* (pp. 195–240). Springer International Publishing., DOI: 10.1007/978-3-030-72670-6_10

El-Emam, R. S., Constantin, A., Bhattacharyya, R., Ishaq, H., & Ricotti, M. E. (2024). Nuclear and renewables in multipurpose integrated energy systems : A critical review. *Renewable & Sustainable Energy Reviews*, *192*, 114157. DOI: 10.1016/j.rser.2023.114157

El-Guebaly, L. A. (2010). Fifty years of magnetic fusion research (1958–2008): Brief historical overview and discussion of future trends. *Energies*, *3*(6), 1067–1086. DOI: 10.3390/en30601067

Eljack, F., & Kazi, M. K. (2021). Prospects and challenges of green hydrogen economy via multi-sector global symbiosis in Qatar. *Frontiers in Sustainability*, *1*, 612762. DOI: 10.3389/frsus.2020.612762

Elmqvist, T., Andersson, E., Frantzeskaki, N., McPhearson, T., Olsson, P., Gaffney, O., Takeuchi, K., & Folke, C. (2019). Sustainability and resilience for transformation in the urban century. *Nature Sustainability*, *2*(4), 267–273. DOI: 10.1038/s41893-019-0250-1

Endo, N., Goshome, K., Tetsuhiko, M., Segawa, Y., Shimoda, E., & Nozu, T. (2021). Thermal management and power saving operations for improved energy efficiency within a renewable hydrogen energy system utilizing metal hydride hydrogen storage. *International Journal of Hydrogen Energy*, *46*(1), 262–271. DOI: 10.1016/j.ijhydene.2020.10.002

Ewan, B. C. R., & Allen, R. W. K. (2005). A figure of merit assessment of the routes to hydrogen. *International Journal of Hydrogen Energy*, *30*(8), 809–819. DOI: 10.1016/j.ijhydene.2005.02.003

Fairén, A. G., & Davila, A. F. (2023). Planetary protection: An international concern & responsibility. *Frontiers in Astronomy and Space Sciences, 10*, 1172546. DOI: 10.3389/fspas.2023.1172546

Fang, Y., & Zhang, W. (2019). Enhancing traffic management efficiency using predictive modeling and AI. *Transportation Research Part C, Emerging Technologies, 101*, 108–122. DOI: 10.1016/j.trc.2019.02.008

Faria, F. A., Silva-Costa, A., Gianuca, D., & Bugoni, L. (2016). Cocoi heron (Ardea cocoi) connects estuarine, coastal, limnetic and terrestrial environments: An assessment based on conventional dietary and stable isotope analysis. *Estuaries and Coasts, 39*(5), 1271–1281. DOI: 10.1007/s12237-016-0073-5

Farkas, J., & Mohácsi-Farkas, C. (2011). History and future of food irradiation. *Trends in Food Science & Technology, 22*(2–3), 121–126. DOI: 10.1016/j.tifs.2010.04.002

Farrell, A. E., Keith, D. W., & Corbett, J. J. (2003). A strategy for introducing hydrogen into transportation. *Energy Policy, 31*(13), 1357–1367. DOI: 10.1016/S0301-4215(02)00195-7

Farwell, J. P., & Rohozinski, R. (2011). Stuxnet & the future of cyber war. *Survival, 53*(1), 23–40. DOI: 10.1080/00396338.2011.555586

Fernández-Arias, P., Antón-Sancho, Á., Lampropoulos, G., & Vergara, D. (2024). Emerging trends and challenges in pink hydrogen research. *Energies, 17*(5), 2291. DOI: 10.3390/en17102291

Fernández-Arias, P., Antón-Sancho, Á., Lampropoulos, G., & Vergara, D. (2024). On green hydrogen generation technologies: A bibliometric review. *Applied Sciences (Basel, Switzerland), 14*(5), 2524. DOI: 10.3390/app14062524

Financial Ratio Analysis in Launchpad Learning Labs Pvt Ltd. (2019). *International Journal of Recent Technology and Engineering, 8(2S8)*, 30–35. https://doi.org/DOI: 10.35940/ijrte.B1308.0882S819

Fleten, S.-E., & Pettersen, E. (2005). Constructing Bidding Curves for a Price-Taking Retailer in the Norwegian Electricity Market. *IEEE Transactions on Power Systems, 20*(2), 701–708. DOI: 10.1109/TPWRS.2005.846082

Forsberg, C. W., & Reich, W. J. (1991). Worldwide advanced nuclear power reactors with passive and inherent safety: What, why, how, and who. Oak Ridge National Laboratory., , ORNL/TM-11907.DOI: 10.2172/6365276

Fragkos, P. (2020). Global energy system transformations to 1.5 °C: The impact of revised Intergovernmental Panel on Climate Change carbon budgets. *Energy Technology (Weinheim)*, *8*(9), 2000395. DOI: 10.1002/ente.202000395

Freire, L. O., & Andrade, D. A. (2021). Novel technological developments with impacts on perspectives for mobile nuclear power plants. *World Journal of Nuclear Science and Technology*, *11*(2), 141–158. DOI: 10.4236/wjnst.2021.114011

Fukami, T. (2024). Enhancing Healthcare Accountability for Administrators : Fostering Transparency for Patient Safety and Quality Enhancement. *Cureus*, *16*. DOI: 10.7759/cureus.66007 PMID: 39221336

Garud, Y. S., & Rebak, R. B. (2023). Effect of surface oxides on tritium entrance and permeation in FeCrAl alloys for nuclear fuel cladding: A review. *Corrosion Reviews*, *41*(2), 143–169. DOI: 10.1515/corrrev-2022-0033

Gas-cooled reactor and advanced gas-cooled reactor (AGR). (2025). Energy Encyclopedia. https://www.energyencyclopedia.com/en/nuclear-energy/the-nuclear-reactors/gas-cooled-reactor-gcr-and-advanced-gas-cooled-reactor-agr

Germany's Energiewende—World Nuclear Association. (s. d.). Consulté 30 janvier 2025, à l'adresse https://world-nuclear.org/information-library/energy-and-the-environment/energiewende

Gibson, M. A., & Poston, D. I. (2020). Kilopower Project: The KRUSTY fission power experiment & potential missions. *Nuclear Technology*, *206*(9), 1247–1255. DOI: 10.1080/00295450.2020.1722554

Glasstone, S. (1955). *Principles of nuclear reactor engineering*. D. Van Nostrand Company.

Glasstone, S., & Sesonske, A. (1994). *Nuclear reactor engineering: Reactor systems engineering* (4th ed.). Springer International Publishing AG.

Global energy & CO_2 status report 2019 (2019). IEA. https://www.iea.org/reports/global-energy-co2-status-report-2019

Global energy transformation: A roadmap. to 2050. https://www.irena.org/publications/2019/Apr/Global-energy-transformation-A-roadmap-to-2050-2019Edition

Goh, J., Tjahjono, B., & Brawn, D. (2019). Predictive maintenance: Towards the next generation of maintenance management. *Journal of Industrial Information Integration*, *15*, 1–10.

Gong, B., Yao, T., Lei, P., Harp, J., Nelson, A. T., & Lian, J. (2020). Spark plasma sintering (SPS) densified U3Si2 pellets: Microstructure control and enhanced mechanical and oxidation properties. *Journal of Alloys and Compounds*, *825*, 154022. DOI: 10.1016/j.jallcom.2020.154022

Gong, Q., & Jia, H. (2019). Frameworks for healthcare delivery optimization in smart cities. *Smart Health (Amsterdam, Netherlands)*, *10*, 100131. DOI: 10.1016/j.smhl.2019.100131

González, E., & García, J. (2016). Nuclear fission and fusion: Comparative analysis and perspectives. *Energy Policy*, (94), 51–60.

Goodfellow, I., Bengio, Y., & Courville, A. (2016). *Deep Learning*. MIT Press.

Green, M. A., Ho-Baillie, A., & Snaith, H. J. (2018). The emergence of perovskite solar cells. *Nature Photonics*, *8*(7), 506–514. DOI: 10.1038/nphoton.2014.134

Greiman, V. A. (2023). Nuclear cyber attacks: A study of sabotage & regulation of nuclear facilities. *Proceedings of 18th International Conference on Cyber Warfare & Security*, 1042-1051. https://papers.academic-conferences.org/index.php/iccws/article/download/1042/932

Gross, R., Leach, M., & Bauen, A. (2003). Progress in renewable energy. *Environment International*, *29*(1), 105–122. DOI: 10.1016/S0160-4120(02)00130-7 PMID: 12605943

Guérin, Y., Was, G. S., & Zinkle, S. J. (2009). Materials challenges for advanced nuclear energy systems. *MRS Bulletin*, *34*(1), 10–19. DOI: 10.1557/mrs2009.6

Gungor, G., & Sari, R. (2022). Nuclear power and climate policy integration in developed and developing countries. *Renewable & Sustainable Energy Reviews*, *169*, 112839. DOI: 10.1016/j.rser.2022.112839

Guo, X., Hu, Y., Zhang, Y., & Zhen, L. (2022). Using 137Cs and 210Pbex to investigate the soil erosion moduli of the sandy area of typical grasslands in northern China. *Sustainability (Basel)*, *14*(18), 12137. DOI: 10.3390/su141912137

Gussev, M. N., Sridharan, N., Babu, S. S., & Terrani, K. A. (2021). Influence of neutron irradiation on Al-6061 alloy produced via ultrasonic additive manufacturing. *Journal of Nuclear Materials*, *550*, 152939. DOI: 10.1016/j.jnucmat.2021.152939

Hajipour, M., Ansarifar, G. R., & Yeganeh, M. H. Z. (2025). Assessment and predicting the effect of accident tolerant fuel composition and geometry on neutronic and safety parameters in small modular reactors via artificial neural network and adaptive neuro-fuzzy inference system. *Nuclear Engineering and Design*, *433*, 113837. DOI: 10.1016/j.nucengdes.2025.113837

Harris, W. (2022). *Developments in Safety Culture—Views by the World Nuclear Association and the International System of Occupational Exposure*. 399-403.

Heavy water reactor (PHWR). (2025). Energy Encyclopedia. https://www.energyencyclopedia.com/en/nuclear-energy/the-nuclear-reactors/heavy-water-reactor-phwr

Hemsley, K. E., & Fisher, R. M. (2018). *History of industrial control system cyber incidents*. U.S. Department of Energy, Idaho National Laboratory., DOI: 10.2172/1505628

Hepp, A. F., Palaszewski, B. A., Colozza, A. J., Landis, G. A., Jaworske, D. A., & Kulis, M. J. (2014). In-situ resource utilization for space exploration: resource processing, mission-enabling technologies, and lessons for sustainability on earth and beyond. In *12th International Energy Conversion Engineering Conference* (p. 3761).

Hidayatullah, H., Susyadi, S., & Subki, M. H. (2015). Design and technology development for small modular reactors–Safety expectations, prospects, and impediments of their deployment. *Progress in Nuclear Energy*, *79*, 127–135. DOI: 10.1016/j.pnucene.2014.11.010

Hong, T., Chou, S. K., & Bong, T. Y. (2009). Building simulation: An overview of developments and information sources. *Building and Environment*, *44*(4), 848–858.

Hsiao, C. Y., Ou, Y., Sheng, N., & Wei, X. (2021). Measuring contagion effects of nuclear energy policies and events. *International Journal of Energy Research*, *45*(8), 11510–11525. DOI: 10.1002/er.6421

Huang, Q., Peng, S., Deng, J., Zeng, H., Zhang, Z., Liu, Y., & Yuan, P. (2023). A review of the application of artificial intelligence to nuclear reactors: Where we are and what's next. *Heliyon*, *9*(3), e12738. DOI: 10.1016/j.heliyon.2023.e13883 PMID: 36895398

Hussein, E. M. (2020). Emerging small modular nuclear power reactors: A critical review. *Physics Open*, *5*, 100038. DOI: 10.1016/j.physo.2020.100038

IEA. (2020). *World Energy Outlook*. https://iea.blob.core.windows.net/assets/a72d8abf-de08-4385-8711-b8a062d6124a/WEO2020.pdf

Ingersoll, D. T., & Popov, E. (2020). Emerging small modular nuclear power reactors: A critical review. *Nuclear Engineering and Design*, *367*, 110411. https://www.researchgate.net/publication/344477235_Emerging_small_modular_nuclear_power_reactors_A_critical_review

International Atomic Energy Agency (IAEA). (2017). Uses of ionizing radiation for tangible cultural heritage conservation (IAEA Radiation Technology Series No. 6). IAEA. https://www.iaea.org/publications/11092/uses-of-ionizing-radiation-for-tangible-cultural-heritage-conservation

International Atomic Energy Agency (IAEA). (2022). International Atomic Energy Agency. https://www.iaea.org

International Atomic Energy Agency. (1992). *The Chernobyl accident: Updating of INSAG-1* (Safety Series No. 75-INSAG-7). IAEA.

International Atomic Energy Agency. (2011). *Computer security at nuclear facilities*. IAEA Nuclear Security Series No. 17. https://www.iaea.org/publications/8939/computer-security-at-nuclear-facilities

International Atomic Energy Agency. (2012). Advances in nuclear power process heat applications. IAEA., IAEA-TECDOC-1682.

International Atomic Energy Agency. (2015). The Fukushima Daiichi accident: Report by the Director General. Technical (Vol. 5). IAEA., STI/PUB/1710.

International Atomic Energy Agency. (2020a). *Nuclear power reactors in the world* (Reference Data Series No. 2). IAEA.

International Atomic Energy Agency. (2020b). *Status of fast reactor research and technology development*. IAEA.

International Atomic Energy Agency. (2021). *Liquid metal coolants for fast reactors cooled by sodium, lead, and lead-bismuth eutectic*. IAEA.

International Atomic Energy Agency. (2022). *What is nuclear energy? The science of nuclear power*. https://www.iaea.org/newscenter/news/what-is-nuclear-energy-the-science-of-nuclear-power

ITER. (n.d.). *The Machine*. Retrieved April 15, 2025 from https://www.iter.org/machine

Iwata, H., Okada, K., & Samreth, S. (2010). Empirical study on the environmental Kuznets curve for CO2 in France: The role of nuclear energy. *Energy Policy*, *38*(8), 4057–4063. DOI: 10.1016/j.enpol.2010.03.031

Jalles, J. T. (2020). The impact of financial crises on the environment in developing countries. *Annals of Finance*, *16*(2), 281–306. DOI: 10.1007/s10436-019-00356-x

Jez, J. M., Topp, C. N., McNicol, G., Yu, Z., Berry, Z. C., Emery, N., Soper, F. M., & Yang, W. H. (2021). Tracing plant–environment interactions from organismal to planetary scales using stable isotopes: A mini review. *Emerging Topics in Life Sciences*, *5*(3), 301–316. DOI: 10.1042/ETLS20200277 PMID: 33634828

Jiang, Y., Oron, T. R., Clark, W. T., Bankapur, A. R., D'Andrea, D., Lepore, R., Funk, C. S., Kahanda, I., Verspoor, K. M., Ben-Hur, A., Koo, D. C. E., Penfold-Brown, D., Shasha, D., Youngs, N., Bonneau, R., Lin, A., Sahraeian, S. M. E., Martelli, P. L., Profiti, G., & Radivojac, P. (2016). An expanded evaluation of protein function prediction methods shows an improvement in accuracy. *Genome Biology*, *17*(1), 184. DOI: 10.1186/s13059-016-1037-6 PMID: 27604469

Jin, T., & Kim, J. (2018). What is better for mitigating carbon emissions – Renewable energy or nuclear energy? A panel data analysis. *Renewable & Sustainable Energy Reviews*, *91*(C), 464–471. DOI: 10.1016/j.rser.2018.04.022

Josephs, R. E., Yap, T., Alamooti, M., Omojiba, T., Benarbia, A., Tomomewo, O., & Ouadi, H. (2025). Regulation of Small Modular Reactors (SMRs): innovative strategies and economic insights. *Eng—Advances in Engineering*, *6*(4), 61. DOI: 10.3390/eng6040061

Kabeyi, M. J. B., & Olanrewaju, O. A. (2022). Sustainable energy transition for renewable and low carbon grid electricity generation and supply. *Frontiers in Energy Research*, *9*, 743114. DOI: 10.3389/fenrg.2021.743114

Kang, J., Kim, H., Kim, J., & Cho, H. (2020). Recent advances in floating solar photovoltaic systems: A review. *Renewable & Sustainable Energy Reviews*, *110*, 332–346.

Karabašević, D., Stanujkić, D., & Urošević, S. (2015). The MCDM Model for Personnel Selection Based on SWARA and ARAS Methods. Management (1820-0222), 20(77).

Kemmerzell, J. (2022). Energy Governance in Germany. In Knodt, M., & Kemmerzell, J. (Eds.), *Handbook of Energy Governance in Europe* (pp. 667–708). Springer International Publishing., DOI: 10.1007/978-3-030-43250-8_11

Kempton, W., & Tomić, J. (2005). Vehicle-to-grid power implementation: From stabilizing the grid to supporting large-scale renewable energy. *Journal of Power Sources*, *144*(1), 280–294. DOI: 10.1016/j.jpowsour.2004.12.022

Kenley, C. R., Klingler, R. D., Plowman, C. M., Soto, R., Turk, R. J., Baker, R. L., Close, S. A., McDonnell, V. L., Paul, S. W., Rabideau, L. R., Rao, S. S., & Reilly, B. P. (2009). Job creation due to the nuclear power resurgence in the United States. *Energy Policy*, *37*(11), 4894–4900. DOI: 10.1016/j.enpol.2009.06.045

Kim, J. H., & Yoo, S. H. (2021). Comparison of the economic effects of nuclear power and renewable energy deployment in South Korea. *Renewable & Sustainable Energy Reviews*, *135*, 110236. DOI: 10.1016/j.rser.2020.110236

Klinger, T., Andreeva, T., Bozhenkov, S., Brandt, C., Burhenn, R., Buttenschön, B., & Hollfeld, K. P. (2019). Overview of the first Wendelstein 7-X high-performance operation. *Nuclear Fusion*, *59*(11), 112004. DOI: 10.1088/1741-4326/ab03a7

Krall, L. M., Macfarlane, A. M., & Ewing, R. C. (2022). Nuclear waste from small modular reactors. *Proceedings of the National Academy of Sciences of the United States of America*, *119*(23), e2111833119. DOI: 10.1073/pnas.2111833119 PMID: 35639689

Kruger, W., & Eberhard, A. (2018). Renewable energy auctions in sub-Saharan Africa: Comparing the South African, Ugandan, and Zambian Programs. *WIREs Energy and Environment, 7(4)*. Portico. https://doi.org/DOI: 10.1002/wene.295

Kumar, N., Prasad, R., Shankar, R., & Lyer, K. C. (2009). Technology transfer for rural housing: An interpretive structural modelling approach. *Journal of Advances in Management Research*, *6*(2), 188–205. DOI: 10.1108/09727980911007208

Kushner, D. (2013). The real story of Stuxnet. *IEEE Spectrum*. https://spectrum.ieee.org/the-real-story-of-stuxnet

Langner, R. (2011). Stuxnet: The emergence of a new cyber weapon & its implications. *Journal of Computer Virology & Hacking Techniques*, *7*(4), 291–298. https://www.researchgate.net/publication/241736813_Stuxnet_The_emergence_of_a_new_cyber_weapon_&_its_implications

Laraia, M. (2018). *Nuclear decommissioning*. Springer., DOI: 10.1007/978-3-319-75916-6

Lee, R. M., Assante, M. J., & Conway, T. (2016). *Analysis of cyber-attack on the Ukrainian power grid*. SANS Industrial Control Systems. https://ics.sans.org/media/E-ISAC_SANS_Ukraine_DUC_5.pdf

Lee, J. I. (2024). Review of Small Modular Reactors: Challenges in Safety and Economy to Success. *Korean Journal of Chemical Engineering*, *41*(10), 2761–2780. DOI: 10.1007/s11814-024-00207-0

Lenzen, M. (2008). Life cycle energy and greenhouse gas emissions of nuclear energy : A review. *Energy Conversion and Management*, *49*(8), 2178–2199. DOI: 10.1016/j.enconman.2008.01.033

Lenzen, M. (2008). Renewable and nuclear electricity: Comparison of environmental impacts. *Energy Policy*, *37*(12), 5058–5068.

Leppert, R., & Kennedy, B. (2024). Growing share of Americans favor more nuclear power. *Pew Research Center*. https://www.pewresearch.org/short-reads/2024/08/05/majority-of-americans-support-more-nuclear-power-in-the-country/

Lights, Z., & Institut, E. I. for M. S. on H. R. and S.-K. (2022). The (Missing) Role of Nuclear Energy in the Sustainable Development Goals. *Journal of the European Institut for Multidisciplinary Studies on Human Rights & Science*, *3*(1). Advance online publication. DOI: 10.5281/zenodo.6632131

Lindley, B. A., Fiorina, C., Gregg, R., Franceschini, F., & Parks, G. T. (2015). The effectiveness of full actinide recycles as a nuclear waste management strategy when implemented over a limited timeframe–Part I: Uranium fuel cycle. *Progress in Nuclear Energy*, *85*, 498–510. DOI: 10.1016/j.pnucene.2015.07.020

Liu, L., Guo, H., Dai, L., Liu, M., Xiao, Y., Cong, T., & Gu, H. (2023). The role of nuclear energy in the carbon neutrality goal. *Progress in Nuclear Energy*, *162*, 104772. DOI: 10.1016/j.pnucene.2023.104772

Liu, N., & Xu, Z. (2021). An overview of ARAS method: Theory development, application extension, and future challenge. *International Journal of Intelligent Systems*, *36*(7), 3524–3565. DOI: 10.1002/int.22425

Li, Y., You, E., Wang, T., Xing, D., Xu, J., & Zhang, H. (2024). Developments, challenges and prospects in thermal-hydraulic research on accident tolerant fuel. *Heliyon*, *10*(19), e38999. DOI: 10.1016/j.heliyon.2024.e38999 PMID: 39498276

Li, Z., Wang, X., Chen, H., Zhang, R., Wei, T., Yao, L., & Wang, P. (2023). Review on performance of chromium-coated zirconium and its failure mechanisms. *Frontiers in Nuclear Engineering*, *2*, 1212351. Advance online publication. DOI: 10.3389/fnuen.2023.1212351

Locatelli, G., Bingham, C., & Mancini, M. (2014). Small modular reactors: A comprehensive overview of their economics and strategic aspects. *Progress in Nuclear Energy*, *73*, 75–85. DOI: 10.1016/j.pnucene.2014.01.010

Locatelli, G., Mancini, M., & Todeschini, N. (2013). Generation IV nuclear reactors: Current status and prospects. *Energy Policy*, *61*, 1503–1520. DOI: 10.1016/j.enpol.2013.06.101

Lombardi, M., & Fargnoli, M. (2018). Prioritization of hazards by means of a QFD-based procedure. *International Journal of Safety and Security Engineering*, *8*(2), 342–353. DOI: 10.2495/SAFE-V8-N2-342-353

Long, X., Naminse, E. Y., Du, J., & Zhuang, J. (2015). Nonrenewable energy, renewable energy, carbon dioxide emissions and economic growth in China from 1952 to 2012. *Renewable & Sustainable Energy Reviews*, *52*(C), 680–688. DOI: 10.1016/j.rser.2015.07.176

Loulou, R., Waaub, J.-P., & Zaccour, G. (Eds.). (2005). *Energy and Environment*. Springer US., DOI: 10.1007/b107228

Lu, C., Lu, C., Lange, R. T., Foerster, J., Clune, J., & Ha, D. (2024). The AI scientist: Towards fully automated open-ended scientific discovery. arXiv. https://arxiv.org/abs/2408.06292

Lukner, K., & Sakaki, A. (2013). Lessons from Fukushima: An assessment of the investigations of the nuclear disaster. *Asia-Pacific Journal: Japan Focus, 2013*(19). https://apjjf.org/2013/19/19/Lukner/3959/article.html

Lund, H., & Mathiesen, B. V. (2009). Energy system analysis of 100% renewable energy systems—The case of Denmark in years 2030 and 2050. *Energy*, *34*(5), 524–531. DOI: 10.1016/j.energy.2008.04.003

Luo, B., Huang, G., Chen, J., Zhang, X., & Zhao, K. (2021). A chance-constrained small modular reactor siting model: A case study for the Province of Saskatchewan, Canada. *Renewable & Sustainable Energy Reviews*, *148*, 111320. DOI: 10.1016/j.rser.2021.111320

Mallick, C. B., Aditya, A. R. N., & Bhattacharya, C. (2020). Adaptive load balancing strategy for a distributed computing environment. *Journal of Cloud Computing: Advances, Systems and Applications*, *9*(1), 1–19. DOI: 10.1186/s13677-020-00197-8

Manchanda, V. K. (2023). Thorium as an abundant source of nuclear energy and challenges in separation science. *Radiochimica Acta*, *111*(4), 243–263. DOI: 10.1515/ract-2022-0006

Marx, C., Tetzlaff, D., Hinkelmann, R., & Soulsby, C. (2021). Isotope hydrology and water sources in a heavily urbanized stream. *Hydrological Processes*, *35*(3), e14377. DOI: 10.1002/hyp.14377

Masood, R. (2016). *Assessment of cyber security challenges in nuclear power plants*. George Washington University. https://cspri.seas.gwu.edu/sites/g/files/zaxdzs4366/files/downloads/GW-CSPRI-2016-03%2BMASOOD%2BRahat%2BNuclear%2BPower%2BPlant%2BCybersecurity_0.pdf

Mathew, M. D. (2022). Nuclear energy : A pathway towards mitigation of global warming. *Progress in Nuclear Energy, 143*, 104080. DOI: 10.1016/j.pnucene.2021.104080

McCulloch, M., Falter, J., Trotter, J., & Montagna, P. (2012). Coral resilience to ocean acidification and global warming through pH up-regulation. *Nature Climate Change*, *2*(8), 623–627. DOI: 10.1038/nclimate1473

McDonald, R. I., Mansur, A. V., Ascensão, F., Crossman, K., Elmqvist, T., Gonzalez, A., Güneralp, B., Haase, D., Hamann, M., & Hillel, O.. (2020). Research gaps in knowledge of the impact of urban growth on biodiversity. *Nature Sustainability*, *3*(1), 16–24. DOI: 10.1038/s41893-019-0436-6

Meier, O. (2013). *Technology transfers and non-proliferation: Between control and cooperation*. Routledge. DOI: 10.4324/9780203798973

Menyah, K., & Wolde-Rufael, Y. (2010). CO2 emissions, nuclear energy, renewable energy and economic growth in the US. *Energy Policy*, *38*(6), 2911–2915. DOI: 10.1016/j.enpol.2010.01.024

Mignacca, B., & Locatelli, G. (2019). Economics and finance of Small Modular Reactors: A systematic review and research agenda. *Renewable & Sustainable Energy Reviews*, *118*, 109519. DOI: 10.1016/j.rser.2019.109519

Mikovits, C., Wetterlund, E., Wehrle, S., Baumgartner, J., & Schmidt, J. (2021). Stronger together : Multi-annual variability of hydrogen production supported by wind power in Sweden. *Applied Energy*, *282*, 116082. DOI: 10.1016/j.apenergy.2020.116082

Minic, D. (Ed.). (2012). Hydrogen Energy: Challenges and Perspectives.

Mohan, P. (2023). Cybersecurity & nuclear facilities. In D. Carlton (Ed.), *Cybersecurity in critical infrastructures*(pp. 123-145). Springer. https://link.springer.com/chapter/10.1007/978-3-031-56814-5_7link.springer.com

Mohurle, S., & Patil, M. (2017). A brief study of WannaCry threat: Ransomware attack 2017. *International Journal of Advanced Research in Computer Science*, *8*(5), 1938–1940. https://www.proquest.com/docview/1912631307/abstract/DEF9AE2FF2924E35PQ/

Mondal, K., Martinez, O., & Jain, P. (2024). Advanced manufacturing and digital twin technology for nuclear energy. *Frontiers in Energy Research*, *12*, 1339836. DOI: 10.3389/fenrg.2024.1339836

Montesclaros, J. M. L. (2023). The Future of Small Modular Reactors: Implications for Nuclear Governance Introduction.

Mundo, A. (2020, March 26). Triton malware spearheads latest attacks on industrial systems. *Security Intelligence*. https://www.trellix.com/en-in/blogs/research/triton-malware-spearheads-latest-generation-of-attacks-on-industrial-systems1/

Murray, L. J., Dincă, M., & Long, J. R. (2009). Hydrogen storage in metal–organic frameworks. *Chemical Society Reviews, 38*(5), 1294–1314. DOI: 10.1039/b802256a PMID: 19384439

NASA. (2023). NASA demonstration rocket for agile cislunar operation (DRACO) (NASA Technical Report No. 20230017310). NASA Technical Reports Server. https://ntrs.nasa.gov/citations/20230017310

Nature-Inspired Optimization Algorithms. (2014). *Nature-Inspired Optimization Algorithms*. https://doi.org/DOI: 10.1016/B978-0-12-416743-8.00016-6

Nayak, A. K., & Sinha, R. K. (2007). Role of passive systems in advanced reactors. *Progress in Nuclear Energy, 49*(6), 486–498. DOI: 10.1016/j.pnucene.2007.07.007

Neeraj, B. R. (2023). Cybersecurity in Indian nuclear facilities. *Electronic Journal of Social & Strategic Studies*, 4(3), 314-330. https://doi.org/DOI: 10.47362/EJSSS.2023.4302ejsss.net.in

Nelson, A. T. (2023). Prospects for additive manufacturing of nuclear fuel forms. *Progress in Nuclear Energy, 155*, 104493. DOI: 10.1016/j.pnucene.2022.104493

Nelson, V. (2009). *Wind Energy : Renewable Energy and the Environment*. CRC Press., DOI: 10.1201/9781420075694

Neutron Bytes. (2020). *NuScale completes key milestone with NRC.*https://neutronbytes.com/2020/08/29/nuscale-completes-key-milestone-with-nrc

Newsome, S. D., Martinez del Rio, C., Bearhop, S., & Phillips, D. L. (2007). A niche for isotopic ecology. *Frontiers in Ecology and the Environment, 5*(8), 429–436. DOI: 10.1890/1540-9295(2007)5[429:ANFIE]2.0.CO;2

Nicholas, T. E. G., Davis, T. P., Federici, F., Leland, J., Patel, B. S., Vincent, C., & Ward, S. H. (2021). Re-examining the role of nuclear fusion in a renewables-based energy mix. *Energy Policy, 149*, 112043. DOI: 10.1016/j.enpol.2020.112043

Nicholson, E., Watermeyer, K. E., Rowland, J. A., Sato, C. F., Stevenson, S. L., Andrade, A., Brooks, T. M., Burgess, N. D., Cheng, S.-T., Grantham, H. S., Hill, S. L., Keith, D. A., Maron, M., Metzke, D., Murray, N. J., Nelson, C. R., Obura, D., Plumptre, A., Skowno, A. L., & Watson, J. E. M. (2021). Scientific foundations for an ecosystem goal, milestones and indicators for the post-2020 global biodiversity framework. *Nature Ecology & Evolution*, *5*(10), 1338–1349. DOI: 10.1038/s41559-021-01538-5 PMID: 34400825

Nickel, H., Nabielek, H., Pott, G., & Mehner, A. (2002). Long time experience with the development of HTR fuel elements in Germany. *Nuclear Engineering and Design*, *217*(1–2), 141–151. DOI: 10.1016/S0029-5493(02)00128-0

Nie, J., Jia, T., Pan, L., Zhang, X., & Zou, J. J. (2022). Development of high-energy-density liquid aerospace fuel: A perspective. *Transactions of Tianjin University*, *28*(1), 1–5. DOI: 10.1007/s12209-021-00302-x

Nisbet, E. G., Fisher, R. E., Lowry, D., France, J. L., Allen, G., Bakkaloglu, S., Broderick, T. J., Cain, M., Coleman, M., Fernandez, J., et al. (2020). Methane mitigation: Methods to reduce emissions, on the path to the Paris agreement. Reviews of Geophysics, 58, e2019RG000675. DOI: 10.1029/2019RG000675

Nuclear energy: Fuel cycle. (2025). Energy Encyclopedia. https://www.energyencyclopedia.com/en/nuclear-energy/nuclear-fuel/fuel-cycle

Nuclear Power in France—World Nuclear Association. (s. d.). Consulté 30 janvier 2025, à l'adresse https://world-nuclear.org/information-library/country-profiles/countries-a-f/france

Nuclear Power in the USA - World Nuclear Association. (s. d.). Consulté 30 janvier 2025, à l'adresse https://world-nuclear.org/information-library/country-profiles/countries-t-z/usa-nuclear-power

O'Reilly, C. M., Alin, S. R., Plisnier, P. D., Cohen, A. S., & McKee, B. A. (2003). Climate change decreases aquatic ecosystem productivity of Lake Tanganyika, Africa. *Nature*, *424*(6950), 766–768. DOI: 10.1038/nature01833 PMID: 12917682

OECD Nuclear Energy Agency. (2018). *GIF R&D outlook for Generation IV nuclear energy systems*. Gen IV International Forum.

OECD-NEA. (2024). High-assay low-enriched uranium (HALEU): Drivers & supply chain challenges. OECD Nuclear Energy Agency. https://www.oecd-nea.org/upload/docs/application/pdf/2024-09/nea_publication_2_2_2024-09-18_16-53-42_174.pdf

Ohta, H., & Barrett, B. (2023). Politics of climate change and energy policy in Japan : Is green transformation likely? *Earth System Governance, 17*(100187), 100187. Advance online publication. DOI: 10.1016/j.esg.2023.100187

Okada, S., & Momoshima, N. (1993). Overview of tritium: Characteristics, sources, and problems. *Health Physics, 65*(6), 595–609. DOI: 10.1097/00004032-199312000-00001 PMID: 8244708

ORGANISATION FOR ECONOMIC CO-OPERATION AND DEVELOPMENT & OECD Nuclear Energy Agency. (2021). Small modular reactors: challenges and opportunities. In ORGANISATION FOR ECONOMIC CO-OPERATION AND DEVELOPMENT & OECD Nuclear Energy Agency, *Nuclear Technology Development and Economics* [Report]. https://inis.iaea.org/collection/NCLCollectionStore/_Public/52/041/52041043.pdf?r=1

Otaki, T., & Shaw, R. (2024). Global hydrogen energy: Potentials and challenges. *Energy, Sustainability and Resilience: A Futuristic Vision from Asia*, 127-138.

Ould Amrouche, S., Rekioua, D., Rekioua, T., & Bacha, S. (2016). Overview of energy storage in renewable energy systems. *International Journal of Hydrogen Energy, 41*(45), 20914–20927. DOI: 10.1016/j.ijhydene.2016.06.243

Paisiripas, D., Cho, K., & Park, S. (2024). Integration of Small Modular Reactors with Renewable Energy for Carbon Neutrality: A Case Study of Phuket, Thailand. *Energies, 17*(22), 5565. DOI: 10.3390/en17225565

Panov, A. V., Kuznetsov, V. K., Isamov, N. N., Geshel', I. V., Trapeznikov, A. V., & Korzhavin, A. V. (2021). Assessment of the influence of BN-800 operation on the radioecological situation in the vicinity of Beloyarsk NPP. *Atomic Energy, 129*(5), 297–304. DOI: 10.1007/s10512-021-00751-6

Parisi, C., Ma, Z., Mandelli, D., Anderson, N., & Zhang, H. (2020). Risk-Informed Safety Analysis for Accident Tolerant Fuels. *Nuclear Science and Engineering, 194*(8–9), 748–770. DOI: 10.1080/00295639.2020.1732699

Parsons, D. K. (2017). *Some notes on neutron up-scattering and the Doppler-broadening of high-Z scattering resonances (No. LA-UR-17-28809).* Los Alamos National Laboratory., https://www.osti.gov/servlets/purl/1371457 DOI: 10.2172/1396101

Peakman, A., & Lindley, B. (2023). A review of nuclear electric fission space reactor technologies for powering future space missions. *Progress in Nuclear Energy, 158*, 103123. DOI: 10.1016/j.pnucene.2023.104815

Pereira, M. A. G., Domingos, M., da Silva, E. A., Aragaki, S., Ramon, M., de Camargo, P. B., & Ferreira, M. L. (2022). Isotopic composition (δ13C and δ15N) in the soil-plant system of subtropical urban forests. *The Science of the Total Environment*, *851*, 158052. DOI: 10.1016/j.scitotenv.2022.158052 PMID: 35988596

Pervès, J.-P. (2022). *What are the CO2 emissions of nuclear power?* Polytechnique Insights. https://www.polytechnique-insights.com/en/columns/energy/carbon-footprint-of-nuclear-reactors-in-france/

Petrie, C. M., Linton, K. D., Vasudevamurthy, G., Schappel, D., Seibert, R. L., Carpenter, D., Nelson, A. T., & Terrani, K. A. (2023). Fission gas retention of densely packed uranium carbonitride tristructural-isotropic fuel particles in a 3D printed SiC matrix. *Journal of Nuclear Materials*, *580*, 154419. DOI: 10.1016/j.jnucmat.2023.154419

Podder, A., Roy, P., Ghosh, S., Sarkar, S. K., & Choudhury, R. R. (2025). Citizen Engagement and Smart Governance. In Ahmad, M., Mumtaz, R., & Khan, M. (Eds.), *Leveraging IoT and Machine Learning for Smart Urban Planning* (pp. 35–68). IGI Global Scientific Publishing., DOI: 10.4018/979-8-3693-9030-6.ch002

Pokharel, S. (2007). Kyoto protocol and Nepal's energy sector. *Energy Policy*, *35*(4), 2514–2525. DOI: 10.1016/j.enpol.2006.09.015

Ponta, F. L., Amador, J., & Ponta, O. (2017). *Harvesting ocean energy: Mechanical conversion systems*. Springer.

Poornima, B. (2022). Cyber threats & nuclear security in India. *Asian Journal of International Affairs*, *2*(1), 34–50. DOI: 10.1177/23477970221099748journals.sagepub.com

Poresky, C., Andreades, C., Kendrick, J., & Peterson, P. F. (2017). *Cyber security in nuclear power plants: Insights for advanced nuclear technologies*. University of California., https://www.researchgate.net/publication/321443750_Cyber_Security_in_Nuclear_Power_Plants_Insights_for_Advanced_Nuclear_Technologies

Pouryousefi, M., & Ashuri, T. (2019). Optimizing multi-objective problems with Pareto Frontier in Renewable Energy Systems. *Energy Reports*, *5*, 198–206. DOI: 10.1016/j.egyr.2019.08.043

Prasad, S., Venkatramanan, V., & Singh, A. (2020). Renewable Energy for a Low-Carbon Future: Policy Perspectives. *Sustainable Bioeconomy*, 267–284. https://doi.org/DOI: 10.1007/978-981-15-7321-7_12

Proskuryakova, L. (2023). The contribution of low-carbon energy technologies to climate resilience. *Climate (Basel)*, *11*(12), 231. DOI: 10.3390/cli11120231

Prosvirova, Y. (2018). *International radioactive waste for long-term disposal* (Doctoral dissertation, University of Wien). Wien University.

Rad, M. M., & Khosravi, F. (2021). Machine learning for optimal waste management in smart cities. *Renewable & Sustainable Energy Reviews, 137*, 110622. DOI: 10.1016/j.rser.2020.110622

Rajca, M. (2024). Privacy risks and regulatory challenges in smart grids and renewable energy systems. [iKAR]. *Internetowy Kwartalnik Antymonopolowy i Regulacyjny, 13*(2), 9–25. DOI: 10.7172/2299-5749.IKAR.2.13.1

Ramana, M. V. (2021). Small modular and advanced nuclear reactors: a reality check. In *IEEEAccess*. https://creativecommons.org/licenses/by/4.0/

Rand, D.A.J., & Dell, R.M. (2007). Hydrogen energy: challenges and prospects.

Rasul, M. G., Hazrat, M. A., Sattar, M. A., Jahirul, M. I., & Shearer, M. J. (2022). The future of hydrogen: Challenges on production, storage and applications. *Energy Conversion and Management, 272*, 116326. DOI: 10.1016/j.enconman.2022.116326

Raza, M. K., Alghassab, M., Altamimi, A., Khan, Z. A., Kazmi, S. A., Ali, M., & Diala, U. (2024). Integration of very small modular reactors and renewable energy resources in the microgrid. *Frontiers in Energy Research, 12*, 1365735. Advance online publication. DOI: 10.3389/fenrg.2024.1365735

Rebak, R. B. (2023). Improved and Innovative Accident-Tolerant Nuclear Fuel Materials Considered for Retrofitting Light Water Reactors—A Review. *Corrosion and Materials Degradation, 4*(3), 466–487. DOI: 10.3390/cmd4030024

Reddy, C. K. K., Anisha, P. R., Khan, S., Hanafiah, M. M., Pamulaparty, L., & Mohana, R. M. (Eds.). (2024). *Sustainability in Industry 5.0: Theory and applications*. CRC Press.

Reinders, L. J. (2021). What is Nuclear Fusion? In Reinders, L. J. (Ed.), *Sun in a Bottle?... Pie in the Sky!: The Wishful Thinking of Nuclear Fusion Energy* (pp. 1–11). Springer International Publishing., DOI: 10.1007/978-3-030-74734-3_1

Richter, J. (2014). Small Modular Reactors: The future of nuclear energy? *ResearchGate, 1–4*, 1–4. Advance online publication. DOI: 10.1109/ETHICS.2014.6893449

Roberts, P., Perera, N., Wedage, O., Deraniyagala, S., Perera, J., Eregama, S., Petraglia, M. D., & Lee-Thorp, J. A. (2017). Fruits of the forest: Human stable isotope ecology and rainforest adaptations in Late Pleistocene and Holocene (~36 to 3 ka) Sri Lanka. *Journal of Human Evolution, 106*, 102–118. DOI: 10.1016/j.jhevol.2017.01.015 PMID: 28434535

Rodrigues, E. A., Aquino, A. R., & Carvalho, A. R. (2021). Convergence between the study of ecosystem services and nuclear technology–A necessary approach. *Brazilian Journal of Radiation Sciences*, *9*(1A), 1–18. DOI: 10.15392/bjrs.v9i1A.1567

Rosner, R., & Goldberg, S. Energy Policy Institute at Chicago, The Harris School of Public Policy Studies, Hezir, J. S., & EOP Foundation, Inc. (2011). *Small modular reactors – key to future nuclear power generation in the U.S.*https://www.energy.gov/sites/prod/files/2015/12/f27/ECON-SMRKeytoNuclearPowerDec2011.pdf

Rowinski, M. K., White, T. J., & Zhao, J. (2015). Small and Medium sized Reactors (SMR) : A review of technology. *Renewable & Sustainable Energy Reviews*, *44*, 643–656. DOI: 10.1016/j.rser.2015.01.006

Ru, Y., Kleissl, J., & Martinez, S. (2012). Battery sizing for grid-connected PV systems with fixed minimum charging/discharging time. *2012 American Control Conference (ACC)*, 270–275. https://doi.org/DOI: 10.1109/ACC.2012.6314783

Sachs, J. D., Schmidt-Traub, G., Mazzucato, M., Messner, D., Nakicenovic, N., & Rockström, J. (2019). Six transformations to achieve the sustainable development goals. *Nature Sustainability*, *2*(9), 805–814. DOI: 10.1038/s41893-019-0352-9

Saidi, K., & Omri, A. (2020). Reducing CO2 emissions in OECD countries : Do renewable and nuclear energy matter? *Progress in Nuclear Energy*, *126*, 103425. DOI: 10.1016/j.pnucene.2020.103425

Salman, M., Wang, G., Qin, L., & He, X. (2024). G20 roadmap for carbon neutrality: The role of Paris Agreement, artificial intelligence, and energy transition in changing geopolitical landscape. *Journal of Environmental Management*, *367*, 122080. DOI: 10.1016/j.jenvman.2024.122080 PMID: 39111003

Sam-Aggrey, H. (2015). Opportunities and challenges related to the deployment of small modular reactors (SMRs) in the Northern Territories of Canada. *Journal of Nuclear Engineering*, *44*(2), 23–34.

Samson, J. A., & Vlasov, A. Y. (2024). Nuclear fusion propulsion for spacecraft: Background, applications, & impact. Research Archive *Preprints*. https://research-archive.org/index.php/rars/preprint/view/956

Sánchez-Murillo, R., Esquivel-Hernández, G., Birkel, C., Ortega, L., Sánchez-Guerrero, M., Rojas-Jiménez, L. D., Vargas-Víquez, J., & Castro-Chacón, L. (2020). From mountains to cities: A novel isotope hydrological assessment of a tropical water distribution system. *Isotopes in Environmental and Health Studies*, *56*(6), 606–623. DOI: 10.1080/10256016.2020.1809390 PMID: 32835532

Sanchez-Torrijos, J., Aragón, P., Queral, C., Feria, F., Schubert, A., & Van Uffelen, P. (2023). A review on the thermo-mechanical modelling needs in system codes. *Nuclear Engineering and Design, 406*, 112243. DOI: 10.1016/j.nucengdes.2023.112243

Sayed, E. T., Olabi, A. G., Alami, A. H., Radwan, A., Mdallal, A., Rezk, A., & Abdelkareem, M. A. (2023). Renewable Energy and Energy Storage Systems. *Energies, 16*(3), 3. Advance online publication. DOI: 10.3390/en16031415

Setty, D. S., Kapoor, K., & Saibaba, N. (2017). Nuclear fuel cycle - developments and challenges in fuel fabrication technology in India. *Progress in Nuclear Energy, 101*, 100–117. DOI: 10.1016/j.pnucene.2017.03.005

Shahzad, S., Abbasi, M. A., Ali, H., Iqbal, M., Munir, R., & Kilic, H. (2023). Possibilities, challenges, and future opportunities of microgrids: A review. *Sustainability (Basel), 15*(8), 6366. DOI: 10.3390/su15086366

Shobeiri, E., Genco, F., Hoornweg, D., & Tokuhiro, A. (2023). Small modular reactor deployment and obstacles to be overcome. *Energies, 16*(8), 3468. DOI: 10.3390/en16083468

Siegrist, M., & Visschers, V. H. M. (2013). Acceptance of nuclear power : The Fukushima effect. *Energy Policy, 59*, 112–119. DOI: 10.1016/j.enpol.2012.07.051

Sihombing, V., Nasution, Z., Al Ihsan, M. A., Siregar, M., Munthe, I. R., Siregar, V. M. M., & Asfar, D. A. (2021, June). Additive Ratio Assessment (ARAS) Method for Selecting English Course Branch Locations. [). IOP Publishing.]. *Journal of Physics: Conference Series, 1933*(1), 012070. DOI: 10.1088/1742-6596/1933/1/012070

Silva, T. R., Ferreira, V. P., Lima, M. M. C., & Sial, A. N. (2016). Two stage mantle-derived granitic rocks and the onset of the Brasiliano orogeny: Evidence from Sr, Nd, and O isotopes. *Lithos, 264*, 189–200. DOI: 10.1016/j.lithos.2016.08.030

Singer, P. W. (2015). Stuxnet & its hidden lessons on the ethics of cyberweapons. *Case Western Reserve Journal of International Law, 47*(1), 79–96. https://scholarlycommons.law.case.edu/cgi/viewcontent.cgi?article=1009&context=jil

Singh, M., Jiao, J., Klobasa, M., & Frietsch, R. (2022). Servitization of energy sector: Emerging service business models and startup's participation. *Energies, 15*(7), 2705. DOI: 10.3390/en15072705

Singh, R., Altaee, A., & Gautam, S. (2020). Nanomaterials in the advancement of hydrogen energy storage. *Heliyon, 6*(7), e04487. DOI: 10.1016/j.heliyon.2020.e04487 PMID: 32743097

Soltani, A., & Mohsenian-Rad, H. (2020). Game theory for energy distribution in smart cities. *Energy*, *202*, 117784. DOI: 10.1016/j.energy.2020.117784

Sonar, D. (2021). Renewable energy-based trigeneration systems—technologies, challenges, and opportunities. *Renewable-Energy-Driven Future*, 125–168. https://doi.org/DOI: 10.1016/B978-0-12-820539-6.00004-2

Soto, G. H., & Martinez-Cobas, X. (2024). Nuclear energy generation's impact on the CO2 emissions and ecological footprint among European Union countries. *The Science of the Total Environment*, *945*, 173844. DOI: 10.1016/j.scitotenv.2024.173844 PMID: 38871309

Stanujkic, D., & Jovanovic, R. (2012). Measuring a quality of faculty website using ARAS method.

Stouffer, K., Falco, J., & Scarfone, K. (2011). Guide to industrial control systems (ICS) security. NIST special publication, 800(82), 16-16.

Sturbois, A., Cucherousset, J., De Caceres, M., Desroy, N., Riera, P., Carpentier, A., Quillien, N., Grall, J., Espinasse, B., Cherel, Y., & Schaal, G. (2022). Stable Isotope Trajectory Analysis (SITA): A new approach to quantify and visualize dynamics in stable isotope studies. *Ecological Monographs*, *92*(2), e1501. DOI: 10.1002/ecm.1501

Suman, S. (2018). Hybrid nuclear-renewable energy systems: A review. *Journal of Cleaner Production*, *181*, 166–177. DOI: 10.1016/j.jclepro.2018.01.262

Sun, B., Zhao, H., Dong, X., Teng, C., Zhang, A., Kong, S., Zhou, J., Zhang, X.-C., & Tu, S. T. (2024). Current challenges in the utilization of hydrogen energy-a focused review on the issue of hydrogen-induced damage and embrittlement. *Advances in Applied Energy*, *14*, 100168. DOI: 10.1016/j.adapen.2024.100168

Suzuki, T. (2019). Nuclear Energy Policy after the Fukushima Nuclear Accident : An Analysis of "Polarized Debate" in Japan. In *Energy Policy*. IntechOpen. DOI: 10.5772/intechopen.83435

Szocik, K., Shelhamer, M., Braddock, M., Cucinotta, F. A., Impey, C., Worden, P., Peters, T., Ćirković, M. M., Smith, K. C., Tachibana, K., Reiss, M. J., Norman, Z., Gouw, A. M., & Munévar, G. (2021). Future space missions & human enhancement: Medical & ethical challenges. *Futures*, *133*, 102819. DOI: 10.1016/j.futures.2021.102819

Tan, S., Cheng, S., Wang, K., Liu, X., Cheng, H., & Wang, J. (2023). The development of micro and small modular reactor in the future energy market. *Frontiers in Energy Research*, *11*, 1149127. Advance online publication. DOI: 10.3389/fenrg.2023.1149127

Tantawy, A., Erradi, A., Abdelwahed, S., & Shaban, K. (2020). Model-based risk assessment for cyber-physical systems security. *arXiv preprint arXiv:2005.13738*. https://arxiv.org/abs/2005.13738

Tarascon, J. M., & Armand, M. (2001). Issues and challenges facing rechargeable lithium batteries. *Nature*, *414*(6861), 359–367. DOI: 10.1038/35104644 PMID: 11713543

Thakkar, J. J., Kanda, A., & Deshmukh, S. G. (2011, January). A decision framework for supply chain planning in SMEs: A QFD-ISM-enabled ANP-GP approach. In *Supply Chain Forum* []. Taylor & Francis.]. *International Journal (Toronto, Ont.)*, *12*(4), 62–75.

Thiri, M. A., & Borsi, M. T. (2024). From Fukushima to fossil fuels : Carbon emissions, climate narratives, and grassroots movements in Japan's energy transition. *Energy Research & Social Science*, *112*, 103520. DOI: 10.1016/j.erss.2024.103520

U.S. Environmental Protection Agency. (2016). *Global greenhouse gas emissions data*. https://climatechange.chicago.gov/ghgemissions/global-greenhouse-gas-emissions-data

United Nations Framework Convention on Climate Change. (2015). *The Paris Agreement*. https://unfccc.int/process-and-meetings/the-paris-agreement/the-paris-agreement

United Nations Office for Outer Space Affairs (UNOOSA). (1967). The Outer Space Treaty. https://www.unoosa.org/oosa/en/ourwork/spacelaw/treaties/introouterspacetreaty.html

Veerendra, A. S., Mohamed, M. R. B., & García Márquez, F. P. (2024). Energy management control strategies for energy storage systems of hybrid electric vehicle: A review. *Energy Storage*, *6*(1), e573. DOI: 10.1002/est2.573

Verma, O. P., Manik, G., & Sethi, S. K. (2019). A comprehensive review of renewable energy source on energy optimization of black liquor in MSE using steady and dynamic state modeling, simulation and control. *Renewable & Sustainable Energy Reviews*, *100*, 90–109. DOI: 10.1016/j.rser.2018.10.002

Vijayan, P. K., & Gangwar, S. (2024). Intrinsically Safe Thermohydraulic Designs for SMRs : Design advantages and challenges. *IEEE Electrification Magazine*, *12*(4), 75-83. *IEEE Electrification Magazine*. Advance online publication. DOI: 10.1109/MELE.2024.3473332

Vinoya, C. L., Ubando, A. T., & Culaba, A. B. (2024). Life cycle analysis of a network of small modular reactors. *IOP Conference Series. Earth and Environmental Science, 1372*(1), 012059. DOI: 10.1088/1755-1315/1372/1/012059

Vinoya, C. L., Ubando, A. T., Culaba, A. B., & Chen, W. (2023). State-of-the-Art review of small modular reactors. *Energies, 16*(7), 3224. DOI: 10.3390/en16073224

Visschers, V. H. M., & Wallquist, L. (2013). Nuclear power before and after Fukushima : The relations between acceptance, ambivalence and knowledge. *Journal of Environmental Psychology, 36*, 77–86. DOI: 10.1016/j.jenvp.2013.07.007

Vujić, J., Bergmann, R. M., Škoda, R., & Miletić, M. (2012). Small modular reactors: Simpler, safer, cheaper? *Energy, 45*(1), 288–295. DOI: 10.1016/j.energy.2012.01.078

Walsh, P. P., Murphy, E., & Horan, D. (2020). The role of science, technology and innovation in the UN 2030 agenda. *Technological Forecasting and Social Change, 154*, 119957. DOI: 10.1016/j.techfore.2020.119957

Wang, Q., & Zhao, L. (2023). Data-driven stochastic robust optimization of sustainable utility system. *Renewable & Sustainable Energy Reviews, 188*, 113841. DOI: 10.1016/j.rser.2023.113841

Wayne, M., Farokh, N., John, S., & John, S. (2002). Role of fusion energy in a sustainable global energy strategy. *Energy & Environment, 13*(4-5), 647–665. DOI: 10.1260/095830502320939606

Weisser, D. (2007). A guide to life-cycle greenhouse gas (GHG) emissions from electric supply technologies. *Energy, 32*(9), 1543–1559. DOI: 10.1016/j.energy.2007.01.008

Weller, R., Drozdz, Z., & Kjaersgaard, S. P. (2019). Hotspot cities: Identifying peri-urban conflict zones. *Journal of Landscape Architecture, 14*(1), 8–19. DOI: 10.1080/18626033.2019.1623542

Wessel, F. J., & Rostoker, N. (2000). *Controlled thermonuclear fusion in a staged Z-pinch, August 1, 1993–December 31, 1998*. University of California., DOI: 10.2172/765161

Wesson, J. (2011). *Tokamaks* (4th ed.). Oxford University Press.

Whitehead, D. E., Owens, K., Gammel, D., & Smith, J. (2016). *Ukraine cyber-induced power outage: Analysis & practical mitigation strategies*. Proceedings of 43rd Annual Western Protective Relay Conference. https://selinc.com/api/download/117044/

Wikipedia. (2025). *Sodium-cooled fast reactor*. Retrieved from https://en.wikipedia.org/wiki/Sodium-cooled_fast_reactor

Worku, A. K., Ayele, D. W., Deepak, D. B., Gebreyohannes, A. Y., Agegnehu, S. D., & Kolhe, M. L. (2024). Recent advances and challenges of hydrogen production technologies via renewable energy sources. *Advanced Energy and Sustainability Research*, *5*(5), 2300273. DOI: 10.1002/aesr.202300273

World Energy & Climate Statistics – Yearbook (2024). Enerdata. https://yearbook.enerdata.net/

World Nuclear Association. (2025) *Carbon dioxide emissions from electricity*. Retrieved September 3, 2024, from https://world-nuclear.org/information-library/energy-and-the-environment/carbon-dioxide-emissions-from-electricity

World Nuclear Association. (2025) *Generation IV nuclear reactors*. Retrieved April 30, 2024, from https://www.world-nuclear.org/information-library/current-and-future-generation/generation-iv-nuclear-reactors.aspx

World Nuclear Association. (2025) *Nuclear power reactors*. Retrieved March 21, 2025, from https://world-nuclear.org/information-library/nuclear-fuel-cycle/nuclear-power-reactors/nuclear-power-reactors

World Nuclear Association. (2025) *RBMK reactors*. Retrieved February 15, 2022, from https://world-nuclear.org/information-library/appendices/rbmk-reactors

World Nuclear Association. (2025) *World energy needs and nuclear power*. Retrieved January 6, 2025, from https://world-nuclear.org/information-library/current-and-future-generation/world-energy-needs-and-nuclear-power

World Nuclear Association. (2025). *World Nuclear Symposium 2025*. https://world-nuclear.org/events/world-nuclear-symposium-2025

Wu, D., & Ehsani, M. (2010). The role of energy storage in high-power hybrid electric vehicles. *Proceedings of the IEEE*, *95*(4), 678–689. DOI: 10.1109/jproc.2007.914016

Wu, Y. (2019). *Neutronics of advanced nuclear systems* (Vol. 765). Springer., DOI: 10.1007/978-981-13-6520-1

Yang, J., Sudik, A., Wolverton, C., & Siege, D. J. (2010). High capacity hydrogen storage materials: Attributes for automotive applications and techniques for materials discovery. *Chemical Society Reviews*, *39*(2), 656–675. DOI: 10.1039/B802882F PMID: 20111786

Yang, X., & Khan, I. (2022). Dynamics among economic growth, urbanization, and environmental sustainability in IEA countries : The role of industry value-added. *Environmental Science and Pollution Research International*, *29*(3), 4116–4127. DOI: 10.1007/s11356-021-16000-z PMID: 34402019

Yin, L., Jurewicz, T. B., Larsen, M., Drobnjak, M., Graff, C. C., Lutz, D. R., & Rebak, R. B. (2021). Uniform corrosion of FeCrAl cladding tubing for accident tolerant fuels in light water reactors. *Journal of Nuclear Materials*, *554*, 153090. DOI: 10.1016/j.jnucmat.2021.153090

Yue, M., Lambert, H., Pahon, E., Roche, R., Jemei, S., & Hissel, D. (2021). Hydrogen energy systems: A critical review of technologies, applications, trends and challenges. *Renewable & Sustainable Energy Reviews*, *146*, 111180. DOI: 10.1016/j.rser.2021.111180

Yue, Q., Zhang, M., & Song, Y. (2024). Impact of digital divide on energy poverty across the globe: The mediating role of income inequality. *Energy Policy*, *195*, 114349. DOI: 10.1016/j.enpol.2024.114349

Yun, D., Lu, C., Zhou, Z., Wu, Y., Liu, W., Guo, S., Shi, T., & Stubbins, J. F. (2021). Current state and prospect on the development of advanced nuclear fuel system materials: A review. *Materials Reports: Energy*, *1*(1), 100007. DOI: 10.1016/j.matre.2021.01.002

Zarębski, P., & Katarzyński, D. (2023). Small Modular Reactors (SMRs) as a Solution for Renewable Energy Gaps : Spatial Analysis for Polish Strategy. *Energies*, *16*(18), 18. Advance online publication. DOI: 10.3390/en16186491

Zavadskas, E. K., & Turskis, Z. (2010). A new additive ratio assessment (ARAS) method in multicriteria decision-making. *Technological and Economic Development of Economy*, *16*(2), 159–172. DOI: 10.3846/tede.2010.10

Zhang, M., Huang, Y., & Bao, Y. (2016). The mechanism of shallow submarine landslides triggered by storm surge. *Natural Hazards*, *81*(2), 1373–1383. DOI: 10.1007/s11069-015-2112-0

Zhang, X., & Chien, S. (2018). Machine learning techniques for integrating renewable energy in urban grids. *Renewable Energy*, *119*, 666–675. DOI: 10.1016/j.renene.2017.12.012

Zhang, Y., Gu, C., Yan, X., & Li, F. (2020). Cournot oligopoly game-based local energy trading considering renewable energy uncertainty costs. *Renewable Energy*, *159*, 1117–1127. DOI: 10.1016/j.renene.2020.06.066

Zhou, K., Fu, C., & Yang, S. (2018). Big data driven smart energy management: From big data to big insights. *Renewable & Sustainable Energy Reviews*, *56*, 215–225. DOI: 10.1016/j.rser.2015.11.050

Zou, Y., & Mao, D. (2022). Simulation of freshwater ecosystem service flows under land-use change: A case study of Lianshui River Basin, China. *Sustainability (Basel)*, *14*(6), 3270. DOI: 10.3390/su14063270

About The Contributors

Shrikaant Kulkarni, Ph.D., is currently working as a Research Professor at Sanjivani University, Kopargaon, India and as an Adjunct Professor, at the Faculty of Business, Victorian Institute of Technology, Melbourne, Australia, and Adjunct Professor at Chitkara University, Punjab, India. Dr. Kulkarni has been a senior academician and researcher for 42 years. He has delivered invited lectures and conducted sessions at national and international conferences and faculty development programs. He has guided many major and minor projects in areas like engineering chemistry, green chemistry, nanotechnology, analytical chemistry, catalysis, chemical engineering materials, industrial organization, and management, Advanced AI, and Educational Technology. He has published over 100+ research papers in national and international journals and conferences of repute. He has edited 36 books including four books in Advanced AI, and authored 100+ book chapters in books published by CRC, Springer Nature, Apple Academic Press, Elsevier, Wiley, and IET. He authored four textbooks in engineering chemistry too. Dr Kulkarni possesses MSc, MPhil, and PhD degrees in Chemistry and Master's degrees in Economics, Business Management, and Political Science.Two of his Patents have been published and two more are applied for grant. He was invited by UNESCO for an invited talk on " Green Chemistry Education for Sustainable Development " at the International Conference on Green Chemistry, Bangkok, and was well received by international Jury members and dignitaries from across the world. E-mail: shrikaant.kulkarni@vit.edu.au Mob: +91-9970663353

Susanta Das is a seasoned educator with a proven record of teaching and mentoring diverse students, working at multiple institutions, and providing service to the various initiatives of organizations. He received his Ph.D. and M.A. degrees from Western Michigan University (WMU), USA, and M.Sc. degree from Banaras Hindu University (BHU), India all in Physics. He continued his research as a Marie-Curie post-doctoral fellow at Stockholm University, Stockholm, Sweden on beam diagnostics for the DESIREE (Double ElectroStatic Ion Ring ExpEriment) in the project DITANET (Diagnostic Techniques for particle Accelerators – a Marie-Curie initial training NETwork), at the Indian Institute of Science Education and Research-Kolkata, India on high-pressure physics as a Project Scientist-B, and at the University of Electro-Communications, Tokyo, Japan on ion-surface interactions as a Post-doctoral fellow. Throughout his career, Dr. Das worked and collaborated with many researchers, Ph.D., master, and visiting students from different countries. He further visited many countries to discuss research and an international conference participant (UK, Italy, Germany, Greece, Belgium, Bulgaria, Romania, Brazil etc.). He co-authored several research articles in WoS/Scopus indexed international journals, conference proceedings, and scholarly book chapters published by renowned international publishing houses. Beyond his research endeavors, Dr. Das has a storied history of service to academic and administrative committees, exhibiting his commitment to the growth and development of educational institutions. His experience includes tenure at Central University South Bihar, Sri Sri University, and P.K. University before assuming his current role at Ajeenkya D Y Patil University. He received the Marie-Curie post-doctoral fellowship in Sweden, Science Academies' Summer Research Fellowship in India, Gwen Frostic Doctoral Fellowship, Department Graduate Research and Creative Scholar Award by WMU, and Leo R. Parpart Doctoral Fellowship by Dept. of Physics, WMU, among many others, throughout his academic journey. At present, Dr. Das continues to delve into cutting-edge fields, with a keen interest in nanotechnology, quantum computing, and data science. His multifaceted contributions, spanning teaching, research, and administrative leadership, showcase a dedicated professional who is instrumental in advancing the vision and mission of the institutions he serves.

***

Adil Aknouch is a researcher and professor of nuclear physics at the Higher Institute of Nursing Professions and Health Techniques in Agadir. He holds a PhD in nuclear physics from Ibn Tofail University (2022) and a Master's degree in nuclear engineering from Sidi Mohamed Ben Abdellah University (2018). His research focuses on radiation protection and the application of nuclear technologies in the medical, agricultural, industrial, and energy sectors. An expert in radiation protection, he works to minimize the risks of radiation exposure while maximizing the benefits of nuclear technologies. His scientific publications highlight his significant contributions to nuclear safety as well as the advancement of medical and industrial techniques. His vision is to continue advancing research in nuclear physics while contributing to the safety and effectiveness of nuclear applications for the benefit of society.

Sonali Bhatnagar is a Senior Assistant Professor of physics at Dayalbagh Educational Institute, Agra with her Ph.D. from Punjab University, Chandigarh. Her current area of expertise is in high energy physics and has set up a muon telescope. Apart from this a mini array for studying secondary cosmic rays with temperature and pressure at Agra has been set up as Dayalbagh Educational Air Shower Array (DEASA) in collaboration with T.I.F.R., Mumbai. She was awarded Young Scientist Award FOR WOMEN IN SCIENCE at FMT 2020 Conference, by Department of Physics, School of Applied Sciences, Kalinga Institute of Industrial Technology (KIIT). She has published around 45 peer-reviewed articles.She has completed an MHRD project, two Institute funded projects

Kashvi Chaturvedi is presently enrolled at Ajeenkya DY Patil University to pursue a B.Tech in Artificial Intelligence and Data Science. Her technological skills are demonstrated by her knowledge on languages like Python, C++, C. Her Semiconductors Research Paper examined developments in microchip technology, while her work on the Comparative Analysis of BJT Configurations offered insights into electronic circuits. Kashvi is now researching on Hyperloop's potential to transform high-speed transportation. She recently took part in the IIT Madras-hosted Global Hyperloop Competition 2025, where she offered creative suggestions for this futuristic form of transportation. In addition, she is investigating the possibilities of Small Modular Reactors (SMRs) and Ethical dimensions of AI in sustainable energy solutions.

M'hamed El Mansouri is a professor of biomedical engineering in ISPITS of Casablanca.

Youssef El-Ouardi is a researcher in nuclear physics.

J. Mangaiyarkkarasi is currently working as an Associate Professor of Physics in Nadar Mahajana Sangam S. Vellaichamy Nadar College, Nagamalai, Madurai-625 019, Tamil Nadu, India. She has 24 years of teaching experience. Her major research interest includes synthesis of ceramic materials, powder X-ray analysis and electron density studies. She has published 11 research papers in the refereed International journals and 17 book chapters. She has prepared study materials (UG & PG Physics) for Madurai Kamaraj University Distance Education Course.

Shanthalakshmi Revathy J has a Bachelor of Engineering (B.E.) in Information Technology from Mohamed Sathak Engineering College, Kilakkarai, and a Master of Engineering (M.E.) in Computer Science and Engineering from PSG College of Technology, Coimbatore. His current research focus lies in the dynamic realm of data analytics. He has authored more than 10 papers in prestigious international journals. His has authored 7 book chapters, and 2 patents.

Yadnyesh Khapekar, currently pursuing a B.Tech. in Computer Science and Engineering from Ajeenkya D Y Patil University, has a broad spectrum of research interests, ranging from nanoparticles, and aerodynamics to many others. His research not only focuses on the synthesis but also the characterization of nanoparticles. 'How can one minimize the air resistance for a moving object?' is his focus for studying aerodynamics. He aims to research in these fields and make them more applicable for research purposes. He recently participated at the Global Hyperloop Competition (GHC) at IIT Madras as a member of the team Vac-Vectoor. The event took place from 21st February through 24th February 2025. He contributed significantly to the performance of his team. The team Vac-Vectoor achieved 2nd position in Business and Socio-Economic Impact of Hyperloop in the Innoquest category and Special Mention Certificate in Traction sub-system under Design X category.

Shrikaant Kulkarni has been a senior academician and researcher for over four decades. He has delivered invited lectures and conducted sessions at national and international conferences and faculty development programs. He has guided many major and minor projects in subject areas like engineering chemistry, green chemistry, nanotechnology, analytical chemistry, catalysis, chemical engineering materials, industrial organization, and management. He has published over 100+ research papers in national and international journals and conferences of repute in Green, Analytical, Coordination Chemistry, Green Nanotechnology, Sustainability, Advanced AI, and Educational Technology. He has edited 25+ books, and authored 50+ book chapters in books published by CRC, Springer Nature, Apple Academic Press, Elsevier, Wiley, and IET. He authored four textbooks in engineering chemistry too. Dr Kulkarni possesses MSc, MPhil, and PhD degrees in Chemistry and master's degrees in Economics, Business Management, and Political Science. He has expertise in green chemistry and engineering, analytical chemistry, green nanoscience and nanotechnology, wastewater treatment, advanced areas in Chemical Engineering and Material Science, and Advanced AI. Two of his Patents have been published and two more are applied for grant. He was invited by UNESCO for an invited talk on " Green Chemistry Education for Sustainable Development " at the International Conference on Green Chemistry, Bangkok, and was well received by international Jury members and dignitaries from across the world.

M. D. Mathew is a distinguished academic and researcher, currently serving as the Associate Provost and Dean of Research at Saintgits College of Engineering (Autonomous), Kottayam, Kerala, India. Earlier, he was a research scientist at the

Indira Gandhi Centre for Atomic Research (IGCAR), Kalpakkam, India for more than three decades. With over four decades of experience in materials science and engineering, he has made significant contributions to both academia and industry. His research interests include engineering education, strength of materials, materials development, and nuclear irradiation effects. His work has significantly contributed to high-temperature materials development, particularly for India's nuclear energy program. Dr. Mathew has an extensive publication record, with over 200 articles in peer-reviewed journals, over 100 articles in conference proceedings, and several book chapters. He has been recognized internationally for his scientific contributions.. Notably, he is among the World's Top 2% Scientists in the field of Materials, according to the annual compilation by the US's Stanford University, and academic publisher Elsevier for the years 2020, 2021, 2022, 2023, and 2024. He is also a Program Evaluator for ABET, USA, contributing to the accreditation of engineering programs.

Ranjit Parmar persuaded his M.E. in Electrical Engineering from MS University Baroda, India. He has cleared GPSC with rank 1. He has 25 years of teaching experience. He has published 4 patents. He has been awarded National Teachers Award 2024 by Hon. President of India for excellence in education. He has been also awarded by the Innovation Award 2025 by Gujrat Technological University, Ahmedabad, India. Presently he is leading the Department of Electrical Engineering at Government Polytechnic Junagadh. Under his leadership, department has been accredited twice by the National Board of Accreditation, New Delhi, department has regularly secured a place in top 10 of university including rank 1. He has emphasized skill based education through Gurukul style securing excellent placement in MNCs for the students even for rural background bridging the socio-economic gap.

Subrata Paul is currently a PhD scholar under MAKAUT and working as Assistant Professor and Examination Coordinator at Brainware University, Barasat, Kolkata. He is currently pursuing PhD and had completed his B.E(CSE) from VTU – Belgaum, Karnataka in 2010 and M.Tech(CS) from Berhampur University in the year 2013. His research area includes Social Network Analysis, Computational Intelligence and Cloud Computing. He had several publications at national and international levels, both in journals and conferences including papers in IEEE and Elsevier conference. He had an experience of nearly 4.5 years in teaching undergraduate courses and 1 year in handling post graduate classes. He usually teaches papers like Compiler Design, Formal Languages and Automata Theory, Software Engineering, Computer Network, Artificial Intelligence and Programming in C Language.

Amitava Podder is an Assistant Professor at Brainware University, Barasat In Computer Science and Engineering- AI Department. He is currently pursuing his Ph.D. and has extensive teaching experience in various reputed enginerring colleges and universities. Amitava's research has been published in reputable journals and has been recognized with multiple patents filed with Intellectual Property India and in United Kingdom. He is proficient in Computer networking, Machine learning, Image processing and Internet of Medical Things (IOMT).

Khawla Rachmoune is specialized in biology and biochemistry. She studies radioprotective biomaterials as part of multidisciplinary research projects and is the co-author of two scientific articles published in peer-reviewed journals. Passionate and dedicated, she is particularly interested in the applications of biopolymers in health, environmental protection, and radiation shielding.

Bhanu Sankara Rao is a distinguished metallurgical engineer and academician, currently serving as a Distinguished Professor of Practice at the Indian Institute of Technology Hyderabad. Dr. Rao's expertise encompasses high-temperature materials development, mechanical metallurgy, physical metallurgy, powder metallurgy, welding science and technology, and structure-property correlations. His significant contributions include supporting the development of Fast Breeder Reactor technology in India during his tenure at the Indira Gandhi Centre for Atomic Research. Notably, he played a pivotal role in developing reduced activation ferritic-martensitic steels for the Indian Test Blanket Module of the International Thermonuclear Experimental Reactor. Dr. Rao has held esteemed positions such as the Dean of the School of Engineering Sciences and Technology at the University of Hyderabad, Ministry of Steel Chair Professor, and Pratt & Whitney Chair Professor at the same university. His accolades include the National Metallurgist Award from the Ministry of Steel, the Platinum Medal from the Indian Institute of Metals, and the NASA Appreciation Award for his contributions to aerospace materials research. Dr. Rao is a Fellow of several prestigious organizations, including the Indian National Academy of Engineering, the Indian Academy of Sciences, and ASM International. His extensive publication record features over 400 research papers, reflecting his profound impact on the field of metallurgical engineering and nuclear technology.

Piyal Roy is an accomplished academician in the field of Computer Science & Engineering. He has completed his BSc in Computer Science, MSc in Computer Science and MTech in Computer Science & Engineering. Currently, he is working as an Assistant Professor at Computer Science & Engineering department at Brainware University, India.

Sunil Sankathala is pursuing B. Tech Computer Science and Engineering at Ajeenkya D Y Patil University, with a diverse range of interests that bridge technology, sustainability, and innovation. He academic focus lies in computer science fundamentals like programming, data structures, algorithms, and software development. He actively supplements this with hands-on experience in web development and robotics. Beyond the core curriculum, He has a strong research inclination in emerging and impactful fields. He current research interests include Artificial Intelligence in Healthcare, bioplastics, nano bioplastics, e-waste management, and solid waste management—areas where he believes technology can make a meaningful difference in solving real-world problems. He has been publishing research articles and book chapters in reputed national and international journals and books. He is also part of an ambitious Hyperloop project, which involves interdisciplinary collaboration and cutting-edge engineering aimed at revolutionizing transportation. Working on this project has enhanced his skills in teamwork, innovation, and systems thinking. In the tech domain, he has developed proficiency in web development using technologies such as HTML, CSS, JavaScript, and React. He interests in robotics further reflects building smart, responsive systems that interact with the physical world. Throughout his course, he has helped strengthen problem-solving and collaborative skills. He proficient in languages such as Python, C, C++, MySQL, Ubuntu, Machine Learning and familiar with frameworks and tools like React, Git. He actively participates in hackathons, coding competitions, and tech events to challenge himself and learn from peers. He is also a quick learner and highly adaptable, with an eagerness to stay updated with the latest trends in the tech world. He aspires to become a skilled software engineer and contribute to building innovative solutions that have a positive impact on society. The long-term goal is to work in a dynamic environment where knowledge, grow professionally, and make meaningful contributions to the field of technology. Driven by a spirit of exploration and sustainability, he aims to contribute to technological solutions that address both human and environmental needs. The long-term goal is to become a versatile engineer and researcher, working at the intersection of computer science, environmental innovation, and human well-being.

Atharva Saraf is currently pursuing a B. Tech degree in Computer Science Engineering at Ajeenkya D Y Patil University. Throughout his academic journey, he has developed a strong foundation in various programming languages, including C, C++, and Python. His interests and focus areas lie in algorithm-based programs, artificial intelligence (with a special emphasis on neural networks), and cybersecurity. He has successfully undertaken several projects that reflect his proficiency and passion in these domains. For instance, he has developed a PC-Tuner application aimed at optimizing system performance by employing various methods such as defragmentation. Additionally, he has created an automated backup utility capable of securely backing up selected files, folders, and stores them in compressed backup folder in system itself it can also store backups on external hard disk and cloud storages. Beyond project work, he has contributed to the academic community through research articles and book chapters. Notable among these is his co-authored chapter on federated learning techniques, exploring their applications and advancements in the field of machine learning. His academic experiences have honed his skills in problem-solving, critical thinking, and collaborative teamwork. He is eager to further expand his knowledge and expertise in computer science, particularly in areas that intersect with his interests in algorithms, artificial intelligence, and their practical applications.

Suchismita Satapathy is an associate professor in the School of Mechanical Sciences, KIIT University, Bhubaneswar, India. She has published more than 130 articles in national and international journals and conferences. She has also published many books and e-books for academic and research purposes. Her areas of interest include production operation management, operation research, acoustics, sustainability, and supply chain management. She has filled three Indian patents, among them two published. She has more than 15 years of teaching and research experience. She has guided many Ph.D. (three completed, three ongoing), MTech (20), and BTech (49) students. Suchismita has published books such as Production operation management (Stadium Press) and MCDM methods for waste management with CRC Press, Innovation, Technology, and Knowledge Management with Springer, Soft Computing and Optimization Techniques for Sustainable Agriculture by DEGRUYTER, The PDCA cycle for industrial improvement, applied case studies by Springer and many of her projects are ongoing. Scopus Author ID: 55085975400

Slimane Semghouli has obtained his undergraduate degree in Nuclear Physcis (2001) from Faculty of Sciences, University Mohammed V of Rabat. I worked as a teacher at the Health Careers Training Institute in Agadir between 2002 and 2007. Subsequently, I participated as a medical physicist in the implementation of radiotherapy techniques at the regional oncology center of Agadir. Since 2013 until today, I have been a professor of Biophysics and head of the health techniques field at the Higher Institute of Nursing and Health Techniques Professions in Agadir.

Anupama Sen is an Assistant Professor in the Computer Science and Engineering- Aritificial Intelligence department at Brainware University. With expertise in artificial intelligence and a passion for fostering innovation, she is dedicated to advancing knowledge and mentoring students in the field of technology.

Aditya Shrivastav is currently studying B.Tech. in Computer Engineering at the School of Engineering, Ajeenkya DY Patil University. His research interests are centered around nanoparticle biosynthesis, exploring ionic liquids' use in various applications, and applying Artificial Intelligence to different fields. Aditya is particularly focused on understanding how these technologies can be utilized together to solve practical problems. He aims to advance research in these areas and contribute to innovative solutions that address real-world challenges. He recently participated at the Global Hyperloop Competition (GHC) at IIT Madras as a member of the team Vac-Vectoor. The event took place from 21st February through 24th February 2025. He contributed significantly to the performance of his team. The team Vac-Vectoor achieved 2nd position in Business and Socio-Economic Impact of Hyperloop in the Innoquest category and Special Mention Certificate in Traction sub-system under Design X category.

Halima Soussi is a geophysicist and geoscientist specializing in tectonics, seismic imaging, and lithospheric processes. She holds a Ph.D. in Geophysics from Mohammed V University- Mohammadia Engineering School. Her research focuses on understanding Earth's structure and geohazards through integrated geophysical methods. She has published extensively and is actively engaged in promoting diversity in the geosciences.

Mohammed Talbi, born in 1985, is a medical physicist, radiation protection expert, and professor of medical physics at Ibn Tofail University in Kénitra. He holds a Bachelor's degree in Radiophysics, Radiobiology, and Radiation Protection, a specialized Master's degree in Medical Physics from Hassan First University of Settat, and a Ph.D. in Applied Physics from Moulay Ismail University of Meknes, completed in 2022. A recognized expert by the IAEA, Dr. Talbi specializes in medical physics, with a particular focus on medical imaging. His research includes radiation dose assessment in Morocco in collaboration with AMSSNuR, and he actively contributes to IAEA-led evaluations of radiation protection in radiology programs across Africa.

Shaurya Talewar is pursuing his B.Tech. in computer science at the School of Engineering of Ajeenkya D Y Patil University, Pune, India. His research interests are cybersecurity, data analysis, and artificial intelligence.

Index

A

Advanced Reactor Designs 143, 147
AI 15, 26, 53, 54, 55, 57, 58, 115, 116, 122, 170, 190, 195, 198, 201, 202, 204, 205, 211, 215, 223, 229, 230, 233, 246, 255

B

Boiling Water Reactors 63, 66, 68, 87, 109, 168

C

Circular Economy 26, 56, 57, 148, 189
Climate Adaptation 95, 98, 128, 198, 202
CPS 218, 221

E

Ecosystem Services 195, 212
Energy-as-a-Service 55, 57
Energy Storage 7, 22, 23, 25, 28, 30, 32, 35, 36, 38, 39, 40, 41, 42, 43, 44, 50, 51, 52, 54, 57, 60, 61, 80, 86, 95, 97, 102, 107, 156, 190, 280

F

Fast Reactors 13, 63, 67, 73, 74, 75, 76, 77, 87, 89, 92, 99, 114, 115, 141, 143, 144, 147, 148, 164, 166
Fission Energy 117
Fuel Efficiency 17, 74, 80, 99, 113, 130, 139, 148, 242

G

Generation IV Reactors 63, 96, 99, 100, 113, 132
Global Warming 2, 8, 9, 20, 22, 63, 64, 66, 77, 85, 87, 89, 92, 93, 94, 113, 120, 123, 128, 161, 194, 211, 265
Greenhouse Gases 3, 9, 12, 16, 17, 36, 50, 63, 66, 79, 92, 98, 103, 107, 113, 117, 128, 154, 264
Grid Integration 27, 36, 194

H

Hybrid Systems 37, 38, 39, 51, 98, 102, 113, 169, 184

I

ICS (Industrial Control Systems) 218
Isotopic Analysis 195, 208
ITER 12, 14, 84, 89, 117, 118, 120, 122, 123

K

Kilopower Reactor 240, 244, 250, 251, 254, 255, 261

M

Multi-Objective Optimization 26, 32, 44, 46, 47, 57

N

Nuclear Fission 63, 64, 65, 66, 67, 84, 85, 87, 88, 91, 117, 120, 128, 194, 240, 244, 250, 254, 261
Nuclear Fuel Cycle 12, 74, 76, 131, 152, 172, 176
Nuclear Fusion 2, 3, 12, 14, 16, 22, 23, 65, 84, 85, 87, 91, 92, 120, 121, 123, 124, 199, 203, 254, 257, 259, 260
Nuclear Licensing 194
Nuclear Reactor 11, 16, 66, 88, 91, 92, 97, 120, 121, 128, 130, 131, 159, 175, 194, 245, 247, 250, 251, 255, 260, 261
Nuclear Technology 15, 64, 66, 80, 109, 112, 130, 132, 154, 175, 179, 186, 192, 195, 197, 198, 199, 200, 202, 203, 208, 212, 240, 241, 242, 243, 244, 246, 247, 248, 249, 250, 253, 254, 256, 257, 258

O

Optimization 25, 26, 27, 28, 29, 30, 31, 32, 33, 34, 38, 39, 40, 41, 43, 44, 45, 46, 47, 48, 49, 51, 52, 53, 54, 55, 56, 57, 58, 59, 60, 123, 132, 148, 177, 204, 205, 206

P

Passive Safety Features 77, 80, 93, 97, 100, 104, 109, 112, 128, 153, 156, 173, 199
Predictivc Maintenance 48, 49, 55, 57, 58, 199, 203
Predictive Modeling 48, 58, 195, 203, 208
Pressurized Water Reactors 63, 66, 67, 87

Q

QFD 263, 265, 266, 278, 279, 280

R

Reducing GHG Emission 194
Regulatory Frameworks 12, 52, 57, 86, 114, 122, 162, 194, 198
Renewable Energy 4, 7, 8, 9, 10, 16, 19, 20, 21, 22, 23, 24, 25, 26, 27, 28, 30, 31, 32, 33, 34, 35, 36, 37, 38, 39, 40, 42, 43, 44, 45, 48, 50, 51, 52, 53, 54, 55, 56, 57, 58, 59, 60, 61, 63, 80, 85, 86, 93, 94, 95, 97, 99, 102, 103, 107, 111, 112, 113, 114, 115, 117, 119, 123, 124, 126, 127, 145, 153, 154, 155, 156, 162, 169, 170, 177, 180, 181, 188, 189, 191, 193, 209, 212, 257, 264, 267, 268, 280

S

Small Modular Reactors 2, 3, 17, 21, 24, 63, 65, 80, 85, 87, 93, 96, 98, 99, 104, 107, 113, 121, 124, 125, 126, 131, 132, 145, 147, 151, 153, 192, 193, 194, 196, 199, 261
Smart Grids 28, 31, 40, 41, 54, 57, 102, 112, 199, 202, 203, 212
SMR 14, 19, 23, 80, 81, 106, 123, 142, 145, 155, 157, 158, 160, 161, 162, 163, 165, 166, 167, 169, 170, 172, 173, 174, 175, 176, 177, 178, 179, 181, 182, 183, 184, 191, 193, 194
Sustainability 2, 13, 23, 26, 27, 39, 40, 53, 55, 56, 57, 66, 74, 75, 80, 82, 87, 95, 97, 98, 109, 118, 126, 129, 130, 131, 133, 139, 141, 142, 143, 144, 145, 146, 147, 148, 159, 161, 186, 189, 190, 191, 195, 202, 203, 208, 209, 210, 211, 212, 213, 258, 278, 279, 280
Sustainable Energy 18, 20, 21, 23, 25, 26, 27, 28, 30, 31, 32, 34, 37, 50, 51, 52, 55, 56, 57, 58, 59, 60, 61, 65, 92, 95, 96, 97, 99, 100, 124, 125, 140, 153, 169, 189, 190, 192, 194, 198, 199, 202, 203, 263, 280

T

Thermal Reactors 63, 67, 72, 73, 74, 76, 80, 82, 87, 91, 92

U

Uranium Silicide 131, 135, 136, 137, 139, 144, 146, 148, 149, 164

www.ingramcontent.com/pod-product-compliance
Lightning Source LLC
LaVergne TN
LVHW061549290926
849162LV00004B/953
* 9 7 9 8 3 3 7 3 0 6 8 5 8 *